NACE CORROSION ENGINEER'S REFERENCE BOOK

Second Edition

R.S. TRESEDER — Editor

R. BABOIAN and C.G. MUNGER — Co-Editors

Published by NACE
1440 South Creek Drive, Houston, TX 77084-4906

Published by
National Association of Corrosion Engineers
1440 South Creek Drive
Houston, TX 77084-4906

Library of Congress Catalog Card Number: 79-67175
ISBN 0-915567-67-9

This book is published on the express condition that no company or individual who supplied information for the book shall be liable in connection with anything contained herein or in connection with any use made of its contents.

PREFACE

The second edition of this book is an extensive revision of the first edition published in 1980. Many new tables have been added. A large number of the original tables were updated, and some were deleted in favor of newer material. The number of tables relating to electrochemistry, cathodic protection, and protective coatings has been increased considerably.

NACE wishes to thank the many sources of information and graphic materials who have given permission for their use in this book. These sources are identified in footnotes following the individual tables and graphs.

TABLE OF CONTENTS

2 CONTENTS

ATMOSPHERIC CORROSION

SEAWATER AND COOLING WATER CORROSION

CATHODIC PROTECTION

PROCESS AND OIL INDUSTRIES CORROSION

4 CONTENTS

METALLIC MATERIALS

NONMETALLIC MATERIALS

PROTECTIVE COATINGS

NACE Glossary
of
Corrosion-Related Terms

Acrylic: Resin polymerized from acrylic acid, methacrylic acid, esters of these acids, or acrylonitrile.

Active: The negative direction of electrode potential. Also used to describe a metal that is corroding without significant influence of reaction product.

Aeration Cell (Oxygen Cell): See Differential Aeration Cell.

Alkyd: Resin used in coatings. Reaction products of polyhydric alcohols and polybasic acids.

Alligatoring: Pronounced wide cracking over the entire surface of a coating having the appearance of alligator hide.

Anaerobic: Free of air or uncombined oxygen.

Anion: A negatively charged ion that migrates through the electrolyte toward the anode under the influence of a potential gradient.

Anode: The electrode of an electrolyte cell at which oxidation occurs. (Electrons flow away from the anode in the external circuit. It is usually at the electrode that corrosion occurs and metal ions enter solution.)

Anode Corrosion Efficiency: The ratio of the actual corrosion (weight loss) of an anode to the theoretical corrosion (weight loss) calculated by Faraday's law from the quantity of electricity that has passed.

Anodic Inhibitor: A chemical substance or mixture that prevents or reduces the rate of the anodic or oxidation reaction.

Anodic Polarization: The change of the electrode potential in the noble (positive) direction due to current flow. (See Polarization.)

Anodic Protection: Polarization to a more oxidizing potential to achieve a reduced corrosion rate by the promotion of passivity.

Anodizing: Oxide coating formed on a metal surface (generally aluminum) by an electrolytic process.

Anolyte: The electrolyte adjacent to the anode of an electrolytic cell.

Antifouling: Intended to prevent fouling of underwater structures (such as the bottoms of ships).

Auger Spectroscopy: Analytical technique in which the sample surface is irradiated with low energy electrons and the energy spectrum of electrons emitted from the surface is measured.

Austenite: A face-centered cubic crystalline phase of iron-base alloys.

Auxiliary Electrode: An electrode commonly used in polarization studies to pass current to or from a test electrode. It is usually made from a noncorroding material.

Backfill: Material placed in the drilled hole to fill space around anodes, vent pipe, and buried components of a cathodic protection system.

Beach Marks: A term used to describe the characteristic fracture markings produced by fatigue crack propagation. Known also as clamshell, conchoidal, and arrest marks.

Bituminous Coating: Coal tar- or asphalt-based coating.

Blow Down: (1) In conjunction with cathodic protection, the injection of air or water under high pressure through a tube to the anode area for the purpose of purging the annular space and possibly correcting high resistance caused by gas blocking. (2) In conjunction with boilers or cooling towers, the process of discharging a significant portion of the aqueous solution in order to remove accumulated salts, deposits, and other impurities.

Blushing: Whitening and loss of gloss of a usually organic coating caused by moisture; blooming.

Calcareous Coating or Deposit: A layer consisting of a mixture of calcium carbonate and magnesium hydroxide deposited on surfaces being cathodically protected because of the increased pH adjacent to the protected surface.

Case Hardening: Hardening a ferrous alloy so that the outer portion, or case, is made substantially harder than the inner portion, or core. Typical processes are carburizing, cyaniding, carbon-nitriding, nitriding, induction hardening, and flame hardening.

Cathode: The electrode of an electrolytic cell at which reduction is the principal reaction. (Electrons flow toward the cathode in the external circuit.) Typical cathodic processes are cations taking up electrons and being discharged, oxygen being reduced, and the reduction of an element or group of elements from a higher to a lower valence state.

Cathodic Corrosion: Corrosion resulting from a cathodic condition of a structure usually caused by the reaction of an amphoteric metal with the alkaline products of electrolysis.

Cathodic Disbondment: The destruction of adhesion between a coating and its substrate by products of a cathodic reaction.

Cathodic Inhibitor: A chemical substance or mixture that prevents or reduces the rate of the cathodic or reduction reaction.

Cathodic Polarization: The change of the electrode potential in the active (negative) direction caused by current flow. (See Polarization.)

Cathodic Protection: Reduction of corrosion rate by shifting the corrosion potential of the electrode toward a less oxidizing potential by applying an external electromotive force.

Catholyte: The electrolyte adjacent to the cathode of an electrolytic cell.

Cation: A positively charged ion that migrates through the electrolyte toward the cathode under the influence of a potential gradient.

Caustic Embrittlement: An obsolete historical term denoting stress corrosion cracking of steel exposed to alkaline solutions.

Cavitation: The formation and rapid collapse within a liquid of cavities or bubbles that contain vapor or gas or both.

Cavitation Damage: Damage of a material associated with collapse of cavities in the liquid at a solid/liquid interface under conditions of severe turbulent flow.

Cell: Electrochemical system consisting of an anode and a cathode immersed in an electrolyte. The anode and cathode may be separate metals or dissimilar areas on the same metal. The cell includes the external circuit that permits the flow of electrons from the anode toward the cathode. (See Electrochemical Cell.)

Chalking: The development of loose removable powder at the surface of an organic coating usually caused by weathering.

Checking: The development of slight breaks in a coating; these breaks do not penetrate to the underlying surface.

Chemical Conversion Coating: A protective or decorative coating produced *in situ* by chemical reaction of a metal with a chosen environment.

Chevron Pattern: A V-shaped pattern on a fatigue or brittle fracture surface. The pattern can also be one of straight radial lines on cylindrical specimens.

Concentration Cell: An electrolytic cell, the electromotive force of which is caused by a difference in concentration of some component in the electrolyte. (This difference leads to the formation of discrete cathode and anode regions.)

Concentration Polarization: That portion of the polarization of a cell produced by concentration changes resulting from passage of current through the electrolyte.

Contact Corrosion: A term mostly used in Europe to describe galvanic corrosion between dissimilar metals.

Continuity Bond: A metallic connection that provides electrical continuity between metal structures.

Corrosion: The deterioration of a material, usually a metal, by reaction with its environment.

Corrosion Fatigue: Fatigue-type cracking of metal caused by repeated or fluctuating stresses in a corrosive environment characterized by shorter life than would be encountered as a result of either the repeated or fluctuating stress alone or the corrosive environment alone.

Corrosion Inhibitor: See Inhibitor.

Corrosion Potential (E_{corr}): The potential of a corroding surface in an electrolyte, relative to a reference electrode. Also called rest potential, open circuit potential, freely corroding potential.

Corrosion Rate: The rate at which corrosion proceeds, expressed as either weight loss or penetration per unit time.

Corrosion Resistance: Ability of a metal to withstand corrosion in a given corrosion system.

Corrosivity: Tendency of an environment to cause corrosion in a given corrosion system.

Counter Electrode: See Auxiliary Electrode.

Couple: See Galvanic Corrosion.

Cracking (of Coating): Breaks in a coating that extend through to the underlying surface.

Crazing: A network of checks or cracks appearing on the surface.

Creep: Time-dependent strain occurring under stress.

Crevice Corrosion: A form of localized corrosion occurring at locations where easy access to the bulk environment is prevented, such as at the mating surfaces of metals or assemblies of metal and nonmetal.

Critical Humidity: The relative humidity above which the atmospheric corrosion rate of some metals increases sharply.

Critical Pitting Potential (E_{cp}, E_p, E_{pp}): The lowest value of oxidizing potential at which pits nucleate and grow. It is dependent on the test method used.

Current Density: The current flowing to or from a unit area of an electrode surface.

Current Efficiency: The ratio of the electrochemical equivalent current density for a specific reaction to the total applied current density.

Dealloying: The selective corrosion of one or more components of a solid solution alloy (also known as parting).

Decomposition Potential (or Voltage): The potential of a metal surface necessary to decompose the electrolyte of a cell or a component thereof.

Deep Groundbed: One or more anodes installed vertically at a nominal depth of 50 feet or more below the earth's surface in a drilled hole for the purpose of supplying cathodic protection for an underground or submerged metallic structure.

Depolarization: The removal of factors resisting the flow of current in a cell.

Deposit Attack: Corrosion occurring under or around a discontinous deposit on a metallic surface (also called poultice corrosion).

Dezincification: A corrosion phenomenon resulting in the selective removal of zinc from copper-zinc alloys. (This is the most common form of dealloying.)

Dielectric Shield: In a cathodic protection system, an electrically nonconductive material, such as a coating, plastic sheet, or pipe, that is placed between an anode and an adjacent cathode to avoid current wastage and to improve current distribution, usually on the cathode.

Differential Aeration Cell: An electrolytic cell, the electromotive force of which is due to a difference in air (oxygen) concentration at one electrode as compared with that at another electrode of the same material.

Diffusion Limited Current Density: The current density, often referred to as limiting current density, that corresponds to the maximum transfer rate that a particular species can sustain because of the limitation of diffusion.

Disbondment: The destruction of adhesion between a coating and the surface coated.

Double Layer: The interface between an electrode or a suspended particle and an electrolyte created by charge-charge interaction leading to an alignment of oppositely charged ions at the surface of the electrode or particle. The simplest model is represented by a parallel plate condenser.

Drainage: Conduction of electric current from an underground metallic structure by means of a metallic conductor. (1) Forced Drainage: Drainage applied to underground metallic structures by means of an applied electromotive force or sacrificial anode. (2) Natural Drainage: Drainage from an underground structure to a more negative (more anodic) structure, such as the negative bus of a trolley substation.

Drying Oil: An oil capable of conversion from a liquid to a solid by slow reaction with oxygen in the air.

Electrical Isolation: The condition of being electrically separated from other metallic structures or the environment.

Electrochemical Cell: An electrochemical system consisting of an anode and a cathode in metallic contact and immersed in an electrolyte. (The anode and cathode may be different metals or dissimilar areas on the same metal surface.)

Electrochemical Equivalent: The weight of an element or group of elements oxidized or reduced at 100 percent efficiency by the passage of a unit quantity of electricity. Usually expressed as grams per coulomb.

Electrochemical Potential: The partial derivative of the total electrochemical free energy of a constituent with respect to the number of moles of this constituent where all factors are kept constant. It is analogous to the chemical potential of a constituent except that it includes the electric as well as chemical contributions to the free energy. The potential of an electrode in an electrolyte relative to a reference electrode measured under open circuit conditions.

Electrode: An electronic conductor used to establish electrical contact with an electrolytic part of a circuit.

Electrode Potential: The potential of an electrode in an electrolyte as measured against a reference electrode. (The electrode potential does not include any resistance losses in potential in either the solution or the external circuit. It represents the reversible work to move a unit charge from the electrode surface through the solution to the reference electrode.)

Electrokinetic Potential: This potential, sometimes called zeta potential, is a potential difference in the solution caused by residual, unbalanced charge distribution in the adjoining solution producing a double layer. The electrokinetic potential is different from the electrode potential in that it occurs exclusively in the solution phase, i.e., it represents the reversible work necessary to bring a unit charge from infinity in the solution up to the interface in question but not through the interface.

Electrolysis: The process that produces a chemical change in an electrolyte resulting from the passage of electricity.

Electrolyte: A chemical substance or mixture, usually liquid, containing ions that migrate in an electric field.

Electrolytic Cleaning: A process for removing soil, scale, or corrosion products from a metal surfae by subjecting it as an electrode to an electric current in an electrolytic bath.

Electromotive Force Series (Emf Series): A list of elements arranged according to their standard electrode potentials, the sign being positive for elements whose potentials are cathodic to hydrogen and negative for those anodic to hydrogen.

Ellipsometry: An optical analytical technique employing plane polarized light to study thin films.

Embrittlement: Loss of ductility of a material resulting from a chemical or physical change.

Endurance Limit: The maximum stress that a material can withstand for an infinitely large number of fatigue cycles. (See Fatigue Strength.)

Environment: The surroundings or conditions (physical, chemical, mechanical) in which a material exists.

Environmental Cracking: Brittle fracture of a normally ductile material in which the corrosive effect of the environment is a causative factor. Environmental cracking is a general term that includes all of the terms listed below. The definitions of these terms are listed elsewhere in the glossary.

Corrosion Fatigue
High Temperature Hydrogen Attack
Hydrogen Blistering
Hydrogen Embrittlement
Hydrogen Induced Cracking
Liquid Metal Cracking
Stress Corrosion Cracking
Sulfide Stress Cracking

The following terms have been used in the past in connection with environmental cracking but are now obsolete and should not be used:

Caustic Embrittlement
Delayed Fracture
Liquid Metal Embrittlement
Season Cracking
Static Fatigue
Stepwise Cracking
Sulfide Corrosion Cracking
Sulfide Stress Corrosion Cracking

Epoxy: Resin formed by the reaction of bisphenol and epichlorohydrin.

Equilibrium (Reversible) Potential: The potential of an electrode in an electrolytic solution when the forward rate of a given reaction is exactly equal to the reverse rate. The electrode potential with reference to a standard equilibrium, as defined by the Nernst equation.

Erosion: Destruction of materials by the abrasive action of moving fluids, usually accelerated by the presence of solid particles.

Erosion Corrosion: A corrosion reaction accelerated by the relative movement of the corrosive fluid and the metal surface.

Exchange Current: When an electrode reaches dynamic equilibrium in a solution, the rate of anodic dissolution balances the rate of cathodic plating. The rate at which either positive or negative charges are entering or leaving the surface at this point in known as the exchange current.

Exfoliation Corrosion: Localized subsurface corrosion in zones parallel to the surface that result in thin layers of uncorroded metal resembling the pages of a book.

External Circuit: The wires, connectors, measuring devices, current sources, etc., that are used to bring about or measure the desired electrical conditions within the test cell. It is this portion of the cell through which electrons travel.

Fatigue: The phenomenon leading to fracture under repeated or fluctuating stresses having a maximum value less than the tensile strength of the material.

Fatigue Strength: The stress to which a material can be subjected for a specified number of fatigue cycles.

Ferrite: A body-centered cubic crystalline phase of iron-base alloys.

Filiform Corrosion: Corrosion that occurs under some coatings in the form of randomly distributed threadlike filaments.

Film: A thin, not necessarily visible layer of material.

Foreign Structure: Any metallic structure that is not intended as part of a cathodic protection system of interest.

Fouling: An accumulation of deposits. This term includes accumulation and growth of marine organisms on a submerged metal surface and also includes the accumulation of deposits (usually inorganic) on heat exchanger tubing.

Fractography: Descriptive treatment (photographs) of fracture, especially in metals.

Fracture Mechanics: A quantitative analysis for evaluating structural reliability in terms of applied stress, crack length, and specimen geometry.

Free Machining: The machining characteristics of an alloy to which an ingredient has been introduced to give small broken chips, lower power consumption, better surface finish, and longer tool life.

Fretting Corrosion: Deterioration at the interface of two contacting surfaces under load, accelerated by relative motion between them of sufficient amplitude to produce slip.

Furan: Resin formed from reactions involving furfuryl alcohol alone or in combination with other constituents.

Galvanic Anode: A metal which, because of its relative position in the galvanic series, provides sacrificial protection to metals that are more noble in the series, when coupled in an electrolyte.

Galvanic Corrosion: Corrosion associated with the current resulting from the electrical coupling of dissimilar electrodes in an electrolyte.

Galvanic Couple: A pair of dissimilar conductors, commonly metals, in electrical contact in electrolyte. (See Galvanic Corrosion.)

Galvanic Current: The electric current that flows between metals or conductive nonmetals in a galvanic couple.

Galvanic Series: A list of metals arranged according to their corrosion potentials in a given environment.

Galvanostatic: A constant current technique of applying current to a specimen in an electrolyte. Also called intentiostatic.

General Corrosion: A form of deterioration that is distributed more or less uniformly over a surface.

Graphitic Corrosion: Deterioration of gray cast iron in which the metallic constituents are selectively leached or converted to corrosion products leaving the graphite intact.

Graphitization: A metallurgical term describing the formation of graphite in iron or steel, usually from decomposition of iron carbide at elevated temperatures. Not recommended as a term to describe Graphitic Corrosion.

Ground Bed: A buried item, such as junk steel or graphite rods, that serves as the anode for the cathodic protection of pipelines or other buried structures.

Heat Affected Zone (HAZ): That portion of the base metal that was not melted during brazing, cutting, or welding, but whose microstructure and properties were altered by the heat of these processes.

High Temperature Hydrogen Attack: A loss of strength and ductility of steel by high temperature reaction of absorbed hydrogen with carbides in the steel resulting in decarburization and internal fissuring.

Holiday: Any discontinuity or bare spot in a coated surface.

Hydrogen Blistering: Subsurface voids produced in a metal by hydrogen absorption in (usually) low strength alloys with resulting surface bulges.

Hydrogen Embrittlement (HE): A loss of ductility of a metal resulting from absorption of hydrogen.

Hydrogen Induced Cracking (HIC): A form of hydrogen blistering in which stepwise internal cracks are created that can affect the integrity of the metal.

Hydrogen Overvoltage: Overvoltage associated with the liberation of hydrogen gas.

Hydrogen Stress Cracking: A cracking process that results from the presence of hydrogen in a metal in combination with tensile stress. It occurs most frequently with high strength alloys.

Impingement Attack: Corrosion associated with turbulent flow of liquid and may be accelerated by entrained gas bubbles. (See also Erosion Corrosion.)

Impressed Current: Direct current supplied by a device employing a power source external to the electrode system of a cathodic protection installation.

Inclusion: A nonmetallic phase such as an oxide, sulfide, or silicate particle in a metal.

Inhibitor: A chemical substance or combination of substances that, when present in the environment, prevents or reduces corrosion without significant reaction with the components of the environment.

Inorganic Zinc Rich Paint: Coating containing a zinc powder pigment in an inorganic vehicle.

Intentiostatic: See Galvanostatic.

Intercrystalline Corrosion: See Intergranular Corrosion.

Interdendritic Corrosion: Corrosive attack of cast metals that progresses preferentially along interdendritic paths.

Intergranular Corrosion: Preferential corrosion at or along the grain boundaries of a metal. Also called intercrystalline corrosion.

Intergranular Stress Corrosion Cracking (IGSCC): Stress corrosion cracking in which the cracking occurs along grain boundaries.

Internal Oxidation: The formation of isolated particles of corrosion products beneath the metal surface.

Intumescence: The swelling or bubbling of a coating usually because of heating (term currently used in space and fire protection applications).

Ion: An electrically charged atom or group of atoms.

Iron Rot: Deterioration of wood in contact with iron-based alloys.

K_{ISCC}: Abbreviation for the critical value of the plane strain stress intensity factor that will produce crack propagation by stress corrosion cracking of a given material in a given environment.

Knife-Line Attack: Intergranular corrosion of an alloy, usually stabilized stainless steel, along a line adjoining or in contact with a weld after heating into the sensitization temperature range.

Lamellar Corrosion: See Exfoliation Corrosion.

Langelier Index: A calculated saturation index for calcium carbonate that is useful in predicting scaling behavior of natural water.

Liquid Metal Cracking: Cracking of a metal caused by contact with a liquid metal.

Long-Line Current: Current flowing through the earth between an anodic and a cathodic area that returns along an underground metallic structure.

Luggin Probe: A small tube or capillary filled with electrolyte, terminating close to the metal surface under study, and used to provide an ionically conducting path without diffusion between an electrode under study and a reference electrode.

Martensite: Metastable body-centered phase of iron super-saturated with carbon, produced from austenite by shear transformation during quenching or deformation.

Metal Dusting: The catastrophic deterioration of metals in carbonaceous gases at elevated temperatures.

Metallizing: The coating of a surface with a thin metal layer by spraying, hot dipping, or vacuum deposition.

Mill Scale: The heavy oxide layer formed during hot fabrication or heat treatment of metals.

Mixed Potential: A potential resulting from two or more electrochemical reactions occurring simultaneously on one metal surface.

Nernst Equation: An equation that expresses the exact electromotive force of a cell in terms of the activities of products and reactants of the cell.

Nernst Layer and Nernst Thickness: The diffusion layer or the hypothetical thickness of this layer as given by the theory of Nernst. It is defined by:

$$i_d = n \, F \, D \, \frac{C_o - C}{\delta}$$

where, i_d = the diffusion limited current density, D = the diffusion coefficient, C_o = the concentration at the electrode surface, and δ = the Nernst thickness (0.5 mm in many cases of unstirred aqueous electrolytes).

Noble: The positive direction of electrode potential, thus resembling noble metals such as gold and platinum.

Noble Metal: A metal that occurs commonly in nature in the free state. Also a metal or alloy whose corrosion products are formed with a low negative or a positive free energy change.

Noble Potential: A potential more cathodic (positive) than the standard hydrogen potential.

Normalizing: Heating a ferrous alloy to a suitable temperature above the transformation range (austenizing), holding at temperature for a suitable time, and then cooling in still air to a temperature substantially below the transformation range.

Open-Circuit Potential: The potential of an electrode measured with respect to a reference electrode or another electrode when no current flows to or from it. (See Corrosion Potential.)

Organic Zinc Rich Paint: Coating containing zinc powder pigment and an organic resin.

Overvoltage: The change in potential of an electrode from its equilibrium or steady state value when current is applied.

Oxidation: Loss of electrons by a constituent of a chemical reaction. (Also refers to the corrosion of a metal that is exposed to an oxidizing gas at elevated temperatures.)

Oxygen Concentration Cell: See Differential Aeration Cell.

Parting: See Dealloying.

Passivation: A reduction of the anodic reaction rate of an electrode involved in corrosion.

Passive: A metal corroding under the control of a surface reaction product.

Passive-Active Cell: A cell, the electromotive force of which is caused by the potential difference between a metal in an active state and the same metal in a passive state.

Passivity: The state of being passive.

Patina: The coating, usually green, which forms on the surface of metals such as copper and copper alloys exposed to the atmosphere. Also used to describe the appearance of a weathered surface of any metal.

pH: A measure of hydrogen ion activity defined by

$$pH = \log_{10} \frac{1}{a_{H^+}}$$

where a_{H^+} = hydrogen ion activity = the molar concentration of hydrogen ions multiplied by the mean ion activity coefficient.

Pickle: A solution or process used to loosen or remove corrosion products such as scale or tarnish.

Pits, Pitting: Localized corrosion of a metal surface that is confined to a small area and takes the form of cavities.

Pitting Factor: The ratio of the depth of the deepest pit resulting from corrosion divided by the average penetration as calculated from weight loss.

Polarization: The deviation from the open circuit potential of an electrode resulting from the passage of current.

Polarization Admittance: The reciprocal of polarization resistance (di/dE).

Polarization Curve: A plot of current density versus electrode potential for a specific electrode-electrolyte combination.

Polarization Resistance: The slope (dE/di) at the corrosion potential of a potential (E)-current density (i) curve. Also used to describe the method of measuring corrosion rates using this slope.

Polyester: Resin formed by condensation of polybasic and monobasic acids with polyhydric alcohols.

Potential-pH Diagram (also known as Pourbaix Diagram): A graphic method of representing the regions of thermodynamic stability of species for metal-electrolyte systems.

Potentiodynamic (Potentiokinetic): The technique for varying the potential of an electrode in a continuous manner at a preset rate.

Potentiostat: An instrument for automatically maintaining an electrode at a constant potential or controlled potential with respect to a reference electrode.

Potentiostatic: The technique for maintaining a constant electrode potential.

Poultice Corrosion: See Deposit Attack.

Pourbaix Diagram: See Potential-pH Diagram.

Precipitation Hardening: Hardening caused by the precipitation of a constituent from a supersaturated solid solution.

Primer: The first coat of paint applied to a surface. Formulated to have good bonding and wetting characteristics: may or may not contain inhibitive pigments.

Primary Passive Potential (Passivation Potential): The potential corresponding to the maximum active current density (critical anodic current density) of an electrode that exhibites active-passive corrosion behavior.

Profile: Anchor pattern on a surface produced by abrasive blasting or acid treatment.

Redox Potential: The equilibrium electrode potential for a reversible oxidation-reduction reaction in a given electrolyte.

Reduction: Gain of electrons by a constituent of a chemical reaction.

Reference Electrode: A reversible electrode used for measuring the potentials of other electrodes.

Relative Humidity: The ratio, expressed as a percentage, of the amount of water vapor present in a given volume of air at a given temperature to the amount required to saturate the air at that temperature.

Rest Potential: See Corrosion Potential.

Riser: That section of pipeline extending from the ocean floor up the platform. Also the vertical tube in a steam generator convection bank that circulates water and steam upwards.

Rust: Corrosion product consisting primarily of hydrated iron oxide; a term properly applied only to iron and ferrous alloys.

Sacrificial Protection: Reduction of corrosion of a metal in an electrolyte by galvanically coupling it to a more anodic metal; a form of cathodic protection.

Scaling: (1) The formation at high temperatures of thick corrosion product layers on a metal surface. (2) The deposition of water insoluble constituents on a metal surface.

Scanning Electron Microscope (SEM): An electron optical device that images topographical details with maximum contrast and depth of field by the detection, amplification, and display of secondary electrons.

Season Cracking: An obsolete historical term usually applied to stress corrosion cracking of brass.

Sensitizing Heat Treatment: A heat treatment, whether accidental, intentional, or incidental (as during welding), which causes precipitation of constituents at grain boundaries, often causing the alloy to become susceptible to intergranular corrosion or intergranular stress corrosion cracking.

Sigma Phase: An extremely brittle Fe-Cr phase in Fe-Ni-Cr alloys which can form at elevated temperatures.

Slip: A deformation process involving shear motion of a specific set of crystallographic planes.

Slow Strain Rate Technique: An experimental technique for evaluating susceptibility to stress corrosion cracking. It involves pulling the specimen to failure in uniaxial tension at a controlled slow strain rate while the specimen is in the test environment and examining the specimen for evidence of stress corrosion cracking.

Slushing Compound: An obsolete term describing oil or grease coatings used to provide temporary protection against atmospheric corrosion.

Solution Heat Treatment: Heating a metal to a suitable temperature and holding at that temperature long enough for one or more constituents to enter into solid solution, then cooling rapidly enough to retain the constituents in solution.

Spalling: The spontaneous chipping, fragmentation, or separation of a surface or surface coating.

Standard Electrode Potential: The reversible potential for an electrode process when all products and reactions are at unit activity on a scale in which the potential for the standard hydrogen half-cell is zero.

Stray Current: Current flowing through paths other than the intended circuit.

Stray Current Corrosion: Corrosion resulting from direct current flow through paths other than the intended circuit. For example, by any extraneous current in the earth.

Stress Corrosion Cracking (SCC): Cracking of a metal produced by the combined action of corrosion and tensile stress (residual or applied).

Subsurface Corrosion: See Internal Oxidation.

Sulfidation: The reaction of a metal or alloy with a sulfur-containing species to produce a sulfur compound that forms on or beneath the surface of the metal or alloy.

Sulfide Stress Cracking (SSC): Brittle failure by cracking under the combined action of tensile stress and corrosion in the presence of water and hydrogen sulfide.

Tafel Line, Tafel Slope, Tafel Diagram: When an electrode is polarized, it frequently will yield a current/potential relationship over a region that can be approximated by:

$$\eta = \pm\, \beta \log \frac{i}{i_o}$$

where η = change in open circuit potential, i = the current density, β and i_o = constants. The constant (β) is also known as the Tafel slope. If this behavior is observed, a plot on semilogarithmic coordinates is known as the Tafel line and the overall diagram is termed a Tafel diagram.

Tarnish: Surface discoloration of a metal resulting from formation of a thin film of corrosion product.

Thermal Spraying: A group of processes by which finely divided metallic or nonmetallic materials are deposited in a molten or semimolten condition to form a coating. (The coating material may be in the form of powder, ceramic rod, wire, or molten material.)

Thermogalvanic Corrosion: Corrosion resulting from an electrochemical cell caused by a thermal gradient.

Throwing Power: The relationship between the current density at a point on a surface and its distance from the counter electrode. The greater the ratio of the surface resistivity shown by the electrode reaction to the volume resistivity of the electrolyte, the better is the throwing power of the process.

Transpassive: The noble region of potential where an electrode exhibits a higher than passive current density.

Tuberculation: The formation of localized corrosion products scattered over the surface in the form of knoblike mounds called tubercles.

Underfilm Corrosion: Corrosion that occurs under organic films in the form of randomly distributed threadlike filaments or spots. In many cases this is identical to filiform corrosion. (See Filiform Corrosion.)

Voids: A term generally applied to paints to describe holidays, holes, and skips in the film. Also used to describe shrinkage in castings or welds.

Wash Primer: A thin, inhibiting paint, usually chromate pigmented with a polyvinyl butyrate binder.

Weld Decay: Not a preferred term. Intergranular corrosion, usually of stainless steel or certain nickel-base alloys, that occurs as the result of sensitization in the heat affected zone during the welding operation.

Working Electrode: The test or specimen electrode in an electrochemical cell.

Glossary
of
Corrosion-Related Acronyms

ABS	Acrylonitrile-butadiene-styrene plastics
AC	Air cooled
AE	Acoustic emission
AES	Auger electron spectroscopy
ANN	Annealed
AUSS	Austenitic stainless steel
AVT	All volatile treatment for BFW
BFW	Boiler feed water
BWR	Boiling water reactor
CAB	Cellulose acetate-butyrate
CCI	Crevice corrosion index
CCT	Critical crevice corrosion temperature
CD	Current density
CDA	Corrosion data acquisition
CF	Corrosion fatigue
CH	Cold work hardened
CHA	Cold work hardened, aged
CN	Concentric neutral
CP	Cathodic protection
CPP	Critical pitting potential
CPT	Critical pitting temperature
CPVC	Chlorinated poly(vinyl chloride)
CR	Cold rolled
CRA	Corrosion resistant alloy
CS	Carbon steel
CSE	Copper/copper sulfate electrode
CW	Cooling water
DCB	Double cantilever beam test
DIMA	Direct imaging mass analyzer
DSS	Duplex stainless steel
DTA	Differential thermal analysis
DW	Distilled water
EC	Environmental cracking
EDXA	Energy dispersive X-ray analysis
EIS	Electrochemical impedance spectroscopy
ELN	Electrochemical noise technique
EPMA	Electron beam microprobe analysis
EPDM	Ethylene propylene elastomer
EPR	Electrochemical potentiokinetic reactivation
ER	Electrical resistance

ESCA	Electron spectroscopy for chemical analysis
EW	Electric welded
FBC	Fluidized bed combustion
FBE	Fusion-bonded epoxy coating
FC	Furnace cooled
FCGR	Fatigue crack growth rate
FEP	Fluorinated ethylene propylene polymer
FGD	Flue gas desulfurization
FPM	Fluorocarbon elastomers
FRP	Fiber-reinforced plastic
FSS	Ferritic stainless steel
GMAW	Gas metal arc welding
GTAW	Gas tungsten arc welding
HAZ	Heat-affected zone
HB	Brinell hardness number
HE	Hydrogen embrittlement
HIC	Hydrogen induced cracking
HK	Knoop hardness number
HLLW	High level liquid waste (nuclear)
HPW	High purity water
HR	Hot rolled
HRA	Hot rolled, aged
HRB	Rockwell B hardness number
HRC	Rockwell C hardness number
HSC	Hydrogen stress cracking
HSLA	High strength low alloy steel
HTR	Heat treatment
HVN	Vickers hardness number
IGC	Intergranular corrosion
IGSCC	Intergranular stress corrosion cracking
IMMA	Ion microprobe mass analyzer
IOZ	Inorganic zinc coating
ISS	Ion scattering spectroscopy
ISWS	Illinois State Water Survey tester
KIC	Critical stress intensity
LAS	Low alloy steel
LMC	Liquid metal cracking
LSI	Langelier saturation index
MAS	Maraging steels
MCA	Multiple crevice assembly
MIC	Microbial induced corrosion
MSS	Martensitic stainless steel
MT	Magnetic particle inspection
NG	Nuclear grade
NHE	Normal hydrogen electrode
NMR	Nuclear magnetic resonance
NPS	Nominal pipe size
NT	Normalized and tempered
OCTG	Oil country tubular goods

OQ	Oil quenched
OTEC	Ocean thermal energy conversion
OZ	Organic zinc coating
PC	Polycarbonate
PD	Pit depth
PE	Polyethylene
PFA	Perfluoro(alkoxy-alkane) copolymer
PHSS	Precipitation hardenable stainless steel
PPC	Polymer modified Portland cement
PP	Polypropylene
PR	Polarization resistance
PT	Dye penetrant survey
PTA	Polythionic acids
PTFE	Polytetrafluoroethylene
PU	Polyurethane
PVC	Poly(vinyl chloride)
PVDC	Poly(vinylidene chloride)
PVDF	Poly(vinylidene fluoride)
PWHT	Post weld heat treatment
PWR	Pressurized water reactor
QT	Quenched and tempered
RH	Relative humidity
RSI	Ryzner saturation index
RT	X-ray or gamma ray survey
RTP	Reinforced thermoset plastics
RX	Recrystallized
SAM	Scanning Auger microscopy
SAW	Submerged arc welding
SBR	Styrene-butadiene rubber
SCC	Stress corrosion cracking
SCE	Saturated calomel electrode
SEM	Scanning electron microscopy
SIMS	Secondary ion mass spectroscopy
SMAW	Shielded metal arc welding
SMLS	Seamless pipe or tubing
SMYS	Specified minimum yield strength
SRA	Stress relief anneal
SRB	Sulfate-reducing bacteria
SRC	Solvent-refined coal
S/N	Fatigue test
SRE	Scanning reference electrode
SS	Stainless steel
SSC	Sulfide stress cracking
SSMS	Spark sources mass spectroscopy
SSR	Slow strain rate test
SSW	Substitute seawater

STA	Solution treated and aged
STEM	Scanning transmission electron microscopy
STQ	Solution treated and quenched
SW	Seawater
TEM	Transmission electron microscopy
TFE	Tetrafluoroethylene
TS	Tensile strength
TTS	Temperature, time, sensitization diagram
URD	Underground residential distribution systems
UT	Ultrasonic survey
UV	Ultraviolet spectroscopy
VCI	Volatile corrosion inhibitor
WFMT	Wet fluorescent magnetic particle inspection
WQ	Water quenched
WOL	Wedge-opening load test
XPS	X-ray photoelectron spectroscopy
XRD	X-ray diffraction
YS	Yield strength
ZRP	Zinc rich paint

INTERNATIONAL SYSTEM OF UNITS (SI)

The International System of Units (SI for short) is a modernized version of the metric system. It is built upon seven base units and two supplementary units. Derived units are related to base and supplementary units by formulas in the righthand column. Symbols for units with specific names are given in parentheses. This information is adapted from the revised "Metric Practice Guide," ASTM Standard E380. Factors for converting U.S. customary units to SI units are given in the table entitled "General Conversion Factors."

Quantity	Unit	Formula
Base Units		
length	metre (m)	
mass	kilogram (kg)	
time	second (s)	
electric current	ampere (A)	
thermodynamic temperature	kelvin (K)	
amount of substance	mole (mol)	
luminous intensity	candela (cd)	
Supplementary Units		
plane angle	radian (rad)	
solid angle	steradian (sr)	
Derived Units		
acceleration	metre per second squared	m/s^2
activity (of a radioactive source)	disintegration per second	(disintegration)/s
angular acceleration	radian per second squared	rad/s^2
angular velocity	radian per second	rad/s
area	square metre	m^2
density	kilogram per cubic metre	kg/m^3
electric capacitance	farad (F)	A·s/V
electric conductance	stemens (S)	A/V
electric field strength	volt per metre	V/m
electric inductance	henry (H)	V·s/A
electric potential difference	volt (V)	W/A
electric resistance	ohm (Ω)	V/A
electromotive force	volt (V)	W/A
energy	joule (J)	N·m
entropy	joule per kelvin	j/K
force	newton (N)	$kg·m/s^2$
frequency	hertz (Hz)	(cycle)/s
illuminance	lux (lx)	lm/m^2
luminance	candela per square metre	cd/m^2
luminous flux	lumen (lm)	cd·sr
magnetic field strength	ampere per metre	A/m
magnetic flux	weber (Wb)	V·s
magnetic flux density	tesla (T)	Wb/m^2
magnetomotive force	ampere (A)	—
power	watt (W)	J/s
pressure	pascal (Pa)	N/m^2
quantity of electricity	coulomb (C)	A·s
quantity of heat	joule (J)	N·m
radiant intensity	watt per steradian	W/sr
specific heat	joule per kilogram-kelvin	J/kg·K

Continued on Next Page

INTERNATIONAL SYSTEM OF UNITS (SI) (Cont)

Quantity	Unit	Formula
stress	pascal (Pa)	N/m^2
thermal conductivity	watt per metre-kelvin	$W/m \cdot K$
velocity	metre per second	m/s
viscosity, dynamic	pascal-second	Pa·s
viscosity, kinematic	square metre per second	m^2/s
voltage	volt (V)	W/A
volume	cubic metre	m^3
wavenumber	reciprocal metre	(wave)/m
work	joule (J)	N·m

Multiplication Factors	Prefix	SI Symbol
1 000 000 000 000 = 10^{12}	tera	T
1 000 000 000 = 10^9	giga	G
1 000 000 = 10^6	mega	M
1 000 = 10^3	kilo	k
100 = 10^2	hecto*	h
10 = 10^1	deka*	da
0.1 = 10^{-1}	deci*	d
0.01 = 10^{-2}	centi*	c
0.001 = 10^{-3}	milli	m
0.000 001 = 10^{-6}	micro	μ
0.000 000 001 = 10^{-9}	nano	n
0.000 000 000 001 = 10^{-12}	pico	p
0.000 000 000 000 001 = 10^{-15}	femto	f
0.000 000 000 000 000 001 = 10^{-18}	atto	a

*To be avoided where possible

Source: ASM. *Metals Progress Databook*, p. 183. 1974.

GENERAL CONVERSION FACTORS

Unit	Conversion to	Multiply by	Reciprocal
Linear Measure			
mil (0.001 inch)	micrometre	25.4	0.03937
mil (0.001 inch)	millimetre	0.0254	39.37
inch	millimetre	25.4	0.03937
foot	metre	0.3048	3.281
yard	metre	0.9144	1.0936
mile	kilometre	1.6093	0.6214
nautical mile	kilometre	1.8532	0.5396
Square Measure			
square inch	square millimetre	645.2	0.00155
square inch	square centimetre	6.452	0.155
square foot	square metre	0.0929	10.764
square yard	square metre	0.8361	1.196
acre	hectare	0.4047	2.471
acre	square metre	4047.	0.0002471
acre	square foot	43560.	0.00002296
square mile	acre	640.	0.001562
square mile	square kilometre	2.590	0.3863
Volume			
cubic inch	cubic centimetre	16.387	0.06102
cubic foot	cubic metre	0.02832	35.31
cubic foot	gallon (U.S.)	7.48	0.1337
cubic foot	litre	28.32	0.03531
cubic yard	cubic metre	0.7646	1.3079
ounce (U.S., liq.)	cubic centimetre	29.57	0.03382
quart (U.S. liq.)	litre	0.9464	1.0566
gallon (U.S.)	gallon (Imperial)	0.8327	1.2009
gallon (U.S.)	litre	3.785	0.2642
barrel (U.S. Petroleum)	gallon (U.S.)	42.	0.0238
barrel (U.S. Petroleum)	litre	158.98	0.00629
Mass			
grain	milligram	64.8	0.01543
ounce (avoirdupois)	gram	28.35	0.03527
pound (avoirdupois)	kilogram	0.4536	2.205
short ton	metric ton	0.9072	1.1023
long ton	metric ton	1.0161	0.9842
Pressure or Stress			
atmosphere	mm Hg @ 0°C	760.	0.001316
atmosphere	pound force per inch^2	14.696	0.06805
atmosphere	bar	1.013	0.9872
atmosphere	megapascal (MPa)	0.1013	9.872
torr (mm Hg)	pascal	133.32	0.007501
inch of water	pascal	248.8	0.004019
foot of water	pound force per inch^2	0.4335	2.307
dyne per centimetre^2	pascal	0.1000	10.00

Continued on Next Page

GENERAL CONVERSION FACTORS (Cont)

Unit	Conversion to	Multiply by	Reciprocal
Pressure or Stress (Cont)			
pound force per inch^2 (psi)	kilopascal (kPa)	6.895	0.1450
kip per inch^2 (ksi)	megapascal (MPa)	6.895	0.1450
pound force per inch^2	bar	0.06895	14.50
kip per inch^2	kilogram per millimetre^2	0.7031	1.4223
Work, Heat, and Energy			
British thermal unit (Btu)	joule	1055.	0.0009479
foot pound - force	joule	1.356	0.7375
calorie	joule	4.187	0.2389
Btu	foot pound - force	778.	0.001285
kilocalorie	Btu	3.968	0.252
Btu	kilogram metre	107.56	0.009297
Btu per hour	watt	0.2929	3.414
watthour	joule	3600.	0.0002778
horse power	kilowatt	0.7457	1.341
Thermal Properties			
(Btu per foot^2, hour, °F) per inch	(kilocalorie per metre^2, hour, °C) per metre	0.1240	8.064
(Btu per foot^2, hour, °F) per inch	watt per metre, K	0.144	6.944
Btu per foot, hour, °F	kilocalorie per metre^2, hour, °C	4.882	0.2048
Btu per foot^2, hour, °F	watt per metre^2, K	5.674	0.1762
Btu per foot^2	kilocalorie per metre^2	2.712	0.3687
Btu per foot^2	joule per metre^2	11360.	0.00008803
Miscellaneous			
pound per foot^3	kilogram per metre^3	16.02	0.06242
pound per gallon (U.S.)	gram per litre	119.8	0.00835
grains per 100 foot^3	milligram per metre^3	22.88	0.0437
ounces per foot^2	gram per metre^2	305.2	0.003277
pound mole (gas)	cubic foot (STP)	359.	0.00279
gram mole (gas)	litre (STP)	22.4	0.0446
day	minute	1440.	0.000694
week	hour	168.	0.00595
year	hour	8766.	0.0001141
U.S. bag cement	kilogram	42.63	0.02346
gallon (U.S.) per bag cement	litre per kilogram	0.0888	11.26
ksi (inch)$^{1\,2}$	megapascal (metre)$^{1\,2}$	1.0989	0.9100
cubic foot of water (60°F)	pound of water	62.37	0.01603
board foot	cubic metre	0.00236	423.7
milliampere per foot^2	milliampere per metre^2	10.76	0.0929
gallons (U.S.) per minute	metre^3 per day	5.451	0.1835
pound - force	newton	4.448	0.2248

TEMPERATURE CONVERSIONS
Celsius — Fahrenheit

The central figures in red refer to the temperatures either in degrees Celsius or degrees Fahrenheit which require conversion. The corresponding temperatures in degrees Fahrenheit or degrees Celsius will be found to the right or left respectively.

$$°C = 5/9 \, (°F - 32°) \qquad °F = 9/5 \, (°C) + 32°$$

°C	°F
-273	-459
-262	-440
-251	-420
-240	-400
-229	-380
-218	-360
-207	-340
-196	-320
-184	-300
-173	-280

°C	°C or °F	°F
-162	-260	-436
-151	-240	-400
-140	-220	-364
-129	-200	-328
-123	-190	-310
-118	-180	-292
-112	-170	-274
-107	-160	-256
-101	-150	-238
-96	-140	-220
-90	-130	-202
-84	-120	-184
-79	-110	-166
-76	-105	-157
-73.3	-100	-148
-71.0	-95	-139
-67.8	-90	-130
-65.0	-85	-121
-62.2	-80	-112
-59.3	-75	-103
-56.7	-70	-94
-53.9	-65	-85
-51.1	-60	-76
-48.3	-55	-67
-45.5	-50	-58
-42.8	-45	-49
-40.0	-40	-40
-37.2	-35	-31
-34.4	-30	-22
-31.7	-25	-13
-28.9	-20	-4
-26.1	-15	5
-23.3	-10	14
-20.6	-5	23
-17.8	0	32
-17.2	1	34
-16.7	2	36
-16.1	3	37
-15.6	4	39
-15	5	41
-14.4	6	43
-13.9	7	45
-13.3	8	46
-12.9	9	48
-12.2	10	50
-11.7	11	52
-11.1	12	54
-10.6	13	55
-10	14	57
-9.4	15	59
-8.9	16	61
-8.3	17	63
-7.8	18	64
-7.2	19	66
-6.7	20	68
-6.1	21	70
-5.6	22	72
-5	23	73
-4.4	24	75
-3.9	25	77
-3.3	26	79
-2.8	27	81
-2.2	28	82
-1.7	29	84
-1.1	30	86
-0.6	31	88
0	32	90
0.6	33	91
1.1	34	93
1.7	35	95
2.2	36	97
2.8	37	99
3.3	38	100
3.9	39	102
4.4	40	104
5	41	106
5.6	42	108
6.1	43	109
6.7	44	111
7.2	45	113
7.8	46	115
8.3	47	117
8.9	48	118
9.4	49	120
10	50	122
10.6	51	124
11.1	52	126
11.7	53	127
12.2	54	129
12.8	55	131
13.3	56	133
13.9	57	135
14.4	58	136
15	59	138
15.6	60	140
16.1	61	142
16.7	62	144
17.2	63	145
17.8	64	147
18.3	65	149
18.9	66	151
19.4	67	153
20	68	154
20.6	69	156
21.1	70	158
21.7	71	160
22.2	72	162
22.8	73	163
23.3	74	165
23.9	75	167
24.4	76	169
25	77	171
25.6	78	172
26.1	79	174
26.7	80	176
27.2	81	178
27.8	82	180
28.3	83	181
28.9	84	183
29.4	85	185
30	86	187
30.6	87	189
31.1	88	190
31.7	89	192
32.2	90	194
32.8	91	196
33.3	92	198
33.9	93	199
34.4	94	201
35	95	203
35.6	96	205
36.1	97	207
36.7	98	208
37.2	99	210
37.8	100	212
41	105	221
43	110	230
46	115	239
49	120	248
52	125	257
54	130	266
57	135	275
60	140	284
63	145	293
66	150	302
68	155	311
71	160	320
74	165	329
77	170	338
79	175	347
82	180	356
85	185	365
88	190	374
91	195	383
93	200	392
99	210	410
104	220	428
110	230	446
115	240	464
121	250	482
127	260	500
132	270	518
138	280	536
143	290	554
149	300	572
154	310	590
160	320	608
165	330	626
171	340	644
177	350	662
182	360	680
188	370	698
193	380	716
199	390	734
204	400	752
210	410	770
215	420	788
221	430	806
226	440	824
232	450	842
238	460	860
243	470	878
249	480	896
254	490	914
260	500	932
265	510	950

Continued on Next Page

(Cont)

In each row the centre column is the temperature to be converted: read as °F it gives the °C value at left; read as °C it gives the °F value at right.

°C	Temp	°F
271	520	968
276	530	986
282	540	1004
288	550	1022
293	560	1040
299	570	1058
304	580	1076
310	590	1094
315	600	1112
321	610	1130
326	620	1148
332	630	1166
338	640	1184
343	650	1202
349	660	1220
354	670	1238
360	680	1256
365	690	1274
371	700	1292
376	710	1310
382	720	1328
387	730	1346
393	740	1364
399	750	1382
404	760	1400
410	770	1418
415	780	1436
421	790	1454
426	800	1471
432	810	1490
438	820	1508
443	830	1526
449	840	1544
454	850	1562
460	860	1580
465	870	1598
471	880	1616
476	890	1634
482	900	1652
487	910	1670
493	920	1688
498	930	1706
504	940	1724
510	950	1743
515	960	1760
520	970	1778
526	980	1796
532	990	1814
538	1000	1832
543	1010	1850
549	1020	1868
554	1030	1886
560	1040	1904
565	1050	1922
571	1060	1940
576	1070	1958
582	1080	1976
587	1090	1994
593	1100	2012
598	1110	2030
604	1120	2048
610	1130	2066
615	1140	2084
620	1150	2102
626	1160	2120
631	1170	2138
637	1180	2156
642	1190	2174
648	1200	2192
653	1210	2210
660	1220	2228
666	1230	2246
671	1240	2264
677	1250	2282
682	1260	2300
688	1270	2318
693	1280	2336
699	1290	2354
704	1300	2372
710	1310	2390
716	1320	2408
721	1330	2426
727	1340	2444
732	1350	2462
738	1360	2480
743	1370	2498
749	1380	2516
754	1390	2534
760	1400	2552
766	1410	2570
771	1420	2588
777	1430	2606
782	1440	2624
788	1450	2642
793	1460	2660
799	1470	2678
804	1480	2696
810	1490	2714
816	1500	2732
821	1510	2750
827	1520	2768
832	1530	2786
838	1540	2804
843	1550	2822
849	1560	2840
854	1570	2858
860	1580	2876
866	1590	2894
871	1600	2912
877	1610	2930
882	1620	2948
888	1630	2966
893	1640	2984
899	1650	3002
904	1660	3020
910	1670	3038
916	1680	3056
921	1690	3074
927	1700	3092
932	1710	3110
938	1720	3128
943	1730	3146
949	1740	3164
954	1750	3182
960	1760	3200
966	1770	3218
971	1780	3236
977	1790	3254
982	1800	3272
988	1810	3290
993	1820	3308
999	1830	3326
1004	1840	3344
1010	1850	3362
1016	1860	3380
1021	1870	3398
1027	1880	3416
1032	1890	3434
1038	1900	3452
1043	1910	3470
1049	1920	3488
1054	1930	3506
1060	1940	3524
1066	1950	3542
1071	1960	3560
1077	1970	3578
1082	1980	3596
1088	1990	3614
1093	2000	3632
1099	2010	3650
1104	2020	3668
1110	2030	3686
1116	2040	3704
1121	2050	3722
1127	2060	3740
1132	2070	3758
1138	2080	3776
1143	2090	3794
1149	2100	3812
1154	2110	3830
1160	2120	3848
1166	2130	3866
1171	2140	3884
1177	2150	3902
1182	2160	3920
1188	2170	3938
1193	2180	3956
1199	2190	3974
1204	2200	3992
1210	2210	4010
1216	2220	4028
1221	2230	4046
1227	2240	4064
1232	2250	4082
1238	2260	4100
1243	2270	4118
1249	2280	4136
1254	2290	4154
1260	2300	4172
1266	2310	4190
1271	2320	4208
1277	2330	4226
1282	2340	4244
1288	2350	4262
1293	2360	4280
1299	2370	4298
1304	2380	4316
1310	2390	4334
1316	2400	4352
1321	2410	4370
1327	2420	4388
1332	2430	4406
1338	2440	4424
1343	2450	4442
1349	2460	4460
1354	2470	4478
1360	2480	4496
1366	2490	4514
1371	2500	4532
1377	2510	4550
1382	2520	4568
1388	2530	4586
1393	2540	4604
1399	2550	4622
1404	2560	4640
1410	2570	4658
1416	2580	4676
1421	2590	4694
1427	2600	4712
1432	2610	4730
1438	2620	4748
1443	2630	4766
1449	2640	4784
1454	2650	4802
1460	2660	4820
1466	2670	4838
1471	2680	4856
1477	2690	4874
1482	2700	4892
1488	2710	4910
1493	2720	4928
1499	2730	4946
1504	2740	4964
1510	2750	4982
1516	2760	5000
1521	2770	5018
1527	2780	5036
1532	2790	5054
1538	2800	5072
1543	2810	5090
1549	2820	5108
1554	2830	5126
1560	2840	5144
1566	2850	5162
1571	2860	5180
1577	2870	5198
1582	2880	5216
1588	2890	5234
1593	2900	5252
1599	2910	5270
1604	2920	5288
1610	2930	5306
1616	2940	5324
1621	2950	5342
1627	2960	5360
1632	2970	5378
1638	2980	5396
1643	2990	5414
1649	3000	5432

ENGLISH / METRIC (SI) STRESS CONVERSION FACTORS

Look up stress to be converted in red column. If in Ksi (1,000 psi), read kg/mm² and MPa in righthand column. If in kg/mm² and MPa in righthand column. If in kg/mm², read ksi in lefthand column.

Note: 1 MPa (megapascal) = 1 MN per m² (meganewton per square metre).

ksi	kg/mm²	MPa	#	ksi	kg/mm²	MPa	#	ksi	kg/mm²	MPa	#	ksi	kg/mm²	MPa	#
—	—	—	0	35.55	17.58	172.4	25	71.10	35.16	344.7	50	106.65	52.74	517.1	75
1.42	0.70	6.89	1	36.97	18.28	179.3	26	72.52	35.86	351.6	51	108.07	53.45	524.0	76
2.84	1.41	13.79	2	38.39	18.99	186.2	27	73.94	36.57	358.5	52	109.49	54.15	530.9	77
4.27	2.11	20.68	3	39.82	19.69	193.1	28	75.37	37.27	365.4	53	110.92	54.85	537.8	78
5.69	2.81	27.57	4	41.24	20.39	199.9	29	76.79	37.97	372.3	54	112.34	55.56	544.7	79
7.11	3.52	34.47	5	42.66	21.10	206.8	30	78.21	38.68	379.2	55	113.76	56.26	551.6	80
8.53	4.22	41.37	6	44.08	21.80	213.7	31	79.63	39.38	386.1	56	115.18	56.96	558.5	81
9.95	4.92	48.26	7	45.50	22.50	220.6	32	81.05	40.08	393.0	57	116.60	57.67	565.4	82
11.38	5.63	55.16	8	46.93	23.21	227.5	33	82.48	40.79	399.9	58	118.03	58.37	572.3	83
12.80	6.33	62.05	9	48.35	23.91	234.4	34	83.90	41.49	406.8	59	119.45	59.07	579.2	84
14.22	7.03	68.95	10	49.77	24.61	241.3	35	85.32	42.19	413.7	60	120.87	59.77	586.1	85
15.64	7.74	75.84	11	51.19	25.32	248.2	36	86.74	42.90	420.6	61	122.29	60.48	593.0	86
17.06	8.44	82.74	12	52.61	26.02	255.1	37	88.16	43.60	427.5	62	123.71	61.18	599.8	87
18.49	9.14	89.63	13	54.04	26.72	262.0	38	89.59	44.30	434.4	63	125.13	61.88	606.7	88
19.91	9.85	96.53	14	55.46	27.43	268.9	39	91.01	45.01	441.3	64	126.55	62.59	613.6	89
21.33	10.55	103.4	15	56.88	28.13	275.8	40	92.43	45.71	448.2	65	127.98	63.29	620.5	90
22.75	11.25	110.3	16	58.30	28.83	282.7	41	93.85	46.41	455.1	66	129.40	63.99	627.4	91
24.17	11.95	117.2	17	59.72	29.54	289.6	42	95.27	47.12	462.0	67	130.82	64.70	634.3	92
25.60	12.66	124.1	18	61.15	30.24	296.5	43	96.70	47.82	468.8	68	132.25	65.40	641.2	93
27.02	13.36	131.0	19	62.57	30.94	303.4	44	98.12	48.52	475.7	69	133.67	66.10	648.1	94
28.44	14.06	137.9	20	63.99	31.65	310.3	45	99.54	49.23	482.6	70	135.09	66.81	655.0	95
29.86	14.77	144.8	21	65.41	32.35	317.2	46	100.96	49.93	489.5	71	136.51	67.51	661.9	96
31.28	15.47	151.7	22	66.83	33.05	324.1	47	102.38	50.63	496.4	72	137.93	68.21	668.8	97
32.71	16.17	158.6	23	68.26	33.76	331.0	48	103.81	51.34	503.3	73	139.36	68.92	675.7	98
34.13	16.88	165.5	24	69.68	34.46	337.8	49	105.23	52.04	510.2	74	140.78	69.62	682.6	99

ENGLISH / METRIC (SI) STRESS CONVERSION FACTORS (Cont)

ksi	kg/mm²		MPa
142.20	70.32	100	689.5
143.62	71.03	101	696.4
145.04	71.73	102	703.3
146.47	72.43	103	710.2
147.89	73.14	104	717.1
149.31	73.84	105	724.0
150.73	74.54	106	730.8
152.15	75.25	107	737.7
153.58	75.95	108	744.6
155.00	76.65	109	751.5
156.42	77.36	110	758.4
157.84	78.06	111	765.3
159.27	78.76	112	772.2
160.69	79.47	113	779.1
162.11	80.17	114	786.0
163.53	80.87	115	792.9
164.95	81.58	116	799.8
166.38	82.28	117	806.7
167.80	82.98	118	813.6
169.22	83.68	119	820.5
170.64	84.39	120	827.4
172.06	85.09	121	834.3
173.48	85.79	122	841.2
174.91	86.50	123	848.1
176.33	87.20	124	855.0
177.75	87.90	125	861.8
179.17	88.61	126	868.7
180.60	89.31	127	875.6
182.02	90.01	128	882.5
183.44	90.72	129	889.4
184.86	91.42	130	896.3
186.28	92.12	131	903.2
187.70	92.83	132	910.1
189.13	93.53	133	917.0
190.55	94.23	134	923.9
191.97	94.94	135	930.8
193.39	95.64	136	937.7
194.81	96.34	137	944.6
196.24	97.05	138	951.5
197.66	97.75	139	958.4
199.08	98.45	140	965.3
200.50	99.16	141	972.2
201.92	99.86	142	979.1
203.35	100.56	143	986.0
204.77	101.27	144	992.9
206.19	101.97	145	999.7
207.61	102.67	146	1.007
209.03	103.38	147	1.014
210.46	104.08	148	1.020
211.88	104.78	149	1.027
213.30	105.49	150	1.034
214.72	106.19	151	1.041
216.14	106.89	152	1.048
217.57	107.59	153	1.054
218.99	108.30	154	1.062
220.41	109.00	155	1.069
221.83	109.70	156	1.076
223.25	110.41	157	1.082
224.68	111.11	158	1.089
226.10	111.81	159	1.096
227.52	112.52	160	1.103
228.94	113.22	161	1.110
230.36	113.92	162	1.117
231.79	114.63	163	1.124
233.21	115.33	164	1.131
234.63	116.03	165	1.138
236.05	116.74	166	1.145
237.47	117.44	167	1.151
238.90	118.14	168	1.158
240.32	118.85	169	1.165
241.74	119.55	170	1.172
243.16	120.25	171	1.179
244.58	120.96	172	1.186
246.01	121.66	173	1.193
247.43	122.36	174	1.200
248.85	123.07	175	1.207
250.27	123.77	176	1.213
251.69	124.47	177	1.220
253.12	125.18	178	1.227
254.54	125.88	179	1.234
255.96	126.58	180	1.241
257.38	127.29	181	1.248
258.80	127.99	182	1.255
260.23	128.69	183	1.262
261.65	129.39	184	1.269
263.07	130.10	185	1.276
264.49	130.80	186	1.282
265.91	131.50	187	1.289
267.34	132.21	188	1.296
268.76	132.91	189	1.303
270.18	133.61	190	1.310
271.60	134.32	191	1.317
273.02	135.02	192	1.324
274.45	135.72	193	1.331
275.87	136.43	194	1.338
277.29	137.13	195	1.344
278.71	137.83	196	1.351
280.13	138.54	197	1.358
281.56	139.24	198	1.365
282.98	139.94	199	1.372
284.40	140.65	200	1.379

APPROXIMATE EQUIVALENT HARDNESS NUMBERS
AND TENSILE STRENGTHS FOR STEEL

Brinell hardness No. 3000 kg load, 10-mm ball	Vickers hardness No.	Rockwell hardness No. B scale, 100-kg load, 1/16-in. diam. ball	C scale, 150 kg load, Brale Indenter	Knoop hardness No., 500 g load and greater	Shore Sclero-scope hardness No.	Tensile strength (approx.)	
						ksi	MPa
(745)	840	—	65.3	852	91	—	—
(712)	783	—	63.4	808	—	—	—
(682)	737	—	61.7	768	84	—	—
(653)	697	—	60.0	732	81	—	—
627	667	—	58.7	703	79	347	2392
601	640	—	57.3	677	77	328	2261
578	615	—	56.0	652	75	313	2158
555	591	—	54.7	626	73	298	2055
534	569	—	53.5	604	71	288	1986
514	547	—	52.1	579	70	273	1882
—	539	—	51.6	571	—	269	1855
495	528	—	51.0	558	68	263	1818
—	516	—	50.3	545	—	257	1782
477	508	—	49.6	537	66	252	1737
—	495	—	48.8	523	—	244	1682
461	491	—	48.5	518	65	242	1669
—	474	—	47.2	499	—	231	1593
444	472	—	47.1	496	63	229	1579
429	455	—	45.7	476	61	220	1517
415	440	—	44.5	459	59	212	1462
401	425	—	43.1	441	58	202	1393
388	410	—	41.8	423	56	193	1331
375	396	—	40.4	407	54	184	1269
363	383	—	39.1	392	52	177	1220
352	372	—	37.9	379	51	172	1186
341	360	—	36.6	367	50	164	1131
331	350	—	35.5	356	48	159	1096
321	339	—	34.3	345	47	154	1062
311	328	—	33.1	336	46	149	1027
302	319	—	32.1	327	45	146	1007
293	309	—	30.9	318	43	142	979
285	301	—	29.9	310	42	138	952
277	292	—	28.8	302	41	134	924
269	284	—	27.6	294	40	131	903
262	276	—	26.6	286	39	127	876
255	269	—	25.4	279	38	123	848
248	261	—	24.2	272	37	120	827
241	253	100.0	22.8	265	36	116	800
235	247	99.0	21.7	259	35	114	786
229	241	98.2	20.5	253	94	111	765

Continued on Next Page

APPROXIMATE EQUIVALENT HARDNESS NUMBERS
AND TENSILE STRENGTHS FOR STEEL (Cont)

Brinell hardness No. 3000 kg load, 10-mm ball	Vickers hardness No.	Rockwell hardness No.		Knoop hardness No., 500 g load and greater	Shore Sclero-scope hardness No.	Tensile strength (approx.)	
		B scale, 100-kg load, 1/16-in. diam. ball	C scale, 150 kg load, Brale Indenter			ksi	MPa
223	234	97.3	—	247	—	107	738
217	228	96.4	—	242	33	105	724
212	222	95.5	—	237	32	102	703
207	218	94.6	—	232	31	100	690
201	212	93.7	—	227	—	98	676
197	207	92.8	—	222	30	95	655
192	202	91.9	—	217	29	93	641
187	196	90.9	—	212	—	90	621
183	192	90.0	—	207	28	89	614
179	188	89.0	—	202	27	87	600
174	182	88.0	—	198	—	85	586
170	178	87.0	—	194	26	83	572
167	175	86.0	—	190	—	81	559
163	171	85.0	—	186	25	79	545
159	167	83.9	—	182	—	78	538
156	163	82.9	—	178	24	76	524
152	159	81.9	—	174	—	75	517
149	156	80.8	—	170	23	73	503
146	153	79.7	—	166	—	72	496
143	150	78.6	—	163	22	71	490
137	143	76.4	—	157	21	67	462
131	137	74.2	—	151	—	65	448
126	132	72.0	—	145	20	63	434
121	127	69.8	—	140	19	60	414
116	122	67.6	—	135	18	58	400
111	117	65.4	—	131	17	56	386

Source: Metals Handbook, Desk Edition, p. 1-61, ASM 1985.

COMMON GAGE SERIES USED FOR SHEET THICKNESS

Name	Acronym	Identical With
American Wire Gage	AWG	B&S
Birmingham Wire Gage	BWG	Stubs Iron Wire Gage[1]
Brown and Sharp	B&S	AWG
Galvanized Iron	GSG	
Standard Wire Gage (British)	SWG	Imperial St., British Std.
Manufacturer's Standard (U.S.)	MSG	
U.S. Standard Plate	USG	
Zinc (American Zinc Gage)	AZG	

[1]But Not Stubs Steel Wire Gage.

SHEET GAGE—THICKNESS CONVERSIONS (INCHES)

Gage No.	Al (U.S.) Copper Brass B&S AWG	Galv. Iron GSG	Al (U.K.) SWG	Sheet MSG	Stainless Steel Strip USG	Zinc BWG	AZG
1	0.289		0.300		0.281		
2	0.258		0.276		0.266		
3	0.229		0.252	0.239	0.250		0.006
4	0.204		0.232	0.224	0.234		0.008
5	0.182		0.212	0.209	0.219		0.010
6	0.162		0.192	0.194	0.203		0.012
7	0.144		0.176	0.179	0.188	0.180	0.014
8	0.128	0.168	0.160	0.164	0.172	0.165	0.016
9	0.114	0.153	0.144	0.149	0.156	0.148	0.018
10	0.102	0.138	0.128	0.134	0.141	0.134	0.020
11	0.091	0.125	0.116	0.120	0.125	0.120	0.024
12	0.081	0.110	0.104	0.105	0.109	0.109	0.028
13	0.072	0.095	0.092	0.090	0.094	0.095	0.032
14	0.064	0.080	0.080	0.075	0.078	0.083	0.036
15	0.057	0.071	0.072	0.067	0.070	0.072	0.040
16	0.051	0.064	0.064	0.060	0.062	0.065	0.045
17	0.045	0.058	0.056	0.054	0.056	0.058	0.050
18	0.040	0.052	0.048	0.048	0.050	0.049	0.055
19	0.036	0.046	0.040	0.042	0.044	0.042	0.060
20	0.032	0.040	0.036	0.036	0.038	0.035	0.070
21	0.028	0.037	0.032	0.033	0.034	0.032	0.080
22	0.025	0.034	0.028	0.030	0.031	0.028	0.090
23	0.023	0.031	0.024	0.027	0.028	0.025	0.100
24	0.020	0.028	0.022	0.024	0.025	0.022	0.125
25	0.018	0.025	0.020	0.021	0.022	0.020	0.250

Source: *Materials Performance*, Vol. 14, No. 12, p. 75 (12/75).

SHEET GAGE—THICKNESS CONVERSIONS (mm)

Gage No.	Al (U.S.) Copper Brass B&S AWG	Galv. Iron GSG	Al (U.K.) SWG	Steel MSG	Stainless Steel		Zinc AZG
					Sheet USG	Strip BWG	
1	7.34		7.62		7.14		
2	6.55		7.71		6.75		
3	5.82		6.40	6.07	6.35		0.15
4	5.18		5.89	5.69	5.95		0.20
5	4.62		5.38	5.31	5.56		0.25
6	4.11		4.88	4.93	5.16		0.30
7	3.66		4.47	4.55	4.76	4.57	0.36
8	3.25	4.27	4.06	4.17	4.37	4.19	0.41
9	2.90	3.89	3.66	3.78	3.97	3.76	0.46
10	2.59	3.50	3.25	3.40	3.57	3.40	0.51
11	2.31	3.18	2.95	3.05	3.18	3.05	0.61
12	2.06	2.79	2.64	2.67	2.78	2.77	0.71
13	1.83	2.41	2.34	2.29	2.38	2.41	0.81
14	1.63	2.03	2.03	1.90	1.98	2.11	0.91
15	1.45	1.80	1.83	1.70	1.79	1.83	1.02
16	1.30	1.63	1.63	1.52	1.59	1.65	1.14
17	1.14	1.47	1.42	1.37	1.42	1.47	1.27
18	1.02	1.32	1.22	1.22	1.27	1.24	1.40
19	0.91	1.17	1.02	1.07	1.11	1.07	1.52
20	0.81	1.02	0.91	0.91	0.95	0.89	1.78
21	0.71	0.94	0.81	0.84	0.87	0.81	2.03
22	0.64	0.86	0.71	0.76	0.79	0.71	2.29
23	0.58	0.79	0.61	0.69	0.71	0.64	2.54
24	0.51	0.71	0.56	0.61	0.64	0.56	3.18
25	0.46	0.64	0.51	0.53	0.56	0.51	6.35

Source: *Materials Performance*, Vol. 14, No. 12, p. 75 (12/75).

METRIC AND DECIMAL EQUIVALENTS
OF FRACTIONS OF AN INCH

inches	mm	inches	mm
1/64	0.015	0.3968	33/64	0.516	13.0966
1/32	0.031	0.7937	17/32	0.531	13.4934
3/64	0.047	1.1906	35/64	0.547	13.8903
1/16	0.063	1.5876	9/16	0.563	14.2872
5/64	0.078	1.9843	37/64	0.578	14.6841
3/32	0.094	2.3812	19/32	0.594	15.0809
7/64	0.109	2.7780	39/64	0.609	15.4778
1/8	0.125	3.1749	5/8	0.625	15.8747
9/64	0.141	3.5718	41/64	0.641	16.2715
5/32	0.156	3.9686	21/32	0.656	16.6684
11/64	0.172	4.3655	43/64	0.672	17.0653
3/16	0.188	4.7624	11/16	0.688	17.4621
13/64	0.203	5.1592	45/64	0.703	17.8590
7/32	0.219	5.5561	23/32	0.719	18.2559
15/64	0.234	5.9530	47/64	0.734	18.6527
1/4	0.250	6.3498	3/4	0.750	19.0496
17/64	0.266	6.7467	49/64	0.766	19.4465
9/32	0.281	7.1436	25/32	0.781	19.8433
19/64	0.297	7.5404	51/64	0.797	20.2402
5/16	0.313	7.9373	13/16	0.813	20.6371
21/64	0.328	8.3342	53/64	0.828	21.0339
11/32	0.344	8.7310	27/32	0.844	21.4308
13/64	0.359	9.1279	55/64	0.859	21.8277
3/8	0.375	9.5248	7/8	0.875	22.2245
25/64	0.391	9.9216	57/64	0.891	22.6214
13/32	0.406	10.3185	29/32	0.906	23.0183
27/64	0.422	10.7154	59/64	0.922	23.4151
7/16	0.438	11.1122	15/16	0.938	23.8120
29/64	0.453	11.5091	61/64	0.953	24.2089
15/32	0.469	11.9060	31/32	0.969	24.6057
31/64	0.484	12.3029	63/64	0.984	25.0026
1/2	0.500	12.6997	—	1.000	25.3995

PHYSICAL PROPERTIES OF GASES AND LIQUIDS

Name	Formula	Molecular Weight	Density g/L	Melting Point °C	Boiling Point °C	Auto-Ignition Point °C	Explosive limits percent by vol. in air	
							Lower	Upper
Acetylene	C_2H_2	26.04	1.173	−81	−83.6 subl.	335	2.5	80.0
Air			1.2929					
Ammonia	NH_3	17.03	0.7710	−77.7	−33.4	780	16.0	27.0
Argon	A	39.94	1.784	−189.2	−185.7			
Butane-n	C_4H_{10}	58.12	2.703	−138	−0.6	430	1.6	8.5
Butane-i	C_4H_{10}	58.12	2.637	−159	−11.7			
Butylene-n	C_4H_8	56.10	2.591	−185	−6.3		1.7	9.0
Carbon dioxide	CO_2	44.01	1.977	−57 5 atm.	−78.5 subl.			
Carbon monoxide	CO	28.01	1.250	−207	−191	650	12.5	74.2
Chlorine	Cl_2	70.91	3.214	−101	−34			
Ethane	C_2H_6	30.07	1.356	−172	−88.6	510	3.1	15.0
Ethylene	C_2H_4	28.05	1.261	−169	−103.7	543	3.0	34.0
Helium	He	4.003	0.1785	−272	−268.9			
Heptane-n	C_7H_{16}	100.20	0.684 g/cm³	−90.6	98.4	233	1.0	6.0
Hexane-n	C_6H_{14}	86.17	0.6594 g/cm³	−95.3	68.7	248	1.2	6.9
Hydrogen	H_2	2.016	0.0899	−259.2	−252.8	580	4.1	74.2
Hydrogen chloride	HCl	36.47	1.639	−112	−84			
Hydrogen fluoride	HF	20.01	0.921	−92.3	19.5			
Hydrogen sulfide	H_2S	34.08	1.539	−84	−62		4.3	45.5
Methane	CH_4	16.04	0.7168	−182.5	−161.5	538	5.3	13.9
Nitrogen	N_2	28.016	1.2506	−209.9	−195.8			
Octane-n	C_8H_{18}	114.23	0.7025 g/cm³	−56.8	125.7	232	0.8	3.2
Oxygen	O_2	32.00	1.4290	−218.4	−183.0			
Pentane-n	C_5H_{12}	72.15	0.026 g/cm³	−131	36.2	310	1.4	8.0
Propane	C_3H_8	44.09	2.020	−189	−44.5	465	2.4	9.5
Propylene	C_3H_6	42.05	1.915	−184	−48	458	2.0	11.1
Sulfur dioxide	SO_2	64.06	2.926	−75.7	−10.0			

Density of gases in g/L at 0°C and 760 mm Hg.
Density of liquids in g/cm³ at 20°/4°C.

PHYSICAL PROPERTIES OF ELEMENTS

	Symbol	Atomic Weight	Density g/cm^2 20°C	Valencies	Melting Point °C	Crystal Structure ***
Aluminum	Al	26.98	2.70	3	660	1
Antimony	Sb	121.75	6.68	3/5	630	5
Argon	A	39.948	1.784*	0	−189.2	1
Arsenic	As	74.92	5.73	3/5	814	5
Barium	Ba	137.34	3.5	2	725	2
Beryllium	Be	9.01	1.85	2	1280	3
Bismuth	Bi	208.98	9.80	3/5	271	5
Boron	B	10.81	2.3	3	2300	—
Bromine	Br	79.91	3.12	1/3/5/7	−7.2	6
Cadmium	Cd	112.40	8.65	2	321	3
Calcium	Ca	40.08	1.55	2	842	1
Carbon	C	12.01	2.25	2/3/4	3550	4
Chlorine	Cl	35.45	1.56**	1/3/5/7	−103	7
Chromium	Cr	52.00	7.2	2/3/6	1890	2
Cobalt	Co	58.93	8.9	2/3	1495	3
Copper	Cu	63.54	8.92	1/2	1083	1
Fluorine	F	19.00	1.69*	1	223	—
Cold	Au	196.97	19.32	1/3	1063	1
Helium	He	4.003	0.177*	0	−272.2	—
Hydrogen	H	1.008	0.090*	1	−259.2	4
Iodine	I	126.90	4.93	1/3/5/7	113.5	6
Iron	Fe	55.85	7.87	2/3/6	1535	2
Lead	Pb	207.19	11.35	2/4	327.4	1
Lithium	Li	6.94	0.53	1	186	2
Magnesium	Mg	24.31	1.74	2	651	3
Manganese	Mn	54.94	7.2	2/3/4/6/7	1260	10
Mercury	Hg	200.59	13.55	1/2	−38.9	5
Molybdenum	Mo	95.94	10.2	2/3/4/5/6	2620	2
Nickel	Ni	58.71	8.90	2/3	1455	1
Niobium	Nb	92.91	8.55	3/5	2500	2
Nitrogen	N	14.007	1.25*	3/5	−209.9	4
Oxygen	O	15.9994	1.429*	2	−218.4	10
Phosphorus	P	30.98	1.82	3/5	44.1	10
Platinum	Pt	195.09	21.37	2/4	1773	1
Potassium	K	39.10	0.87	1	62.3	2
Rhodium	Rh	102.91	12.5	1/2/3/4	1966	1
Selenium	Se	78.96	4.8	2/4/6	220	4
Silicon	Si	28.09	2.42	4	1420	8
Silver	Ag	107.87	10.50	1	960.5	1
Sodium	Na	22.99	0.97	1	97.5	2

Continued on Next Page

PHYSICAL PROPERTIES OF ELEMENTS (Cont)

	Symbol	Atomic Weight	Density g/cm^2 20°C	Valencies	Melting Point °C	Crystal Structure ***
Sulfur	S	32.06	2.07	2/4/6	119	9
Tantalum	Ta	180.95	16.6	3/5	2996	2
Tin	Sn	118.69	7.31	2/4	231.9	7
Titanium	Ti	47.90	4.5	2/3/4	1800	3
Tungsten	W	183.85	19.3	2/4/5/6	3370	2
Vandium	V	50.94	5.96	2/3/4/5	1710	2
Zinc	Zn	65.73	7.14	2	419.5	3
Zirconium	Zr	91.22	6.4	4	1857	3

*kg/Nm^3
**Liquid at boiling point −37°C
***at 20°C

Crystal structures:
1 Face-centered cubic
2 Body-centered cubic
3 Close packed hexagonal
4 Hexagonal
5 Rhombohedral

6 Orthorhombic
7 Tetragonal
8 Diamond cubic
9 Face-centered orthorhombic
10 Cubic (complex)

PROPERTIES OF DRY SATURATED STEAM
(English Units)

Temp. F	Pressure psia	Pressure psig	Specific Volume ft³/lb.	Specific Enthalpy Btu/lb.
32	0.08859	—	3305	1075.5
35	0.09991	—	2948	1076.8
40	0.12163	—	2446	1079.0
45	0.14744	—	2037.8	1081.2
50	0.17796	—	1704.8	1083.4
60	0.2561	—	1207.6	1087.7
70	0.3629	—	868.4	1092.1
80	0.5068	—	633.3	1096.4
90	0.6981	—	468.1	1100.8
100	0.9492	—	350.4	1105.1
110	1.2750	—	265.4	1109.3
120	1.6927	—	203.26	1113.6
130	2.2230	—	157.33	1117.8
140	2.8892	—	123.00	1122.0
150	3.718	—	97.07	1126.1
160	4.741	—	77.29	1130.2
170	5.993	—	62.06	1134.2
180	7.511	—	50.22	1138.2
190	9.340	—	40.96	1142.1
200	11.526	—	33.64	1146.0
210	14.123	—	27.82	1149.7
212	14.696	0.000	26.80	1150.5
220	17.186	2.490	23.15	1153.4
230	20.779	6.083	19.381	1157.1
240	24.968	10.272	16.321	1160.6
250	29.825	15.129	13.819	1164.0
260	35.427	20.731	11.762	1167.4
270	41.856	27.160	10.060	1170.6
280	49.200	34.524	8.644	1173.8
290	57.550	42.854	7.460	1176.8
300	67.005	52.309	6.466	1179.7
310	77.67	62.97	5.626	1182.5
320	89.64	74.94	4.914	1185.2
340	117.99	103.29	3.788	1190.1
360	153.01	138.31	2.957	1194.4
380	195.73	181.03	2.335	1198.0
400	247.26	232.56	1.8630	1201.0
420	308.78	274.08	1.4997	1203.1
440	381.54	366.84	1.2169	1204.4
460	466.9	452.2	0.9942	1204.8
480	566.2	551.5	0.8172	1204.1

Continued on Next Page

PROPERTIES OF DRY SATURATED STEAM (Cont)
(English Units)

Temp. F	Pressure psia	Pressure psig	Specific Volume ft^3/lb.	Specific Enthalpy Btu/lb.
500	680.9	662.2	0.6749	1202.2
520	812.5	797.8	0.5596	1199.0
540	962.8	948.1	0.4651	1194.3
560	1133.4	1118.1	0.3871	1187.7
580	1326.2	1311.5	0.3222	1179.0
600	1543.2	1528.5	0.2675	1167.7
620	1786.9	1772.2	0.2208	1153.2
640	2059.9	2045.2	0.1802	1133.7
660	2365.7	2351.0	0.1443	1107.0
680	2708.6	2693.9	0.1112	1068.5
700	3094.3	3079.6	0.0752	995.2
705.5	3208.2	3193.5	0.0508	906.0

Source: Babcock & Wilcox, *Steam*, p. 2-3.

PROPERTIES OF DRY SATURATED STEAM (SI UNITS)

Abs. Pressure bar	Temp. °C	Specific Volume dm³/kg	Specific Enthalpy kJ/kg
0.01	7.0	129,209	2514
0.025	21.1	54,256	2540
0.05	32.9	28,194	2562
0.075	40.3	19,239	2575
0.10	45.8	14,675	2585
0.15	54.0	10,023	2599
0.20	60.1	7,650	2610
0.25	65.0	6,205	2618
0.30	69.1	5,229	2625
0.40	75.9	3,993	2637
0.50	81.3	3,240	2646
0.75	91.8	2,217	2663
1.0	99.6	1,694	2675
1.5	111.4	1,159	2693
2.0	120.2	885	2706
2.5	127.4	718	2716
3.0	133.5	606	2725
3.5	138.9	524	2732
4.0	143.6	462	2738
5.0	151.8	375	2748
6.0	158.8	315	2756
7.0	164.9	273	2762
8.0	170.4	240	2768
9.0	175.4	215	2772
10.0	179.9	194.3	2776
12.5	189.8	156.9	2784
15.0	198.3	131.7	2790
17.5	205.7	113.4	2794
20.0	212.4	99.5	2797
22.5	218.4	88.7	2799
25.0	223.9	79.9	2801
27.5	229.0	72.7	2802
30.0	233.8	66.6	2802
32.5	238.3	61.5	2802
35.0	242.5	57.0	2802
37.5	246.5	53.2	2801
40.0	250.3	49.7	2800
45.0	257.4	44.0	2798
50.0	263.9	39.4	2794
55.0	269.9	35.6	2790

Continued on Next Page

PROPERTIES OF DRY SATURATED STEAM (SI UNITS) (Cont)

Abs. Pressure bar	Temp. °C	Specific Volume dm³/kg	Specific Enthalpy kJ/kg
60.0	275.6	32.4	2785
65.0	280.8	29.7	2780
70.0	285.8	27.4	2774
75.0	290.5	25.3	2767
80.0	295.0	23.5	2760
85.0	299.2	21.9	2753
90.0	303.2	20.5	2745
95.0	307.2	19.21	2736
100.0	311.0	18.04	2728
110.0	318.0	16.01	2709
120.0	324.6	14.28	2689
130.0	330.8	12.80	2667
140.0	336.6	11.50	2642
150.0	342.1	10.34	2615
160.0	347.3	9.31	2585
170.0	352.3	8.37	2552
180.0	357.0	7.50	2514
190.0	361.4	6.68	2471
200.0	365.7	5.88	2418
210.0	369.8	5.02	2348
220.0	373.7	3.73	2196
221.2	374.2	3.17	2107

Source: A. Parrish, *Mechanical Engineers Handbook*, pp. 2-86, 2-93. Reproduced with permission of ASME.

VAPOR PRESSURE OF WATER BELOW 100°C

Temperature °C	°F	Pressure mm Hg	Millibar	Temperature °C	°F	Pressure mm Hg	Millibar
−15	5	1.4	1.9	38	100	49.7	66.3
−14	7	1.6	2.1	39	102	52.4	69.8
−13	9	1.7	2.3	40	104	55.3	73.7
−12	10	1.8	2.4	41	106	58.3	77.7
−11	12	2.0	2.7	42	108	61.5	81.2
−10	14	2.1	2.8	43	109	64.8	86.4
−9	16	2.3	3.1	44	111	68.3	91.0
−8	18	2.5	3.3	45	113	71.9	95.8
−7	19	2.7	3.6	46	115	75.7	101.
−6	21	2.9	3.9	47	117	79.6	106.
−5	23	3.2	4.3	48	118	83.7	112.
−4	25	3.4	4.5	49	120	88.0	117.
−3	27	3.7	4.9	50	122	92.5	123.
−2	28	4.0	5.3	51	124	97.2	130.
−1	30	4.3	5.7	52	126	102.	136.
0	32	4.6	6.1	53	127	107.	143.
1	34	4.9	6.5	54	129	113.	151.
2	36	5.3	7.1	55	131	118.	157.
3	37	5.7	7.6	56	133	124.	165.
4	39	6.1	8.1	57	135	130.	173.
5	41	6.5	8.7	58	136	136.	181.
6.	43	7.0	9.3	59	138	143.	191.
7	45	7.5	10.0	60	140	149.	199
8	46	8.0	10.7	61	142	156.	208.
9	48	8.6	11.5	62	144	164.	219.
10	50	9.2	12.3	63	145	171.	228.
11	52	9.8	13.1	64	147	179.	239.
12	54	10.5	14.0	65	149	187.	249.
13	55	11.2	14.9	66	151	196.	261.
14	57	12.0	16.0	67	153	205.	273.
15	59	12.8	17.1	68	154	214.	285.
16	61	13.6	18.1	69	156	224.	299.
17	63	14.5	19.3	70	158	234.	312.
18	64	15.5	20.7	71	160	244.	325.
19	66	16.5	22.0	72	162	255.	340.
20	68	17.5	23.3	73	163	266.	355.
21	70	18.6	24.8	74	165	277.	369.
22	72	19.8	26.4	75	167	289.	385.
23	73	21.1	28.1	76	169	301.	401.
24	75	22.4	29.9	77	171	314.	419.
25	77	23.8	31.7	78	172	327.	436.
26	79	25.2	33.6	79	174	341.	455.
27	81	26.7	35.6	80	176	355.	473.
28	82	28.3	37.7	81	178	370.	493.
29	84	30.0	40.0	82	180	385.	513.
30	86	31.8	42.4	83	181	401.	535.
31	88	33.7	44.9	84	183	417.	556.
32	90	35.7	47.6	85	185	434.	579.
33	91	37.7	50.3	86	187	451.	601.
34	93	39.9	53.2	87	189	469.	625.
35	95	42.2	56.3	88	190	487.	649.
36	97	44.6	59.5	89	192	506.	674.
37	99	47.1	62.8	90	194	526.	701.

Continued on Next Page

VAPOR PRESSURE OF WATER BELOW 100°C (Cont)

Temperature		Pressure		Temperature		Pressure	
°C	°F	mm Hg	Millibar	°C	°F	mm Hg	Millibar
91	196	546.	728.	96	205	658.	877.
92	198	567.	756.	97	207	682.	909.
93	199	589.	785.	98	208	707.	942.
94	201	617.	822.	99	210	733.	977.
95	203	634.	845.	100	212	760.	1013.

Source: Condensed from *CRC Handbook*, 55th Edition, p. D-159.

DEW POINT OF MOIST AIR

The temperature drop required for condensation to occur at a specified air temperature and relative humidity is given in the table below. The temperature drops are mean values for the indicated air temperature ranges.

RH %	Air Temperature °C		RH %	Air Temperature °F	
	0-20	20-35		32-68	68-95
55	9	10	55	16	18
60	7	9	60	13	15
65	6	7	65	11	13
70	5	6	70	9	11
75	4	5	75	8	9
80	3	4	80	6	7
85	2	3	85	4	5
90	1.6	1.8	90	3	3
92	1.2	1.4	92	2.2	2.5
95	0.8	0.9	95	1.4	1.6
98	0.3	0.3	98	0.5	0.5

Example: At 30°C (86°F) and 80% RH, a temperature drop of 4°C (7°F) would result in condensation.

Dew point temperatures of moist air as a function of air temperature and relative humidity are tabulated on the following four pages.

DEW POINT OF MOIST AIR °C

Air Temperature °C

RH %	0	2	4	6	8	10	12	14	16	18	20	22	24
1	−50	−49	−47	−46	−45	−44	−42	−41	−40	−39	−38	−36	−35
3	−40	−39	−37	−36	−34	−33	−32	−30	−29	−28	−26	−25	−24
5	−35	−34	−32	−31	−29	−28	−26	−25	−24	−22	−21	−19	−18
7	−32	−30	−29	−27	−26	−24	−23	−21	−20	−18	−17	−15	−14
9	−29	−27	−26	−24	−23	−21	−20	−18	−17	−15	−14	−12	−11
11	−27	−25	−24	−22	−21	−19	−17	−16	−14	−13	−11	−10	−8
13	−25	−23	−22	−20	−19	−17	−15	−14	−12	−11	−9	−8	−6
15	−23	−22	−20	−19	−17	−15	−14	−12	−11	−9	−7	−6	−4
17	−22	−20	−19	−17	−15	−14	−12	−11	−9	−7	−6	−4	−2
19	−21	−19	−17	−16	−14	−12	−11	−9	−8	−6	−4	−3	−1
21	−20	−18	−16	−15	−13	−11	−10	−8	−6	−5	−3	−1	0
23	−19	−17	−15	−14	−12	−10	−8	−7	−5	−3	−2	0	2
25	−18	−16	−14	−13	−11	−9	−7	−6	−4	−2	−1	1	3
27	−17	−15	−13	−12	−10	−8	−6	−5	−3	−1	1	2	4
29	−16	−14	−12	−11	−9	−7	−5	−4	−2	0	2	3	5
31	−15	−13	−12	−10	−8	−6	−5	−3	−1	1	2	4	6
33	−14	−13	−11	−9	−7	−6	−4	−2	0	2	3	5	7
35	−14	−12	−10	−8	−7	−5	−3	−1	1	2	4	6	8
37	−13	−11	−9	−8	−6	−4	−2	0	1	3	5	7	9
39	−12	−10	−9	−7	−5	−3	−1	0	2	4	6	8	9
41	−12	−10	−8	−6	−4	−3	−1	1	3	5	6	8	10
43	−11	−9	−7	−6	−4	−2	0	2	4	5	7	9	11
45	−11	−9	−7	−5	−3	−1	1	2	4	6	8	10	11
47	−10	−8	−6	−5	−3	−1	1	3	5	7	8	10	12
49	−9	−8	−6	−4	−2	0	2	4	5	7	9	11	13
51	−9	−7	−5	−3	−1	0	2	4	6	8	10	12	13
53	−8	−7	−5	−3	−1	1	3	5	7	8	10	12	14
55	−8	−6	−4	−2	−1	1	3	5	7	9	11	13	14
57	−7	−6	−4	−2	0	2	4	6	8	9	11	13	15
59	−7	−5	−3	−1	1	2	4	6	8	10	12	14	16
61	−7	−5	−3	−1	1	3	5	7	9	11	12	14	16
63	−6	−4	−2	0	2	3	5	7	9	11	13	15	17
65	−6	−4	−2	0	2	4	6	8	10	11	13	15	17
67	−5	−3	−2	0	2	4	6	8	10	12	14	16	18
69	−5	−3	−1	1	3	5	7	8	10	12	14	16	18
71	−5	−3	−1	1	3	5	7	9	11	13	15	17	19
73	−4	−2	0	2	4	6	7	9	11	13	15	17	19
75	−4	−2	0	2	4	6	8	10	12	14	15	17	19
77	−4	−2	0	2	4	6	8	10	12	14	16	18	20
79	−3	−1	1	3	5	7	8	10	12	14	16	18	20
81	−3	−1	1	3	5	7	9	11	13	15	17	19	21
83	−3	−1	1	3	5	7	9	11	13	15	17	19	21
85	−2	0	2	4	6	8	10	12	14	16	18	19	21
87	−2	0	2	4	6	8	10	12	14	16	18	20	22
89	−2	0	2	4	6	8	10	12	14	16	18	20	22
91	−1	1	3	5	7	9	11	13	15	17	19	21	23
93	−1	1	3	5	7	9	11	13	15	17	19	21	23
95	−1	1	3	5	7	9	11	13	15	17	19	21	23
97	0	2	4	6	8	10	12	14	16	18	20	22	24
99	0	2	4	6	8	10	12	14	16	18	20	22	24

Continued on Next Page

DEW POINT OF MOIST AIR °C (Cont)

Air Temperature °C

RH %	26	28	30	32	34	36	38	40	42	44	46	48	50
1	−34	−33	−32	−30	−29	−28	−27	−26	−25	−24	−22	−21	−20
3	−22	−21	−20	−18	−17	−16	−14	−13	−12	−11	−9	−8	−7
5	−16	−15	−14	−12	−11	−10	−8	−7	−5	−4	−3	−1	0
7	−12	−11	−9	−8	−7	−5	−4	−2	−1	1	2	4	5
9	−9	−8	−6	−5	−3	−2	0	1	3	4	6	7	9
11	−7	−5	−4	−2	−1	1	3	4	6	7	9	10	12
13	−4	−3	−1	0	2	3	5	6	8	10	11	13	14
15	−3	−1	1	2	4	5	7	9	10	12	13	15	16
17	−1	1	2	4	6	7	9	10	12	14	15	17	18
19	1	2	4	6	7	9	10	12	14	15	17	19	20
21	2	4	5	7	9	10	12	14	15	17	19	20	22
23	3	5	7	8	10	12	13	15	17	18	20	22	23
25	5	6	8	10	11	13	15	16	18	20	21	23	25
27	6	7	9	11	12	14	16	18	19	21	23	24	26
29	7	8	10	12	14	15	17	19	20	22	24	25	27
31	8	9	11	13	15	16	18	20	21	23	25	27	28
33	9	10	12	14	16	17	19	21	22	24	26	28	29
35	9	11	13	15	16	18	20	22	23	25	27	29	30
37	10	12	14	16	17	19	21	23	24	26	28	30	31
39	11	13	15	16	18	20	22	24	25	27	29	31	32
41	12	14	15	17	19	21	23	24	26	28	30	31	33
43	13	14	16	18	20	22	23	25	27	29	31	32	34
45	13	15	17	19	20	22	24	26	28	29	31	33	35
47	14	16	17	19	21	23	25	27	28	30	32	34	36
49	15	16	18	20	22	24	26	27	29	31	33	35	36
51	15	17	19	21	23	24	26	28	30	32	34	35	37
53	16	18	20	21	23	25	27	29	31	32	34	36	38
55	16	18	20	22	24	26	27	29	31	33	35	37	38
57	17	19	21	23	24	26	28	30	32	34	36	37	39
59	17	19	21	23	25	27	29	30	32	34	36	38	40
61	18	20	22	24	26	27	29	31	33	35	37	39	41
63	19	20	22	24	26	28	30	32	34	35	37	39	41
65	19	21	23	25	27	28	30	32	34	36	38	40	42
67	19	21	23	25	27	29	31	33	35	37	38	40	42
69	20	22	24	26	28	30	31	33	35	37	39	41	43
71	20	22	24	26	28	30	32	34	36	38	40	42	43
73	21	23	25	27	29	31	32	34	36	38	40	42	44
75	21	23	25	27	29	31	33	35	37	39	41	42	44
77	22	24	26	28	29	31	33	35	37	39	41	43	45
79	22	24	26	28	30	32	34	36	38	40	41	43	45
81	23	25	26	28	30	32	34	36	38	40	42	44	46
83	23	25	27	29	31	33	35	37	39	40	42	44	46
85	23	25	27	29	31	33	35	37	39	41	43	45	47
87	24	26	28	30	32	34	36	37	39	41	43	45	47
89	24	26	28	30	32	34	36	38	40	42	44	46	48
91	25	27	29	31	32	34	36	38	40	42	44	46	48
93	25	27	29	31	33	35	37	39	41	43	45	47	49
95	25	27	29	31	33	35	37	39	41	43	45	47	49
97	26	28	30	32	34	36	38	40	42	44	46	48	50
99	26	28	30	32	34	36	38	40	42	44	46	48	50

DEW POINT OF MOIST AIR °F

Air Temperature °F

RH %	32	35	38	41	44	47	50	53	56	59	62	65	68	71	74
1	−57	−56	−54	−52	−50	−48	−46	−45	−43	−41	−39	−37	−36	−34	−32
3	−40	−38	−36	−34	−32	−30	−28	−25	−23	−21	−19	−17	−15	−13	−11
5	−31	−29	−27	−25	−22	−20	−18	−16	−14	−12	−9	−7	−5	−3	−1
7	−25	−23	−20	−18	−16	−14	−11	−9	−7	−5	−2	0	2	4	6
9	−20	−18	−15	−13	−11	−9	−6	−4	−2	1	3	5	7	10	12
11	−16	−14	−12	−9	−7	−5	−2	0	2	5	7	9	12	14	16
13	−13	−11	−8	−6	−3	−1	1	4	6	8	11	13	16	18	20
15	−10	−8	−5	−3	0	2	4	7	9	12	14	16	19	21	24
17	−8	−5	−3	0	2	5	7	10	12	15	17	19	22	24	27
19	−5	−3	0	2	5	7	10	12	15	17	19	22	24	27	29
21	−3	−1	2	4	7	9	12	14	17	19	22	24	27	29	32
23	−1	1	4	6	9	11	14	16	19	21	24	26	29	31	34
25	0	3	5	8	11	13	16	18	21	23	26	28	31	33	36
27	2	5	7	10	12	15	17	20	23	25	28	30	33	35	38
29	4	6	9	11	14	17	19	22	24	27	29	32	35	37	40
31	5	8	10	13	16	18	21	23	26	29	31	34	36	39	42
33	6	9	12	14	17	20	22	25	27	30	33	35	38	41	43
35	8	10	13	16	18	21	24	26	29	31	34	37	39	42	45
37	9	12	14	17	20	22	25	28	30	33	36	38	41	44	46
39	10	13	15	18	21	23	26	29	32	34	37	40	42	45	48
41	11	14	17	19	22	25	27	30	33	35	38	41	44	46	49
43	12	15	18	20	23	26	29	31	34	37	39	42	45	48	50
45	13	16	19	21	24	27	29	32	35	38	40	43	46	49	51
47	14	17	20	22	25	28	31	33	36	39	42	44	47	50	52
49	15	18	21	24	26	29	32	35	37	40	43	46	48	51	54
51	16	19	22	24	27	30	33	36	38	41	44	47	49	52	55
53	17	20	23	25	28	31	34	37	39	42	45	48	50	53	56
55	18	21	23	26	29	32	35	37	40	43	46	48	51	54	57
57	19	21	24	27	30	33	35	38	41	44	47	50	52	55	58
59	19	22	25	28	31	33	36	39	42	45	48	50	53	56	59
61	20	23	26	29	32	34	37	40	43	46	49	51	54	57	60
63	21	24	27	29	32	35	38	41	44	47	49	52	55	58	61
65	22	25	27	30	33	36	39	42	44	47	50	53	56	59	62
67	22	25	28	31	34	37	40	42	45	48	51	54	57	60	62
69	23	26	29	32	35	37	40	43	46	49	52	55	58	60	63
71	24	27	30	32	35	38	41	44	47	50	53	55	58	61	64
73	24	27	30	33	36	39	42	45	48	51	53	56	59	62	65
75	25	28	31	34	37	39	42	45	48	51	54	57	60	63	66
77	26	29	32	34	37	40	43	46	49	52	55	58	61	64	66
79	26	29	32	35	38	41	44	47	50	52	55	58	61	64	67
81	27	30	33	36	39	42	44	47	50	53	56	59	62	65	68
83	27	30	33	36	39	42	45	48	51	54	57	60	63	66	68
85	28	31	34	37	40	43	46	49	52	55	58	60	64	66	69
87	29	32	34	37	40	43	46	49	52	55	58	61	64	67	70
89	29	32	35	38	41	44	47	50	53	56	59	62	65	68	71
91	30	33	36	39	42	45	48	51	54	56	60	62	65	68	71
93	30	33	36	39	42	45	48	51	54	57	60	63	66	69	72
95	31	34	37	40	43	46	49	52	55	58	61	64	67	70	73
97	31	34	37	40	43	46	49	52	55	58	61	64	67	70	73
99	32	35	38	41	44	47	50	53	56	59	62	65	68	71	74

Continued on Next Page

DEW POINT OF MOIST AIR °F (Cont)

Air Temperature °F

RH %	77	80	83	86	89	92	95	98	101	104	107	110	113	116	119	122
1	−30	−28	27	−25	−23	−21	−20	−18	−16	−14	−13	−11	−9	−7	−6	−4
3	−9	−7	−5	−3	−1	1	2	4	6	8	10	12	14	16	18	20
5	1	3	5	8	10	12	14	16	18	20	22	24	26	28	30	32
7	8	11	13	15	17	19	22	24	26	28	30	32	35	37	39	41
9	14	16	19	21	23	25	28	30	32	34	36	39	41	43	45	47
11	19	21	23	26	28	30	32	35	37	39	41	44	46	48	50	53
13	23	25	27	30	32	34	37	39	41	44	46	48	51	53	55	57
15	26	28	31	33	35	38	40	43	45	47	50	52	54	57	59	61
17	29	31	34	36	39	41	44	46	48	51	53	55	58	60	63	65
19	32	34	37	39	42	44	46	49	51	54	56	59	61	63	66	68
21	34	37	39	42	44	47	49	52	54	56	59	61	64	66	69	71
23	37	39	42	44	47	49	51	54	56	59	61	64	66	69	71	74
25	38	41	44	46	49	51	54	56	59	61	64	66	69	71	74	76
27	41	43	46	48	51	53	56	58	61	64	66	68	71	74	76	79
29	42	45	48	50	53	55	58	60	63	65	68	71	73	76	78	81
31	44	47	49	52	55	57	60	62	65	67	70	73	75	78	80	83
33	46	48	51	54	56	59	61	64	67	69	72	74	77	80	82	85
35	47	50	53	55	58	60	63	66	68	71	74	76	79	81	84	87
37	49	52	54	57	60	62	65	67	70	73	75	78	81	83	86	89
39	50	53	56	58	61	64	66	69	72	74	77	80	82	85	87	90
41	52	54	57	60	62	65	68	70	73	76	78	81	84	86	89	92
43	53	56	58	61	64	66	69	72	75	77	80	83	85	88	91	93
45	54	57	60	62	65	58	70	73	76	78	81	84	87	89	92	95
47	55	58	61	63	66	69	72	74	77	80	82	85	88	91	93	96
49	56	59	62	65	68	70	73	76	78	81	84	87	89	92	95	98
51	58	60	63	66	69	71	74	77	80	82	85	88	91	93	96	99
53	59	62	64	67	70	73	75	78	81	84	86	89	92	95	97	100
55	60	62	65	68	71	73	76	79	82	85	87	90	93	96	98	101
57	61	64	66	69	72	75	78	80	83	86	89	91	94	97	100	103
59	62	64	67	70	73	76	78	81	84	87	90	92	95	98	101	104
61	63	66	68	71	74	77	80	82	85	88	91	94	96	99	102	105
63	64	66	69	72	75	78	81	83	86	89	92	95	98	100	103	106
65	64	67	70	73	76	79	82	84	87	90	93	96	98	101	104	107
67	65	68	71	74	77	80	82	85	88	91	94	97	100	102	105	108
69	66	69	72	75	78	80	83	86	89	92	95	98	100	103	106	109
71	67	70	73	76	79	81	84	87	90	93	96	99	101	104	107	110
73	68	71	74	76	79	82	85	88	91	94	97	100	102	105	108	111
75	69	71	74	77	80	83	86	89	92	94	97	100	103	106	109	112
77	69	72	75	78	81	84	87	90	93	96	98	101	104	107	110	113
79	70	73	76	79	82	84	87	90	93	96	99	102	105	108	111	114
81	71	74	77	80	82	85	88	91	94	97	100	103	106	109	112	115
83	71	74	77	80	83	86	89	92	95	98	101	104	107	109	112	115
85	72	75	78	81	84	87	90	93	96	99	102	105	108	110	113	116
87	73	76	79	82	85	88	91	94	96	99	102	105	108	111	114	117
89	73	76	79	82	85	88	91	94	97	100	103	106	109	112	115	118
91	74	77	80	83	86	89	92	95	98	101	104	107	110	113	116	119
93	75	78	81	84	87	90	93	96	99	102	105	108	111	114	117	120
95	76	79	82	85	87	90	93	96	99	102	105	108	111	114	117	120
97	76	79	82	85	88	91	94	97	100	103	106	109	112	115	118	121
99	77	80	83	86	89	92	95	98	101	104	107	110	113	116	119	122

VAPOR PRESSURE VS. TEMPERATURE FOR VOLATILE COMPOUNDS

Temperature °F	°C	Propane psia	bar	Butane psia	bar	Isobutane psia	bar	Pentane psia	bar	Carbon Dioxide psia	bar	Hydrogen Sulfide psia	bar	Sulfur Dioxide psia	bar	Ammonia psia	bar
-70	-56.7	7.4	0.51														
-60	-51.1	9.7	0.67													5.5	0.38
-50	-45.6	12.6	0.87													7.7	0.53
-40	-40.0	16.2	1.12													10.4	0.72
-30	-34.4	20.3	1.40													13.9	0.96
-20	-29.9	25.4	1.75			7.5	0.52			221.	15.24					18.3	1.26
-10	-23.3	31.4	2.17			9.3	0.64			262.	18.07			3.1	0.21	23.7	1.63
0	-17.8	38.2	2.63	7.3	0.50	11.6	0.80			309.	21.3			4.3	0.30	30.4	2.10
+10	-12.2	46.0	3.17	9.2	0.63	14.6	1.01			362.	25.0			5.9	0.41	38.5	2.66
20	-6.7	55.5	3.83	11.6	0.80	18.2	1.26			422.	29.1			7.9	0.54	48.2	3.32
30	-1.1	66.3	4.57	14.4	0.99	22.3	1.54			489.	33.7			10.4	0.72	59.7	4.12
40	+4.4	78.0	5.38	17.7	1.22	26.9	1.86			565.	39.0			13.4	0.92	73.3	5.06
50	10.0	91.8	6.33	21.6	1.49	32.5	2.24			650.	44.8	197.	13.6	17.2	1.19	89.2	6.15
60	15.6	107.1	7.39	26.3	1.81	38.7	2.67			744.	51.3	233.	16.0	21.7	1.50	107.6	7.42
70	21.1	124.0	8.55	31.6	2.18	45.8	3.16			849.	58.6	268.	18.5	27.1	1.87	128.8	8.88
80	26.7	142.8	9.85	37.6	2.59	53.9	3.72			964.	66.5	303.	20.9	33.5	2.31	153.0	10.55
90	32.2	164.0	11.31	44.5	3.07	63.3	4.37					349.	24.1	40.9	2.82	180.6	12.46
100	37.8	187.0	12.90	52.2	3.60	73.7	5.08	15.7	1.08			394.	27.2	49.6	3.42	211.9	14.61
110	43.3	213.0	14.19	60.8	4.19	85.1	5.87	19.1	1.32			448.	30.9	59.7	4.12	247.0	17.03
120	48.9	240.	16.55	70.8	4.88	98.0	6.76	22.4	1.54			502.	34.6	71.2	4.91	286.4	19.75
130	54.4			81.4	5.61	112.0	7.72	25.8	1.78			564.	38.9	84.5	5.83		
140	60.0			92.6	6.39	126.8	8.74	31.5	2.17			630.	43.5				

Source: Condensed from *CRC Handbook*, 55th Edition, pp. E-27-E-31.

APPROXIMATE pH VALUES AT 25°C

Solution	Concentration N	g/L	pH	Solution	Concentration N	g/L	pH
Acids				Bases			
Hydrochloric	1	36.5	0.1	Sodium hydroxide	1	40.01	14.0
	0.1	3.65	1.1		0.1	4.00	13.0
	0.01	0.365	2.0		0.01	0.40	12.0
Sulfuric	1	49.0	0.3	Potassium hydroxide	1	56.1	14.0
	0.1	4.9	1.2		0.1	5.61	13.0
	0.01	0.49	2.1		0.01	0.56	12.0
Sulfurous	0.1	4.1	1.5	Sodium carbonate	0.1	5.3	11.6
Ortho-phosphoric	0.1	3.27	1.5	Sodium bicarbonate	0.1	4.2	8.4
Formic	0.1	4.60	2.3	Trisodium phosphate	0.1	5.47	12.0
Acetic	1	60.05	2.4	Ammonia	1	17.03	11.6
	0.1	6.01	2.9		0.1	1.7	11.1
	0.01	0.60	3.4		0.01	0.17	10.6
Carbonic (saturated)			3.8	Calcium carbonate (saturated)			9.4
Hydrogen sulfide	0.1	3.41	4.1	Calcium hydroxide (saturated)			12.4
Hydrocyanic	0.1	2.70	5.1				

BOILING POINTS VS. CONCENTRATION OF COMMON CORROSIVE MEDIA

BOILING POINT

Concentration, Percent By Weight	Hydrochloric Acid °F	°C	Sulfuric Acid °F	°C	Nitric Acid °F	°C	Phosphoric Acid °F	°C	Acetic Acid °F	°C	Formic Acid °F	°C	Sodium Hydroxide °F	°C
10	219	104	215	102	217	103	212	100	213	101	214	101	218	103
20	230*	110*	219	104	222	106	—		—		215	102	226	108
30	—		226	108	228	109	215	102	—		216	102	241	116
40	—		237	114	234	112	—		—		218	103	262	128
50	—		253	123	242	117	226	108	217	103	—		—	
60	—		284	140	249	121	—		—		222	106	—	
65	—		304	151	251	122	—		—		—		—	
70	—		329	165	250	121	—		—		—		—	
80	—		395	202	—		—		—		—		—	
85	—		437	225	—		316	158	—		222	106	—	
90	—		491	255	—		—		—		—		—	
96	—		554	290	—		—		—		—		—	
98	—		626	330	—		—		—		—		—	
99	—		—		—		—		243	117	—		—	

*Constant Boiling Point Mixture at 20.2 percent concentration.

Source: Cabot Corp., *Corrosion Resistance of Hastelloy Alloys.*

pH-VALUES OF PURE WATER AT DIFFERENT TEMPERATURES

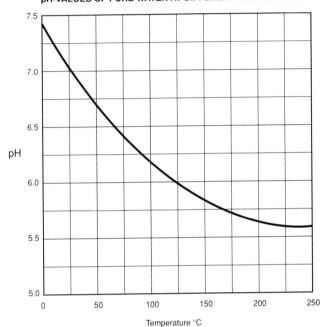

SOLUBILITY OF GASES IN WATER
(Partial pressure of the gas = 760 mm Hg)

| Temperature | | CO_2 | | H_2S | | O_2 | |
°C	°F	cm³/L	g/L	cm³/L	g/L	cm³/L	g/L
0	32	1713	3.36	4670	7.09	48.9	0.070
10	50	1194	2.35	3399	5.16	38.0	0.054
20	68	878	1.72	2582	3.92	31.0	0.044
30	86	665	1.31	2037	3.09	26.1	0.037
40	104	530	1.04	1660	2.52	23.1	0.033
50	122	436	0.86	1392	2.11	20.9	0.030
60	140	359	0.71	1190	1.81	19.5	0.028
70	158	—		1022	1.55	18.3	0.026
80	176	—		917	1.39	17.6	0.025
90	194	—		840	1.28	17.2	0.025
100	212	—		810	1.23	17.0	0.024

SOLUBILITY OF AIR IN WATER AND SOLVENTS
(Air Pressure = 1 Atmosphere)

Temperature		Distilled Water			Sea Water[a]		Ethanol		Iso-Octane	
		Air	O_2		O_2		O_2		O_2	
°C	°F	cm³/L	cm³/L	ppm	cm³/L	ppm	cm³/L	ppm	cm³/L	ppm
0	32.0	29.2	10.2	14.6	7.9	11.0	—		—	
1	33.8	28.4	9.9	14.1	—		—		—	
2	35.6	27.7	9.6	13.7	—		—		—	
3	37.4	27.0	9.4	13.4	—		—		—	
4	39.2	26.3	9.1	13.0	—		—		—	
5	41.0	25.7	8.9	12.7	7.0	9.7	—		—	
6	42.8	25.1	8.7	12.4	—		—		—	
7	44.6	24.5	8.5	12.1	—		—		—	
8	46.4	23.9	8.3	11.9	—		—		—	
9	48.2	23.4	8.1	11.6	—		—		—	
10	50.0	22.8	7.9	11.3	6.3	8.7	—		—	
11	51.8	22.3	7.7	11.0	—		—		—	
12	53.6	21.9	7.5	10.7	—		—		—	
13	55.4	21.4	7.4	10.5	—		—		—	
14	57.2	21.0	7.2	10.3	—		—		—	
15	59.0	20.6	7.0	10.1	5.7	7.9	—		—	
16	60.8	20.1	6.9	9.9	—		—		—	
17	62.6	19.8	6.8	9.7	—		—		—	
18	64.4	19.4	6.6	9.5	—		—		—	
19	66.2	19.0	6.5	9.3	—		—		—	
20	68.0	18.7	6.4	9.1	5.2	7.2	44.	79.	62.	126.
21	69.8	18.3	6.2	8.9	—		—		—	
22	71.6	18.0	6.1	8.7	—		—		—	
23	73.4	17.7	6.0	8.6	—		—		—	
24	75.2	17.4	5.9	8.4	—		—		—	
25	77.0	17.1	5.8	8.3	4.7	6.6	—		—	
26	78.8	16.8	5.7	8.1	—		—		—	
27	80.6	16.5	5.6	8.0	—		—		—	
28	82.4	16.2	5.5	7.9	—		—		—	
29	84.2	15.9	5.4	7.7	—		—		—	
30	86.0	15.6	5.3	7.6	3.9	5.4	—		—	

a) Chlorinity = 20.

SOLUBILITY OF WATER IN HYDROCARBONS

Hydrocarbon	°C	°F	Solubility	
			mg/100g	Gal/1000 Bbl
n-Butane	20	68	6.5	1.6
Isobutane	19	66	6.9	1.7
n-Pentane	15	59	6.1	1.6
	25	77	12.0	3.2
Isopentane	20	68	9.4	2.4
n-Hexane	20	68	11.1	3.1
Cyclohexane	20	68	10.0	3.3
n-Heptane	20	68	12.6	3.6
n-Octane	20	68	14.2	4.2
Benzene	20	68	43.5	16.1
Heptene-1	21	70	104.7	30.8
Butene-1	20	68	39.7	11.1
Gasoline	4	40	6.	1.8
	10	50	7.2	2.1
	16	60	8.2	2.4
	21	70	9.2	2.7
	27	80	10.2	3.0
	32	90	11.3	3.3
	38	100	12.3	3.6
	43	110	13.6	4.0

Source: *Corrosion Inhibitors*, NACE, p. 89, 1973.

THERMOCOUPLE DATA

Thermo-Couple	Cu-Const.	Fe-Const.	Ni Cr-Ni	Pt Rh-Pt
+ pole	Copper	Iron	Nickel-Chromium	Platinum-10% Rhodium
− pole	Constantan		Nickel	Platinum
Measuring Temp. °C	Approximate Thermocouple Voltage in mV			
−200	−5.70	−8.15		
−100	−3.40	−4.60		
0	0	0	0	0
100	4.25	5.37	4.04	0.64
200	9.20	10.95	8.14	1.44
300	14.89	16.55	12.24	2.32
400	20.99	22.15	16.38	3.26
500	27.40	27.84	20.64	4.22
600	34.30	33.66	24.94	5.23
700		39.72	29.15	6.27
800		46.23	33.27	7.34
900		53.15	37.32	8.45
1000			41.32	9.60
1100			45.22	10.77
1200			49.02	11.97
1300				13.17
1400				14.38
1500				15.58
1600				16.76

THERMOCOUPLE	TEMP. LIMITS	
	Average °C	Intermittent °C
Copper/Constantan	400	600
Iron/Constantan	750	1000
Nickel-Chromium/Nickel	1000	1300
Platinum-Rhodium/Platinum	1450	1700

HYPOTHETICAL CATHODIC AND ANODIC POLARIZATION DIAGRAM

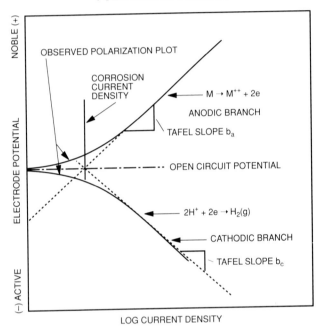

Source: ASTM G 3, Fig. 3 (1988 Edition)

TYPICAL CATHODIC AND ANODIC
POLARIZATION DIAGRAM

POLARIZATION BEHAVIOR OF STEEL
IN 1.0N Na$_2$ SO$_4$

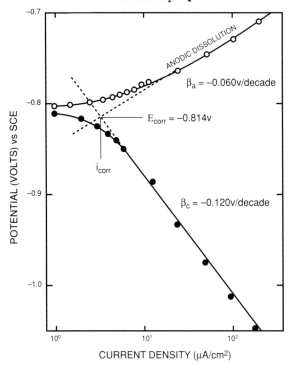

Source: R. Baboian

HYPOTHETICAL CATHODIC AND ANODIC POLARIZATION
PLOTS FOR A PASSIVE ANODE

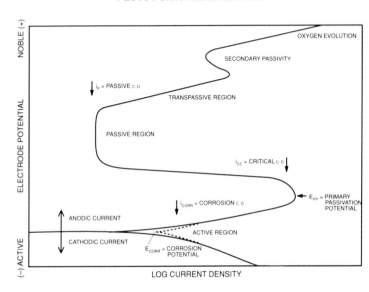

Source: ASTM G 3, Fig. 4 (1988 Edition)

**TYPICAL STANDARD POTENTIOSTATIC ANODIC
POLARIZATION PLOT**

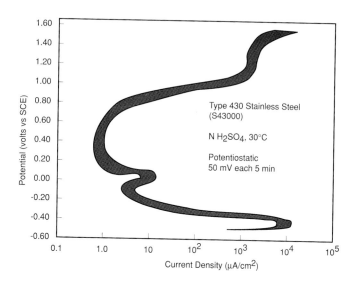

Source: ASTM G 5, Fig. 4 (1988 Edition)

DATA FOR TAFEL EQUATION CALCULATIONS

$$\eta = \beta \log \frac{i}{i_o}$$

Metal	Temper-ature °C	Solution	β volts	i_o A/m^2	η [1 mA/cm^2 (V)]*
		Hydrogen Overvoltage			
Pt (smooth)	20	1N HCl	0.03	10	0.00
	25	0.1N NaOH	0.11	0.68	0.13
Pd	20	0.6N HCl	0.03	2	0.02
Mo	20	1N HCl	0.04	10^{-2}	0.12
Au	20	1N HCl	0.05	10^{-2}	0.15
Ta	20	1N HCl	0.08	10^{-1}	0.16
W	20	5N HCl	0.11	10^{-1}	0.22
Ag	20	0.1N HCl	0.09	5×10^{-3}	0.30
Ni	20	0.1N HCl	0.10	8×10^{-3}	0.31
	20	0.12N NaOH	0.10	4×10^{-3}	0.34
Bi	20	1N HCl	0.10	10^{-3}	0.40
Nb	20	1N HCl	0.10	10^{-3}	0.40
Fe	16	1N HCl	0.15	10^{-2}	0.45
	25	4% NaCl pH 1-4	0.10	10^{-3}	0.40 (Stern)
Cu	20	0.1N HCl	0.12	2×10^{-3}	0.44
	20	0.15N NaOH	0.12	1×10^{-2}	0.36
Sb	20	2N H$_2$SO$_4$	0.10	10^{-5}	0.60
Al	20	2N H$_2$SO$_4$	0.10	10^{-6}	0.70
Be	20	1N HCl	0.12	10^{-5}	0.72
Sn	20	1N HCl	0.15	10^{-4}	0.75
Cd	16	1N HCl	0.20	10^{-3}	0.80
Zn	20	1N H$_2$SO$_4$	0.12	1.6×10^{-7}	0.94
Hg	20	0.1N HCl	0.12	7×10^{-9}	1.10
	20	0.1N H$_2$SO$_4$	0.12	2×10^{-9}	1.16
	20	0.1N NaOH	0.10	3×10^{-11}	1.15
Pb	20	0.01-8N HCl	0.12	2×10^{-9}	1.16
		Oxygen Overvoltage			
Pt (smooth)	20	0.1N H$_2$SO$_4$	0.10	9×10^{-8}	0.81
	20	0.1N NaOH	0.05	4×10^{-9}	0.47
Au	20	0.1N NaOH	0.05	5×10^{-9}	0.47
		Metal Overvoltage (deposition)			
Zn	25	1M ZnSO$_4$	0.12	0.2	0.20 (Bockris)
Cu	25	1M CuSO$_4$	0.12	0.2	0.20 (Bockris)
Fe	25	1M FeSO$_4$	0.12	10^{-4}	0.60 (Bockris)
Ni	25	1M NiSO$_4$	0.12	2×10^{-5}	0.68 (Bockris)

Source: H.H. Uhlig, Corrosion and Corrosion Control, 3rd ed., p. 44, John Wiley & Sons, 1985.

POLARIZATION RESISTANCE METHOD FOR DETERMINING CORROSION RATES

Defining the polarization resistance R_p as

$$R_p = \left(\frac{\partial \phi}{\partial I}\right)_{\phi_{corr}}$$

and combining the constants as

$$B = \frac{b_a b_c}{2.303(b_a + b_c)}$$

the corrosion current I_{corr} can be calculated as

$$I_{corr} = \frac{b_a b_c}{2.303(b_a + b_c)} \left(\frac{\partial I}{\partial \phi}\right)_{\phi_{corr}}$$

$$= \frac{B}{R_p}$$

The dimension of R_p as determined from a potential-current plot is ohms (Ω). In order to obtain a value of R_p, which is independent of the electrode surface and which can be converted into corrosion rates, polarization resistance values should be reported in $\Omega \cdot cm^2$ (e.g., $mV/mA/cm^2$).

See following page for typical values for constant B.

Source: Florian Mansfeld, Advances in Corrosion Science and Technology, Vol. 6, p. 76.

VALUES OF THE CONSTANT B
FOR THE POLARIZATION RESISTANCE METHOD

Corroding System	b_a, mV	b_c, mV	B, mV
Theoretical	30	30	6.5
(Values of B calculated from arbitrary	30	60	9
b_a and b_c values using formula on	30	120	10
previous page; b_a and b_c values	30	180	11
can be interchanged.)	30	∞	13
	60	60	13
	60	90	16
	60	120	17
	60	180	20
	60	∞	26
	90	90	20
	90	120	22
	90	180	26
	90	∞	39
	120	120	26
	120	∞	52
	180	180	39
	180	∞	78
Iron, 4% NaCl, pH 1.5	—	—	17
Iron, 0.5N H_2SO_4, 30 C	—	—	17
Iron, 1N H_2SO_4	—	—	10-20
Iron, 1N HCl	—	—	18-23
Iron, 0.02M citric acid, pH 2.6. 35 C	—	—	12
Carbon steel, seawater	57	∞	25
Carbon steel, 1N Na_2SO_4, H_2, pH 6.3, 30 C	—	—	19
304L SS, 1N H_2SO_4, O_2	inf.	50	22
304 SS, lithiated water, 288 C	85	160	24
304 SS, 3% NaCl, 90 C	inf.	50	22
430 SS, 1N H_2SO_4, H_2, 30 C	—	—	20
600 alloy, lithiated water, 288 C	82	160	24
Al 1199, 1N NaCl, pH 2, 30 C	—	—	44
Aluminum, seawater	45	600	18
Zircaloy 2, lithiated water, 288 C	inf.	186	81
OFHC Copper, 1N NaCl, H_2, pH 6.2, 30 C	—	—	26

Source: Adapted from a collection of literature values compiled by Florian Mansfeld, Advances in Corrosion Science and Technology, Vol. 6, p. 185.

STANDARD REFERENCE POTENTIALS AND CONVERSION TABLE

REFERENCE POTENTIALS

Electrode	Potential (V) @ 25°C		Thermal Temperature Coefficient[a] (mV/°C)
	E'[b]	E''[c]	
$(Pt)/H_2$ ($\alpha = 1$)/H + ($\alpha = 1$) (SHE)	0.000	•••	+0.87
Ag/AgCl/1M KCl	+0.235	•••	+0.25
Ag/AgCl/0.6M Cl⁻ (seawater)	+0.25	•••	•••
Ag/AgCl/0.1M Cl⁻	+0.288	•••	+0.22
Hg/Hg_2Cl_2/sat KCl (SCE)	+0.241	+0.244	+0.22
Hg/Hg_2Cl_2/1M KCl	+0.280	+0.283	+0.59
Hg/Hg_2Cl_2/0.1M KCl	+0.334	+0.336	+0.79
$Cu/CuSO_4$sat	+0.30	•••	+0.90
$Hg/HgSO_4/H_2SO_4$	+0.616	•••	+0.09

[a]To convert from thermal to isothermal temperature coefficients, subtract 0.87 mV/°C. Thus the isothermal temperature coefficient for Ag/AgCl/1M KCl is −0.62 mV/°C.
[b]E' is the standard potential for the half cell corrected for the concentration of the ions.
[c]E'' also includes the liquid junction potentials for a saturated KCl salt bridge.

CONVERSION FACTORS[d]

From (E')	To SHE Scale	To SCE Scale (E')
H_2/H^+	•••	−0.241
Ag/AgCl/1M KCl	+0.235	−0.006
Ag/AgCl/0.6M Cl⁻ (seawater)	+0.25	+0.09
Ag/AgCl/0.1M Cl⁻	+0.288	+0.047
Hg/Hg_2Cl_2/sat KCl (SCE)	+0.241	•••
Hg/Hg_2Cl_2/1M KCl	+0.280	+0.039
Hg/Hg_2Cl_2/0.1M KCl	+0.334	+0.093
$Cu/CuSO_4$ sat	+0.30	+0.06
$Hg/HgSO_4/H_2SO_4$	+0.616	•••

[d]To convert from one scale to another, add the value indicated.

Example:
An electrode potential of +1.000V versus SCE would be (1.000 + 0.241) = +1.241V versus SHE. An electrode potential of −1.000V versus SCE would give (−1.000 + 0.241) = −0.759V versus SHE.

Source: R. Baboian.

ELECTROCHEMICAL SERIES
REDUCTION REACTIONS HAVING E° VALUES MORE POSITIVE THAN THAT
OF THE STANDARD HYDROGEN ELECTRODE

Reaction	E°, V	Reaction	E°, V
$2\,H^+ + 2e \rightleftharpoons H_2$	0.00000	$BiOCl + 2\,H^+ + 3\,e \rightleftharpoons Bi + Cl^- + H_2O$	0.1583
$CuI_2^- + e \rightleftharpoons Cu + 2\,I^-$	0.00	$Bi(Cl)_4^- + 3\,e \rightleftharpoons Bi + 4\,Cl^-$	0.16
$Ge^{4+} + 2\,e \rightleftharpoons Ge^{2+}$	0.00	$Co(OH)_3 + e \rightleftharpoons Co(OH)_2 + OH^-$	0.17
$NO_3^- + H_2O + 2\,e \rightleftharpoons NO_2^- + 2\,OH^-$	0.01	$SO_4^{2-} + 4\,H^+ + 2\,e \rightleftharpoons H_2SO_3 + H_2O$	0.172
$Tl_2O_3 + 3\,H_2O + 4\,e \rightleftharpoons 2\,Tl^+ + 6\,OH^-$	0.02	$SbO^+ + 2\,H^+ + 3\,e \rightleftharpoons Sb + 2\,H_2O$	0.212
$SeO_4^{2-} + H_2O + 2\,e \rightleftharpoons SeO_3^{2-} + 2\,OH^-$	0.05	$AgCl + e \rightleftharpoons Ag + Cl^-$	0.22233
$UO_2^{2+} + e \rightleftharpoons UO_2^+$	0.062	$As_2O_3 + 6\,H^+ + 6\,e \rightleftharpoons 2\,As + 3\,H_2O$	0.234
$Pd(OH)_2 + 2\,e \rightleftharpoons Pd + 2\,OH^-$	0.07	Calomel electrode, saturated NaCl (SSCE)	0.2360
$AgBr + e \rightleftharpoons Ag + Br^-$	0.07133	$Ge^{2+} + 2\,e \rightleftharpoons Ge$	0.24
$S_4O_6^{2-} + 2\,e \rightleftharpoons 2\,S_2O_3^{2-}$	0.08	Calomel electrode, saturated KCl	0.2412
$AgSCN + e \rightleftharpoons Ag + SCN^-$	0.08951	$PbO_2 + H_2O + 2\,e \rightleftharpoons PbO + 2\,OH^-$	0.247
$N_2 + 2\,H_2O + 6\,H^+ + 6\,e \rightleftharpoons 2\,NH_4OH$	0.092	$HAsO_2 + 3\,H^+ + 3_c \rightleftharpoons As + 2\,H_2O$	0.248
$HgO + H_2O + 2\,e \rightleftharpoons Hg + 2\,OH^-$	0.0977	$Ru^{3+} + e \rightleftharpoons Ru^{2+}$	0.2487
$Ir_2O_3 + 3\,H_2O + 6\,e \rightleftharpoons 2\,Ir + 6\,OH^-$	0.098	$ReO_2 + 4\,H^+ + 4\,e \rightleftharpoons Re + 2\,H_2O$	0.2513
$2\,NO + 2\,e \rightleftharpoons N_2O_2^{2-}$	0.10	$IO_3^- + 3\,H_2O + 6\,e \rightleftharpoons I^- + OH^-$	0.26
$[Co(NH_3)_6]^{3+} + e \rightleftharpoons [Co(NH_3)_6]^{2+}$	0.108	$Hg_2Cl_2 + 2\,e \rightleftharpoons 2\,Hg + 2\,Cl^-$	0.26808
$Hg_2O + H_2O + 2\,e \rightleftharpoons 2\,Hg + 2\,OH^-$	0.123	Calomel electrode, molal KCl	0.2800
$Ge^{4+} + 4\,e \rightleftharpoons Ge$	0.124	Calomel electrode, 1 mol/1 KCl (NCE)	0.2801
$Hg_2Br_2 + 2\,e \rightleftharpoons 2\,Hg + 2\,Br^-$	0.13923	$Re^{3+} + 3e \rightleftharpoons Re$	0.300
$Pt(OH)_2 + 2\,e \rightleftharpoons Pt + 2\,OH^-$	0.14	$BiO^+ + 2\,H^+ + 3\,e \rightleftharpoons Bi + H_2O$	0.320
$S + 2H^+ + 2\,e \rightleftharpoons H_2S(aq)$	0.142	$UO_2^{2+} + 4\,H^+ + 2\,e \rightleftharpoons U^{4+} + 2\,H_2O$	0.327
$Np^{4+} + e \rightleftharpoons Np^{3+}$	0.147	$ClO_3^- + H_2O + 2\,e \rightleftharpoons ClO_2^- + 2\,OH^-$	0.33
$Ag_4[Fe(CN)_6] + 4\,e \rightleftharpoons 4\,Ag + [Fe(CN)_6]^{4-}$	0.1478	$2\,HCNO + 2\,H^+ + 2\,e \rightleftharpoons (CN)_2 + 2\,H_2O$	0.330
$Mn(OH)_3 + e \rightleftharpoons Mn(OH)_2 + OH^-$	0.15	Calomel electrode, 0.1 mol/l KCl	0.3337
$2\,NO_2^- + 3\,H_2O + 4\,e \rightleftharpoons N_2O + 6\,OH^-$	0.15	$VO^{2+} + 2\,H^+ + e \rightleftharpoons V^{3+} + H_2O$	0.337
$Sn^{4+} + 2\,e \rightleftharpoons Sn^{2+}$	0.151	$Cu^{2+} + 2\,e \rightleftharpoons Cu$	0.3419
$Sb_2O_3 + 6\,H^+ + 6\,e \rightleftharpoons 2\,Sb + 3\,H_2O$	0.152	$Ag_2O + H_2O + 2\,e \rightleftharpoons 2\,Ag + 2\,OH^-$	0.342

Reaction	E° (V)
$AgIO_3 + e \rightleftharpoons Ag + IO^-3$	0.354
$[Fe(CN)_6]^{3-} + e \rightleftharpoons [Fe(CN)_6]^{4-}$	0.358
$ClO_4^- + H_2O + 2\,e \rightleftharpoons ClO_3^- + 2\,OH^-$	0.36
$Ag_2SeO_3 + 2\,e \rightleftharpoons 2\,Ag + SeO_3^{2-}$	0.3629
$ReO_4^- + 8\,H^+ + 7\,e \rightleftharpoons Re + 4\,H_2O$	0.368
$(CN)_2 + 2\,H^+ + 2\,e \rightleftharpoons 2\,HCN$	0.373
$[Ferricinium]^+ + e \rightleftharpoons ferrocene$	0.400
$Tc^{2+} + 2\,e \rightleftharpoons Tc$	0.400
$O_2 + 2\,H_2O + 4\,e \rightleftharpoons 4\,OH^-$	0.401
$AgOCN + e \rightleftharpoons Ag + OCN^-$	0.41
$[RhCl_6]^{3-} + 3\,e \rightleftharpoons Rh + 6\,Cl^-$	0.431
$Ag_2CrO_4 + 2\,e \rightleftharpoons 2\,Ag + CrO_4^{2-}$	0.4470
$H_2SO_3 + 4\,H^+ + 4\,e \rightleftharpoons S + 3\,H_2O$	0.449
$Ru^{2+} + 2\,e \rightleftharpoons Ru$	0.455
$Ag_2MoO_4 + 2\,e \rightleftharpoons 2\,Ag + MoO_4^{2-}$	0.4573
$Ag_2C_2O_4 + 2\,e \rightleftharpoons 2\,Ag + C_2O_4^{2-}$	0.4647
$Ag_2WO_4 + 2\,e \rightleftharpoons 2\,Ag + WO_4^{2-}$	0.4660
$Ag_2CO_3 + 2\,e \rightleftharpoons 2\,Ag + CO_3^{2-}$	0.47
$TeO_4^- + 8\,H^+ + 7\,e \rightleftharpoons Te + 4\,H_2O$	0.472
$IO^- + H_2O + 2\,e \rightleftharpoons I^- + 2\,OH^-$	0.485
$ReO_4^- + 4\,H^+ + 3\,e \rightleftharpoons ReO_2 + 2\,H_2O$	0.510
$Hg(ac)_2 + 2\,e \rightleftharpoons 2\,Hg + 2\,(ac)^-$	0.51163
$Cu^+ + e \rightleftharpoons Cu$	0.521
$I_2 + 2\,e \rightleftharpoons 2\,I^-$	0.5355
$I_3^- + 2\,e \rightleftharpoons 3\,I^-$	0.536
$AgBrO_3 + e \rightleftharpoons Ag + BrO_3^-$	0.546
$MnO_4^- + e \rightleftharpoons MnO_4^{2-}$	0.558
$H_3AsO_4 + 2\,H^+ + 2\,e \rightleftharpoons HAsO_2 + 2\,H_2O$	0.560
$IO_3^- + 3\,H_2O + 6\,e \rightleftharpoons IO^- + 4\,OH^-$	0.56
$S_2O_6^{2-} + 4\,H^+ + 2\,e \rightleftharpoons 2\,H_2SO_3$	0.564
$AgNO_2 + e \rightleftharpoons Ag + NO_2^-$	0.564
$Te^{4+} + 4\,e \rightleftharpoons Te$	0.568
$Sb_2O_5 + 6\,H^+ + 4\,e \rightleftharpoons 2\,SbO^+ + 3\,H_2O$	0.581
$RuO_4^- + e \rightleftharpoons RuO_4^{2-}$	0.59
$[PdCl_4]^{2-} + 2\,e \rightleftharpoons Pd + 4\,Cl^-$	0.591
$TeO_2 + 4\,H^+ + 4\,e \rightleftharpoons Te + 2\,H_2O$	0.593
$MnO_4^- + 2\,H_2O + 3\,e \rightleftharpoons MnO_2 + 4\,OH^-$	0.595
$Rh^{2+} + 2\,e \rightleftharpoons Rh$	0.600
$Rh^+ + e \rightleftharpoons Rh$	0.600
$MnO_4^- + 2\,H_2O + 2\,e \rightleftharpoons MnO_2 + 4\,OH^-$	0.60
$2\,AgO + H_2O + 2\,e \rightleftharpoons Ag_2O + 2\,OH^-$	0.607
$BrO_3^- + 3\,H_2O + 6\,e \rightleftharpoons Br^- + 6\,OH^-$	0.61
$UO_2^+ + 4\,H^+ + e \rightleftharpoons U^{4+} + 2\,H_2O$	0.612
$Hg_2SO_4 + 2\,e \rightleftharpoons 2\,Hg + SO_4^{2-}$	0.6125
$ClO_3^- + 3\,H_2O + 6\,e \rightleftharpoons Cl^- + 6\,OH^-$	0.62
$Hg_2HPO_4 + 2\,e \rightleftharpoons 2\,Hg + HPO_4^{2-}$	0.6359
$Ag(ac) + e \rightleftharpoons Ag + (ac)^-$	0.643
$Sb_2O_5 \text{ (valentinite)} + 4\,H^+ + 4\,e \rightleftharpoons Sb_2O_3 + 2\,H_2O$	0.649
$Ag_2SO_4 + 2\,e \rightleftharpoons 2\,Ag + SO_4^{2-}$	0.654
$ClO_2^- + H_2O + 2\,e \rightleftharpoons ClO^- + 2\,OH^-$	0.66
$Sb_2O_5 \text{ (senarmontite)} + 4\,H^+ + 4\,e \rightleftharpoons Sb_2O_3 + 2\,H_2O$	0.671
$[PtCl_6]^{2-} + 2\,e \rightleftharpoons [PtCl_4]^{2-} + 2\,Cl^-$	0.68
$O_2 + 2\,H^+ + 2\,e \rightleftharpoons H_2O_2$	0.695
$p\text{-benzoquinone} + 2\,H^+\ 2\,e \rightleftharpoons \text{hydroquinone}$	0.6992
$H_3IO_6 + 2\,e \rightleftharpoons IO_3^- + 3\,OH^-$	0.7
$Ag_2O_3 + H_2O + 2\,e \rightleftharpoons 2\,AgO + 2\,OH^-$	0.739
$[PtCl_4]^{2-} + 2\,e \rightleftharpoons Pt + 4\,Cl^-$	0.755
$Rh^{3+} + 3\,e \rightleftharpoons Rh$	0.758
$ClO_2^- + 2\,H_2O + 4\,e \rightleftharpoons Cl^- + 4\,OH^-$	0.76
$2\,NO + H_2O + 2\,e \rightleftharpoons N_2O + 2\,OH^-$	0.76
$BrO^- + H_2O + 2\,e \rightleftharpoons Br^- + 2\,OH^-$	0.761
$ReO_4^- + 2\,H^+ + e \rightleftharpoons ReO_3 + H_2O$	0.768
$(CNS)_2 + 2\,e \rightleftharpoons 2\,CNS^-$	0.77
$[IrCl_6]^{3-} + 3e \rightleftharpoons Ir + 6\,Cl^-$	0.77
$Fe^{3+} + e \rightleftharpoons Fe^{2+}$	0.771
$Ag(F) + e \rightleftharpoons Ag + F^-$	0.779
$TcO_4^- + 4\,H^+ + 3\,e \rightleftharpoons TcO_2 + 2\,H_2O$	0.782
$Hg_2^{2+} + 2\,e \rightleftharpoons 2\,Hg$	0.7973
$Ag^+ + e \rightleftharpoons Ag$	0.7996
$2\,NO_3^- + 4\,H^+ + 2\,e \rightleftharpoons N_2O_4 + 2\,H_2O$	0.803
$ClO^- + H_2O + 2\,e \rightleftharpoons Cl^- + 2\,OH^-$	0.841
$OsO_4 + 8\,H^+ + 8\,e \rightleftharpoons Os + 4\,H_2O$	0.85

ELECTROCHEMICAL SERIES (Cont)

Reaction	E
$Hg_2^{2+} + 2\,e \rightleftharpoons Hg$	0.851
$AuBr_4^- + 3\,e \rightleftharpoons Au + 4\,Br^-$	0.854
$SiO_2\ (quartz) + 4\,H^+ + 4\,e \rightleftharpoons Si + 2\,H_2O$	0.857
$2\,HNO_2 + 4\,H^+ + 4\,e \rightleftharpoons H_2N_2O_2 + 2\,H_2O$	0.86
$[IrCl_6]^{2-} + e \rightleftharpoons [IrCl_6]^{3-}$	0.8665
$N_2O_4 + 2\,e \rightleftharpoons 2\,NO_2^-$	0.867
$HO_2^- + H_2O + 2\,e \rightleftharpoons 3\,OH^-$	0.878
$2\,Hg^{2+} + 2\,e \rightleftharpoons Hg_2^{2+}$	0.920
$NO_3^- + 3\,H^+ + 2\,e \rightleftharpoons HNO_2 + H_2O$	0.934
$Pd^{2+} + 2\,e \rightleftharpoons Pd$	0.951
$ClO_2\ (aq) + e \rightleftharpoons ClO_2^-$	0.954
$NO_3^- + 4\,H^+ + 3\,e \rightleftharpoons NO + 2\,H_2O$	0.957
$AuBr_2^- + e \rightleftharpoons Au + 2\,Br^-$	0.959
$HNO_2 + H^+ + e \rightleftharpoons NO + H_2O$	0.983
$HIO + H^+ + 2\,e \rightleftharpoons I^- + H_2O$	0.987
$VO_2^+ + 2\,H^+ + e \rightleftharpoons VO^{2+} + H_2O$	0.991
$RuO_4^- + e \rightleftharpoons RuO_4^{2-}$	1.00
$V(OH)_4^+ + 2\,H^+ + e \rightleftharpoons VO^{2+} + 3\,H_2O$	1.00
$AuCl_4^- + 3\,e \rightleftharpoons Au + 4\,Cl^-$	1.002
$Pu^{4+} + e \rightleftharpoons Pu^{3+}$	1.006
$H_6TeO_6 + 2\,H^+ + 2\,e \rightleftharpoons TeO_2 + 4\,H_2O$	1.02
$N_2O_4 + 4\,H^+ + 4\,e \rightleftharpoons 2\,NO + 2\,H_2O$	1.035
$[Fe(phen)_3]^{3+} + e \rightleftharpoons [Fe(phen)_3]^{2+}$ (1 (mol/l H_2SO_4)	1.06
$PuO_2(OH)_2 + H^+ + e \rightleftharpoons PuO_2OH + H_2O$	1.062
$N_2O_4 + 2\,H^+ + 2\,e \rightleftharpoons 2\,HNO_2$	1.065
$Br_2(l) + 2\,e \rightleftharpoons 2\,Br^-$	1.066
$IO_3^- + 6\,H^+ + 6\,e \rightleftharpoons I^- + 3\,H_2O$	1.085
$Br_2(aq) + 2\,e \rightleftharpoons 2\,Br^-$	1.0873
$Pu^{5+} + e \rightleftharpoons Pu^{4+}$	1.099
$Cu^{2+} + 2\,CN^- + e \rightleftharpoons [Cu(CN)2]^- \quad Pt^{2+} + 2\,e \rightleftharpoons Pt$	1.118
$RuO_2 + 4\,H^+ + 2\,e \rightleftharpoons Ru^{2+} + 2\,H_2O$	1.120
$[Fe(phenanthroline)_3]^{3+} + e \rightleftharpoons [Fe(phen)_3]^{2+}$	1.147
$SeO_4^{2-} + 4\,H^+ + 2\,e \rightleftharpoons H_2SeO_3 + H_2O$	1.151
$ClO_3^- + 2\,H^+ + e \rightleftharpoons ClO_2 + H_2O$	1.152
$Ir^{3+} + 3\,e \rightleftharpoons Ir$	1.156
$2\,IO_3^- + 12\,H^+ + 10\,e \rightleftharpoons I_2\ 6\,H_2O$	1.195
$ClO_3^- + 3\,H^+ + 2\,e \rightleftharpoons HClO_2 + H_2O$	1.214
$MnO_2 + 4\,H^+ + 2\,e \rightleftharpoons Mn^{2+} + 2\,H_2O$	1.224
$O_2 + 4\,H^+ + 4\,e \rightleftharpoons 2\,H_2O$	1.229
$Cr_2O_7^{2-} + 14\,H^+ + 6\,e \rightleftharpoons 2\,Cr^{3+} + 7\,H_2O$	1.232
$O_3 + 2\,H^+ + 2\,e \rightleftharpoons O_2 + 2\,OH^-$	1.24
$Tl^{3+} + 2\,e \rightleftharpoons Tl^+$	1.252
$N_2H_5^+ + 3\,H^+ + 2\,e \rightleftharpoons 2\,NH_4^+$	1.275
$ClO_2^- + H^+ + e \rightleftharpoons HClO_2$	1.277
$[PdCl_6]^{2-} + 2\,e \rightleftharpoons [PdCl_4]^{2-} + 2\,Cl^-$	1.288
$2\,HNO_2 + 4\,H^+ + 4\,e \rightleftharpoons N_2O + 3\,H_2O$	1.297
$PuO_2(OH)_2 + 2\,H^+ + 2\,e \rightleftharpoons Pu(OH)_4$	1.325
$HBrO^- + H^+ + 2\,e \rightleftharpoons Br^- + H_2O$	1.331
$HCrO_4^- + 7\,H^+ + 3\,e \rightleftharpoons Cr^{3+} + 4\,H_2O$	1.350
$Cl_2(g) + 2\,e \rightleftharpoons Cl^-$	1.35827
$ClO_4^- + 8\,H^+ + 8\,e \rightleftharpoons Cl^- + 4\,H_2O$	1.389
$ClO_4^- + 8\,H^+ + 7\,e \rightleftharpoons \tfrac{1}{2}Cl_2 + 4\,H_2O$	1.39
$Au^{3+} + 2\,e \rightleftharpoons Au^+$	1.401
$2\,NH_3OH^+ + H^+ + 2\,e \rightleftharpoons N_2H_5^+ + 2\,H_2O$	1.42
$BrO_3^- + 6\,H^+ + 6\,e \rightleftharpoons Br^- + 3\,H_2O$	1.423
$2\,HIO + 2\,H^+ + 2\,e \rightleftharpoons I_2 + 2\,H_2O$	1.439
$Au(OH)_3 + 3\,H^+ + 3\,e \rightleftharpoons Au^- + 3\,H_2O$	1.45
$3\,IO_5^- + 6\,H^+ + 6\,e \rightleftharpoons Cl^- + 3\,H_2O$	1.451
$PbO_2 + 4\,H^+ + 2\,e \rightleftharpoons Pb^{2+} + 2\,H_2O$	1.455
$ClO_3^- + 6\,H^+ + 5\,e \rightleftharpoons \tfrac{1}{2}Cl_2 + 3\,H_2O$	1.47
$BrO_3^- + 6\,H^+ + 5\,e \rightleftharpoons \tfrac{1}{2}Br_2 + 3\,H_2O$	1.482
$HClO + H^+ + e \rightleftharpoons \tfrac{1}{2}Cl_2 + H_2O$	1.482
$HO_2 + H^+ + e \rightleftharpoons H_2O_2$	1.495
$Au^{3+} + 3\,e \rightleftharpoons Au$	1.498
$MnO_4^- + 8\,H^+ + 5\,e \rightleftharpoons Mn^{2+} + 4\,H_2O$	1.507
$Mn^{3+} + e \rightleftharpoons Mn^{2+}$	1.5415
$HClO_2 + 3\,H^+ + 4\,e \rightleftharpoons Cl^- + 2\,H_2O$	1.570
$HBrO + H^+ + e \rightleftharpoons \tfrac{1}{2}Br_2(aq) + H_2O$	1.574
$2\,NO + 2\,H^+ + 2\,e \rightleftharpoons N_2O + H_2O$	1.591
$Bi_2O_4 + 4\,H^+ + 2\,e \rightleftharpoons 2\,BiO^+ + 2\,H_2O$	1.593

Reaction	E°, V
$H_5IO_6 + H^+ + 2e \rightleftharpoons IO_3^- + 3H_2O$	1.601
$Ce^{4+} + e \rightleftharpoons Ce^{3+}$	1.61
$HClO + H^+ + e \rightleftharpoons 1/2\,Cl_2 + H_2O$	1.611
$HClO_2 + 3H^+ + 3e \rightleftharpoons 1/2\,Cl_2 + 2H_2O$	1.628
$HClO_2 + 2H^+ + 2e \rightleftharpoons HClO + H_2O$	1.645
$NiO_2 + 4H^+ + 2e \rightleftharpoons Ni^{2+} + 2H_2O$	1.678
$MnO_4^- + 4H^+ + 3e \rightleftharpoons MnO_2 + 2H_2O$	1.679
$PbO_2 + SO_4^{2-} + 4H^+ + 2e \rightleftharpoons PbSO_4 + 2H_2O$	1.6913
$Au^+ + e \rightleftharpoons Au$	1.692
$CeOH^{3+} + H^+ + e \rightleftharpoons Ce^{3+} + H_2O$	1.715
$N_2O + 2H^+ + 2e \rightleftharpoons N_2 + H_2O$	1.766
$H_2O_2 + 2H^+ + 2e \rightleftharpoons 2H_2O$	1.776
$Co^{3+} + e \rightleftharpoons Co^{2+}$ (2 mol/l H_2SO_4)	1.83
$Ag^{2+} + e \rightleftharpoons Ag^+$	1.980
$S_2O_8^{2-} + 2e \rightleftharpoons 2SO_4^{2-}$	2.010
$OH\cdot + e \rightleftharpoons OH^-$	2.02
$O_3 + 2H^+ + 2e \rightleftharpoons O_2 + H_2O$	2.076
$S_2O_6^{2-} + 2H^+ + 2e \rightleftharpoons 2HSO_4^-$	2.123
$F_2O + 2H^+ + 4e \rightleftharpoons H_2O + 2F^-$	2.153
$FeO_4^{2-} + 8H^+ + 3e \rightleftharpoons Fe^{3+} + 4H_2O$	2.20
$O(g) + 2H^+ + 2e \rightleftharpoons H_2O$	2.421
$H_2N_2O_2 + 2H^+ + 2e \rightleftharpoons N_2 + 2H_2O$	2.65
$F_2 + 2e \rightleftharpoons 2F^-$	2.866
$F_2 + 2H^+ + 2e \rightleftharpoons 2HF$	3.053

REDUCTION REACTIONS HAVING E° VALUES MORE NEGATIVE THAN THAT OF THE STANDARD HYDROGEN ELECTRODE

Reaction	E°, V	Reaction	E°, V
$2H^+ + 2e \rightleftharpoons H_2$	0.00000	$WO_2 + 4H^+ + 4e \rightleftharpoons W + 2H_2O$	-0.119
$AgCN + e \rightleftharpoons Ag + CN^-$	-0.017	$Pb^{2+} + 2e \rightleftharpoons Pb(Hg)$	-0.1205
$2WO_3 + 2H^+ + 2e \rightleftharpoons W_2O_5 + H_2O$	-0.029	$Pb^{2+} + 2e \rightleftharpoons Pb$	-0.1262
$W_2O_5 + 2H^+ + 2e \rightleftharpoons 2WO_2 + H_2O$	-0.031	$CrO_4^- + 4H_2O + 3e \rightleftharpoons Cr(OH)_3 + 5OH^-$	-0.13
$D^+ + e \rightleftharpoons 1/2D_2$	-0.0034	$Sn^{2-} + 2e \rightleftharpoons Sn$	-0.1375
$Ag_2S + 2H^+ + 2e \rightleftharpoons 2Ag + H_2S$	-0.0366	$In^+ + e \rightleftharpoons In$	-0.14
$Fe^{3+} + 3e \rightleftharpoons Fe$	-0.037	$O_2 + 2H_2O + 2e \rightleftharpoons H_2O_2 + 2OH^-$	-0.146
$Hg_2I_2 + 2e \rightleftharpoons 2Hg + 2I^-$	-0.0405	$AgI + e \rightleftharpoons Ag + I^-$	-0.15224
$2D^+ + 2e \rightleftharpoons D_2$	-0.044	$2NO_2 + 2H_2O + 4e \rightleftharpoons N_2O_2^{2-} + 4OH^-$	-0.18
$Tl(OH)_3 + 2e \rightleftharpoons TlOH + 2OH^-$	-0.05	$H_2GeO_3 + 4H^+ + 4e \rightleftharpoons Ge + 3H_2O$	-0.182
$TlOH^{3+} + H^+ + e \rightleftharpoons Tl^{3+} + H_2O$	-0.055	$Co_2 + 2H^+ + 2e \rightleftharpoons HCOOH$	-0.199
$2H_2SO_3 + 2H^+ + 2e \rightleftharpoons HS_2O_4^- + 2H_2O$	-0.056	$Mo^{3+} + 3e \rightleftharpoons Mo$	-0.200
$P(white) + 3H^+ + 3e \rightleftharpoons PH_3(g)$	-0.063	$2SO_4^{2-} + 4H^+ + 2e \rightleftharpoons S_2O_6^{2-} + H_2O$	-0.22
$O_2^- + H_2O + 2e \rightleftharpoons HO_2^- + OH^-$	-0.076	$Cu(OH)_2 + 2e \rightleftharpoons Cu + 2OH^-$	-0.222
$2Cu(OH)_2 + 2e \rightleftharpoons Cu_2O + 2OH^- + H_2O$	-0.080	$CdSO_4 + 2e \rightleftharpoons Cd + SO_4^{2-}$	-0.246
$WO_3 + 6H^+ + 6e \rightleftharpoons W + 3H_2O$	-0.090	$V(OH)_4^+ + 4H^+ + 5e \rightleftharpoons V + 4H_2O$	-0.254
$P(red) + 3H^+ + 3e \rightleftharpoons PH_3(g)$	-0.111	$V^{3+} + 3e \rightleftharpoons V^{2+}$	-0.255
$GeO_2 + 2H^+ + 2e \rightleftharpoons GeO + H_2O$	-0.118	$Ni^{2+} + 2e \rightleftharpoons Ni$	-0.257

ELECTROCHEMICAL SERIES (Cont)

Reaction	E (V)
$PbCl_2 + 2\,e \rightleftharpoons Pb + 2\,Cl^-$	−0.2675
$H_3PO_4 + 2\,H^+ + 2\,e \rightleftharpoons H_3PO_3 + H_2O$	−0.276
$Co^{2+} + 2\,e \rightleftharpoons Co$	−0.28
$PbBr_2 + 2\,e \rightleftharpoons Pb + 2\,Br^-$	−0.284
$Tl^+ + e \rightleftharpoons Tl(Hg)$	−0.3338
$Tl^+ + e \rightleftharpoons Tl$	−0.336
$In^{3+} + 3\,e \rightleftharpoons In$	−0.3382
$TlOH + e \rightleftharpoons Tl + OH^-$	−0.34
$PbF_2 + 2\,e \rightleftharpoons Pb + 2\,F^-$	−0.3444
$PbSO_4 + 2\,e \rightleftharpoons Pb(Hg) + SO_4^{2-}$	−0.3505
$Cd^{2+} + 2\,e \rightleftharpoons Cd(Hg)$	−0.3521
$PbSO_4 + 2\,e \rightleftharpoons Pb + SO_4^{2-}$	−0.3588
$Cu_2O + H_2O + 2\,e \rightleftharpoons 2\,Cu + 2\,OH^-$	−0.360
$Eu^{3+} + e \rightleftharpoons Eu^{2+}$	−0.36
$PbI_2 + 2\,e \rightleftharpoons Pb + 2\,I^-$	−0.365
$SeO_3^{2-} + 3\,H_2O + 4\,e \rightleftharpoons Se + 6\,OH^-$	−0.366
$Ti^{3+} + e \rightleftharpoons Ti^{2+}$	−0.368
$Se + 2\,H^+ + 2\,e \rightleftharpoons H_2Se(aq)$	−0.399
$In^{2+} + e \rightleftharpoons In^+$	−0.40
$Cd^{2+} + 2\,e \rightleftharpoons Cd$	−0.4030
$Cr^{3+} + e \rightleftharpoons Cr^{2+}$	−0.407
$2\,S + 2\,e \rightleftharpoons S_2^{2-}$	−0.42836
$Tl_2SO_4 + 2\,e \rightleftharpoons 2\,Tl + SO_4^{2-}$	−0.4360
$In^{3+} + 2\,e \rightleftharpoons In^+$	−0.443
$Fe^{2+} + 2\,e \rightleftharpoons Fe$	−0.447
$H_3PO_3 + 3\,H^+ + 3\,e \rightleftharpoons P + 3\,H_2O$	−0.454
$Bi_2O_3 + 3\,H_2O + 6\,e \rightleftharpoons 2\,Bi + 6\,OH^-$	−0.46
$NO_2^- + H_2O + e \rightleftharpoons NO + 2\,OH^-$	−0.46
$PbHPO_4 + 2\,e \rightleftharpoons Pb + HPO_4^{2-}$	−0.465
$S + 2\,e \rightleftharpoons S^{2-}$	−0.47627
$S + H_2O + 2\,e \rightleftharpoons HS^- + OH^-$	−0.478
$NiO_2 + 2\,H_2O + 2\,e \rightleftharpoons Ni(OH)_2 + 2\,OH^-$	−0.490
$In^{3+} + e \rightleftharpoons In^{2+}$	−0.49
$H_3PO_3 + 2\,H^+ + 2\,e \rightleftharpoons H_3PO_2 + H_2O$	−0.499
$TiO_2 + 4\,H^+ + 2\,e \rightleftharpoons Ti^{2+} + 2\,H_2O$	−0.502
$H_3PO_2 + H^+ + e \rightleftharpoons P + 2\,H_2O$	−0.508
$Sb + 3\,H^+ + 3\,e \rightleftharpoons SbH_3$	−0.510
$HPbO_2^- + H_2O + 2\,e \rightleftharpoons Pb + 3\,OH^-$	−0.537
$TlCl + e \rightleftharpoons Tl + Cl^-$	−0.5568
$Ga^{3+} + 3\,e \rightleftharpoons Ga$	−0.560
$Fe(OH)_3 + e \rightleftharpoons Fe(OH)_2 + OH^-$	−0.56
$TeO_3^{2-} + 3\,H_2O + 4\,e \rightleftharpoons Te + 6\,OH^-$	−0.57
$2\,SO_3^{2-} + 3\,H_2O + 4\,e \rightleftharpoons S_2O_4^{2-} + 6\,OH^-$	−0.571
$PbO + H_2O + 2\,e \rightleftharpoons Pb + 2\,OH^-$	−0.580
$ReO_4^- + 4\,H_2O + 7\,e \rightleftharpoons Re + 8\,OH^-$	−0.584
$SbO_3^- + H_2O + 2\,e \rightleftharpoons SbO_2^- + 2\,OH^-$	−0.59
$U^{4+} + e \rightleftharpoons U^{3+}$	−0.607
$As + 3\,H^+ + 3\,e \rightleftharpoons AsH_3$	−0.608
$Nb_2O_5 + 10\,H^+ + 10\,e \rightleftharpoons 2\,Nb + 5\,H_2O$	−0.644
$TlBr + e \rightleftharpoons Tl + Br^-$	−0.658
$SbO_2^- + 2\,H_2O + 3\,e \rightleftharpoons Sb + 4\,OH^-$	−0.66
$AsO_2^- + 2\,H_2O + 3\,e \rightleftharpoons As + 4\,OH^-$	−0.68
$Ag_2S + 2\,e \rightleftharpoons 2\,Ag + S^{2-}$	−0.691
$AsO_4^{3-} + 2\,H_2O + 2\,e \rightleftharpoons AsO_2^- + 4\,OH^-$	−0.71
$Ni(OH)_2 + 2\,e \rightleftharpoons Ni + 2\,OH^-$	−0.72
$Co(OH)_2 + 2\,e \rightleftharpoons Co + 2\,OH^-$	−0.73
$H_2SeO_3 + 4\,H^+ + 4\,e \rightleftharpoons Se + 3\,H_2O$	−0.74
$Cr^{3+} + 3\,e \rightleftharpoons Cr$	−0.744
$Ta_2O_5 + 10\,H^+ + 10\,e \rightleftharpoons 2\,Ta + 5\,H_2O$	−0.750
$TlI + e \rightleftharpoons Tl + I^-$	−0.752
$Zn^{2+} + 2\,e \rightleftharpoons Zn$	−0.7618
$Zn^{2+} + 2\,e \rightleftharpoons Zn(Hg)$	−0.7628
$Te + 2\,H^+ + 2\,e \rightleftharpoons H_2Te$	−0.793
$ZnSO_4 \cdot 7H_2O + 2\,e \rightleftharpoons Zn(Hg) + SO_4^{2-}$ (Sat'd $ZnSO_4$)	−0.7993
$Cd(OH)_2 + 2\,e \rightleftharpoons Cd(Hg) + 2\,OH^-$	−0.809
$2\,H_2O + 2\,e \rightleftharpoons H_2 + 2\,OH^-$	−0.8277
$2\,NO_3^- + 2\,H_2O + 2\,e \rightleftharpoons N_2O_4 + 4\,OH^-$	−0.85
$H_3BO_3 + 3\,H^+ + 3\,e \rightleftharpoons B + 3\,H_2O$	−0.8698
$P + 3\,H_2O + 3\,e \rightleftharpoons PH_3(g) + 3\,OH^-$	−0.87
$HSnO_2^- + H_2O + 2\,e \rightleftharpoons Sn + 3\,OH^-$	−0.909
$Cr^{2+} + 2\,e \rightleftharpoons Cr$	−0.913
$Se + 2\,e \rightleftharpoons Se^{2-}$	−0.924

Reaction	E
$SO_4^{2-} + H_2O + 2\,e \rightleftharpoons SO_3^{2-} + 2\,OH^-$	-0.93
$Sn(OH)_6^{2-} + 2\,e \rightleftharpoons HSnO_2^- + 3\,OH^- + H_2O$	-0.93
$NpO_2 + H_2O + H^+ + e \rightleftharpoons Np(OH)_3$	-0.962
$PO_4^{3-} + 2\,H_2O + 2\,e \rightleftharpoons HPO_3^{2-} + 3\,OH^-$	-1.05
$Nb^{3+} + 3\,e \rightleftharpoons Nb$	-1.099
$2\,SO_3^{2-} + 2\,H_2O + 2\,e \rightleftharpoons S_2O_4^{2-} + 4\,OH^-$	-1.12
$Te + 2\,e \rightleftharpoons Te^{2-}$	-1.143
$V^{2+} + 2\,e \rightleftharpoons V$	-1.175
$Mn^{2+} + 2\,e \rightleftharpoons Mn$	-1.185
$CrO_2 + 2\,H_2O + 3\,e \rightleftharpoons Cr + 4\,OH^-$	-1.2
$ZnO_2^{2-} + 2\,H_2O + 2\,e \rightleftharpoons Zn + 4\,OH^-$	-1.215
$H_2GaO_3^- + H_2O + 3\,e \rightleftharpoons Ga + 4\,OH^-$	-1.219
$H_2BO_3^- + 5\,H_2O + 8\,e \rightleftharpoons BH_4^- + 8\,OH^-$	-1.24
$SiF_6^{2-} + 4\,e \rightleftharpoons Si + 6\,F^-$	-1.24
$Ce^{3+} + 3\,e \rightleftharpoons Ce(Hg)$	-1.4373
$UO_2^{2+} + 4\,H^+ + 6\,e \rightleftharpoons U + 2\,H_2O$	-1.444
$Cr(OH)_3 + 3\,e \rightleftharpoons Cr + 3\,OH^-$	-1.48
$HfO_2 + 4\,H^+ + 4\,e \rightleftharpoons Hf + 2\,H_2O$	-1.505
$ZrO_2 + 4\,H^+ + 4\,e \rightleftharpoons Zr + 2\,H_2O$	-1.553
$Mn(OH)_2 + 2\,e \rightleftharpoons Mn + 2\,OH^-$	-1.56
$Ba^{2+} + 2\,e \rightleftharpoons Ba(Hg)$	-1.570
$Ti^{2+} + 2\,e \rightleftharpoons Ti$	-1.630
$HPO_3^{2-} + 2\,H_2O + 2\,e \rightleftharpoons H_2PO_2^- + 3\,OH^-$	-1.65
$Al^{3+} + 3\,e \rightleftharpoons Al$	-1.662
$SiO_3^{2-} + H_2O + 4\,e \rightleftharpoons Si + 6\,OH^-$	-1.697
$HPO_3^{2-} + 2\,H_2O + 3\,e \rightleftharpoons P + 5\,OH^-$	-1.71
$HfO^{2+} + 2\,H^+ + 4\,e \rightleftharpoons Hf + H_2O$	-1.724
$ThO_2 + 4\,H^+ + 4\,e \rightleftharpoons Th + 2\,H_2O$	-1.789
$H_3BO_3 + H_2O + 3\,e \rightleftharpoons B + 4\,OH^-$	-1.79
$Sr^{2+} + 2\,e \rightleftharpoons Sr(Hg)$	-1.793
$U^{3+} + 3\,e \rightleftharpoons U$	-1.798
$H_2PO_2^- + e \rightleftharpoons P + 2\,OH^-$	-1.82
$Be^{2+} + 2\,e \rightleftharpoons Be$	-1.847
$Np^{3+} + 3\,e \rightleftharpoons Np$	-1.856
$Th^{4+} + 4\,e \rightleftharpoons Th$	-1.899
$Pu^{3+} + 3\,e \rightleftharpoons Pu$	-2.031
$AlF_6^{3-} + 3\,e \rightleftharpoons Al + 6\,F^-$	-2.069
$Sc^{3+} + 3\,e \rightleftharpoons Sc$	-2.077
$H_2 + 2\,e \rightleftharpoons 2\,H^-$	-2.23
$H_2AlO_3^- + H_2O + 3\,e \rightleftharpoons Al + 4\,OH^-$	-2.33
$ZrO(OH)_2 + H_2O + 4\,e \rightleftharpoons Zr + 4\,OH^-$	-2.36
$Mg^{2+} + 2\,e \rightleftharpoons Mg$	-2.372
$Y^{3+} + 3\,e \rightleftharpoons Y$	-2.372
$Eu^{3+} + 3\,e \rightleftharpoons Eu$	-2.407
$Nd^{3+} + 3\,e \rightleftharpoons Nd$	-2.431
$Th(OH)_4 + 4\,e \rightleftharpoons Th + 4\,OH^-$	-2.48
$Ce^{3+} + 3\,e \rightleftharpoons Ce$	-2.483
$HfO(OH)_2 + H_2O + 4\,e \rightleftharpoons Hf + 4\,OH^-$	-2.50
$La^{3+} + 3\,e \rightleftharpoons La$	-2.522
$Be_2O_3^{2-} + 3\,H_2O + 4\,e \rightleftharpoons 2\,Be + 6\,OH^-$	-2.63
$Mg(OH)_2 + 2\,e \rightleftharpoons Mg + 2\,OH^-$	-2.690
$Mg^+ + e \rightleftharpoons Mg$	-2.70
$Na^+ + e \rightleftharpoons Na$	-2.71
$Ca^{2+} + 2\,e \rightleftharpoons Ca$	-2.868
$Sr(OH)_2 + 2\,e \rightleftharpoons Sr + 2\,OH^-$	-2.88
$Sr^{2+} + 2\,e \rightleftharpoons Sr$	-2.89
$La(OH)_3 + 3\,e \rightleftharpoons La + 3\,OH^-$	-2.90
$Ba^{2+} + 2\,e \rightleftharpoons Ba$	-2.912
$Cs^+ + e \rightleftharpoons Cs$	-2.92
$K^+ + e \rightleftharpoons K$	-2.931
$Rb^+ + e \rightleftharpoons Rb$	-2.98
$Ba(OH)_2 + 2\,e \rightleftharpoons Ba + 2\,OH^-$	-2.99
$Ca(OH)_2 + 2\,e \rightleftharpoons Ca + 2\,OH^-$	-3.02
$Li^+ + e \rightleftharpoons Li$	-3.0401
$3\,N_2 + 2\,H^+ + 2\,e \rightleftharpoons 2\,NH_3$	-3.09
$Eu^{2+} + 2\,e \rightleftharpoons Eu$	-3.395
$Ca^+ + e \rightleftharpoons Ca$	-3.80
$Sr^+ + e \rightleftharpoons Sr$	-4.10

Source: Handbook of Chemistry and Physics, 65th Edition, pp. D159-D162 (CRC Press).

**TYPICAL POTENTIAL-pH (POURBAIX) DIAGRAM
IRON IN WATER AT 25°C**

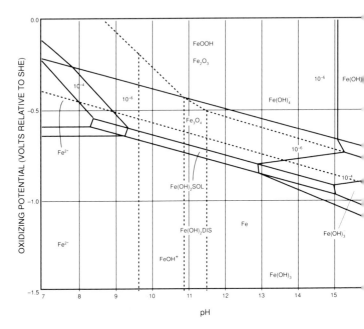

IONIC SPECIES ARE AT ACTIVITIES OF 10^{-4} AND 10^{-6}

Source: D. Inman and D. G. Lovering, Comprehensive Treatise of Electrochemistry,
Vol. 7, Plenum Publishing, 1983

STANDARD ENVIRONMENTS FOR
ENVIRONMENTAL CRACKING TESTS

Standard	Environment	Temperature °C	Materials
NACE TM0177	5.0% NaCl + 0.5% Acetic acid, saturated with H_2S (1 atm.)	21-27	All metals
ASTM G 35	Polythionic acid solution	22-25	Stainless steels Related nickel-chromium-iron alloys
ASTM G 36	45% $MgCl_2$ (boiling)	154-156	Stainless steels Related alloys
ASTM G 37	Mattsson's Solution, pH 7.2 $(CuSO_4 + (NH4)_2SO_4 + NH_4OH)$	18-24	Copper-zinc base alloys
ASTM G 41	NaCl or other salts or synthetic seawater	230-450	All metals
ASTM G 44, G 47	3.5% NaCl (alt. immersion)	26-28	Aluminum alloys Ferrous alloys

SPECIMEN TYPES USED IN ENVIRONMENTAL CRACKING TESTS

Loading System	Specimen Type	Standard	Product Form					Used to Determine				
			Sheet	Bar	Plate	Tube	Wire	$\overline{T}_f$	σ_{th}	K_{Iscc}	da/dt	I_{ICC}
Constant Load	Bent beam	—		X	X			X	X			
	Notched beam	—		X	X			X	X			
	Direct tension	ASTM G 49		X	X	X		X	X	X		
Constant Strain	U-bend	ASTM G 30	X					X				
	Cup	—	X					X				
	C-ring	ASTM G 38	X	X	X	X		X	X			
	Bent beam	ASTM G 39		X	X	X		X	X			
	Direct tension	ASTM G 49		X	X	X		X	X			
	Tuning fork	—		X	X			X				
	Weld bead	—	X					X				
	Rough ground	—	X					X				
	Hairpin	—					X	X				
Slow Strain Rate	Round tensile	ISO 156/WG2		X	X				X	X		X
	Tube					X						X
Sustained-Loading Crack-Growth	DCB	—			X	X				X	X	

T_f time to failure.
σ_{th} threshold stress.
K_{Iscc} threshold stress intensity.
da/dt crack growth rate.
I_{ICC} index of susceptibility.

Source: NACE-OSU Corrosion Course.

PLANNED INTERVAL CORROSION TEST

Used to Evaluate Effect of Time on Corrosion Rate and to Determine if the Time Effect is Due to Changes in Environment Corrosiveness or in Metal Corrodibility

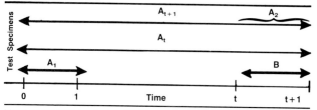

Identical specimens all placed in the same corrosive fluid. Imposed conditions of the test kept constant for entire time $t+1$. Letters A_1, A_t, A_{t+1}, B, represent corrosion damage experienced by each test specimen. A_2 is calculated by subtracting A_t from A_{t+1}.

	Occurrences During Corrosion Test	Criteria
	unchanged	$A_1 = B$
Liquid corrosiveness	decreased	$B < A_1$
	increased	$A_1 < B$
	unchanged	$A_2 = B$
Metal corrodibility	decreased	$A_2 < B$
	increased	$B < A_2$

	Combinations of Situations	
Liquid Corrosiveness	**Metal Corrodibility**	**Criteria**
1. unchanged	unchanged	$A_1 = A_2 = B$
2. unchanged	decreased	$A_2 < A_1 = B$
3. unchanged	increased	$A_1 = B < A_2$
4. decreased	unchanged	$A_2 = B < A_1$
5. decreased	decreased	$A_2 < B < A_1$
6. decreased	increased	$A_1 > B < A_2$
7. increased	unchanged	$A_1 < A_2 = B$
8. increased	decreased	$A_1 < B > A_2$
9. increased	increased	$A_1 < B < A_2$

EXAMPLE

	Interval, days	Wt. Loss, mg	Penetration, mils	Apparent Corrosion Rate, mils/yr.
A_1 0-1	1080	1.69	620
A_t 0-3	1430	2.24	270
A_{t+1} 0-4	1460	2.29	210
B 3-4	70	0.11	40
A_2 . . . . calc. 3-4	30	0.05	18

$$A_2 < B < A_1$$
$$0.05 < 0.11 < 1.69$$

Therefore, liquid markedly decreased in corrosiveness during test, and formation of partially protective scale on the steel was indicated.

Source: *Chemical Engineering Progress*, Vol. 43, p. 318 (1947).

CORROSION RATE CONVERSION FACTORS

$$\text{Mils/year (mpy)} = C \times \frac{\text{weight loss}}{\text{area} \times \text{time}} \times K$$

$$\text{Millimetres/year (mm/y)} = 0.0254 \text{ mpy}$$

Weight Loss	Area	C Factors				
		Hour	Day	Week	Month	Year
mg	cm²	437	18.2	2.59	0.598	0.0498
	dm²	4.37	0.182	0.0259	5.98×10^{-3}	4.98×10^{-4}
	m²	0.0437	1.82×10^{-3}	2.59×10^{-4}	5.98×10^{-5}	4.98×10^{-6}
	in²	67.7	2.82	0.402	0.0927	7.72×10^{-3}
	ft²	0.470	0.0196	2.79×10^{-3}	6.44×10^{-4}	5.36×10^{-5}
g	cm²	437×10^{3}	182×10^{2}	2590	598	49.8
	dm²	4370	182	25.9	5.98	0.498
	m²	43.7	1.82	0.259	0.0598	4.98×10^{-3}
	in²	677×10^{2}	2820	402	92.7	7.72
	ft²	470	19.6	2.79	0.644	0.0536
lb	cm²	198×10^{6}	825×10^{4}	118×10^{4}	271×10^{3}	226×10^{2}
	dm²	198×10^{4}	825×10^{2}	118×10^{2}	2710	226
	m²	198×10^{2}	825	118	27.1	2.26
	in²	307×10^{5}	128×10^{4}	182×10^{3}	420×10^{2}	3500
	ft²	213×10^{3}	8880	1270	292	24.3

K is a density factor.
K = 1.000 for carbon steel.
K factors for other alloys are given on the next page.

Courtesy Aaron Wachter.

EXAMPLE: A 5.0 square inch specimen of copper has a weight loss of 218 mg in a 40 hour corrosion test.

$$\text{mpy} = 67.7 \times \frac{218}{5.0 \times 40} \times 0.88 = 65$$

DENSITIES OF COMMON ALLOYS
(K = ratio of carbon steel density to that of alloy)

UNS	Common Name	Density g/cm³	K	UNS	Common Name	Density g/cm³	K
A91100	Al 1100	2.72	2.89	N06007	G Alloy	8.34	0.94
A93003	Al 3003	2.74	2.87	N06022	C-22 Alloy	8.69	0.90
A95052	Al 5052	2.68	2.93	N06030	G-30 Alloy	8.22	0.96
A96061	Al 6061	2.70	2.91	N06455	C-4 Alloy	8.64	0.91
A97075	Al 7075	2.80	2.81	N06600	600 Alloy	8.47	0.93
C11000	ETP Copper	8.94	0.88	N06601	601 Alloy	8.11	0.97
C22000	Commercial Bronze	8.89	0.88	N06625	625 Alloy	8.44	0.93
C23000	Red Brass	8.75	0.90	N06985	G-3 Alloy	8.30	0.95
C26000	Cartridge Brass	8.53	0.92	N07001	Waspaloy	8.19	0.96
C27000	Yellow Brass	8.39	0.94	N07041	Rene 41	8.25	0.95
C28000	Muntz Metal	8.39	0.94	N07718	718 Alloy	8.19	0.96
C44300	Admiralty brass, As	8.52	0.92	N07750	X-750 Alloy	8.28	0.95
C46500	Naval Brass, As	8.41	0.93	N08020	20Cb-3	8.08	0.97
C51000	Phosphor Bronze A	8.86	0.89	N08024	20Mo-4	8.11	0.97
C52400	Phosphor Bronze D	8.78	0.90	N08026	20Mo-6	8.13	0.97
C61300	Aluminum Bronze, 7%	7.89	1.00	N08028	Sanicro 28	8.0	0.98
C61400	Aluminum Bronze D	7.78	1.01	N08366	AL-6X	8.0	0.98
C63000	Ni-Al Bronze	7.58	1.04	N08800	800 Alloy	7.94	0.99
C65500	High-Silicon Bronze	8.52	0.92	N08825	825 Alloy	8.14	0.97
C67500	Manganese Bronze A	8.36	0.94	N08904	904L Alloy	8.0	0.98
C68700	Aluminum Brass, As	8.33	0.94	N08925	25-6Mo	8.1	0.97
C70600	9-10 Copper-Nickel	8.94	0.88	N09925	925 Alloy	8.05	0.98
C71500	70-30 Copper-Nickel	8.94	0.88	N10003	N Alloy	8.79	0.89
C75200	Nickel Silver	8.73	0.90	N10004	W Alloy	9.03	0.87
C83600	Ounce Metal	8.80	0.89	N10276	C-276 Alloy	8.89	0.88
C86500	Manganese Bronze	8.3	0.95	N10665	B-2 Alloy	9.22	0.85
C90500	Gun Metal	8.72	0.90	R03600	Molybdenum	10.22	0.77
C92200	M Bronze	8.64	0.91	R04210	Niobium	8.57	0.92
C95700	Cast Mn-Ni-Al Bronze	7.53	1.04	R05200	Tantalum	16.60	0.47
C95800	Cast Ni-Al Bronze	7.64	1.03	R50250	Titanium, Gr 1	4.54	1.73
F10006	Gray Cast Iron	7.20	1.09	R50400	Titanium, Gr 2	4.54	1.73
F20000	Malleable Cast Iron	7.27	1.08	R53400	Titanium, Gr 12	4.52	1.74
F32800	Ductile Iron	7.1	1.11	R56400	Titanium, Gr 5	4.43	1.77
F41002	Ni-Resist Type 2	7.3	1.08	R60702	Zr 702	6.53	1.20
F43006	Ductile Ni-Resist, D5	7.68	1.02	S20100	201 SS	7.94	0.99
F47003	Duriron	7.0	1.12	S20200	202 SS	7.94	0.99
G10200	1020 Carbon Steel	7.86	1.00	S30400	304 SS	7.94	0.99
G41300	4130 Steel	7.86	1.00	S30403	304L SS	7.94	0.99
J91150	CA-15 Cast SS	7.61	1.03	S30900	309 SS	7.98	0.98
J91151	CA-15M Cast SS	7.61	1.03	S31000	310 SS	7.98	0.98
J91540	CA-6NM Cast SS	7.7	1.02	S31254	254 SMO	8.0	0.98
J92600	CF-8 Cast SS	7.75	1.01	S31500	3RE60	7.75	1.01
J92800	CF-3MN Cast SS	7.75	1.01	S31600	316 SS	7.98	0.98
J92900	CF-8M Cast SS	7.75	1.01	S31603	316L SS	7.98	0.98
J94204	HK-40 Cast SS	7.75	1.01	S31700	317 SS	7.98	0.98
J95150	CN-7M Cast SS	8.00	0.98	S32100	321 SS	7.94	0.99
K11597	1.25Cr-0.5Mo Steel	7.85	1.00	S32550	Ferralium 255	7.81	1.01
K81340	9Ni Steel	7.86	1.00	S32950	7 Mo Plus	7.75	1.01
L51120	Chemical Lead	11.3	0.70	S34700	347 SS	8.03	0.98
M11311	Mg AZ31B	1.77	4.44	S41000	410 SS	7.70	1.02
N02200	Nickel 200	8.89	0.88	S43000	430 SS	7.72	1.02
N04400	400 Alloy	8.80	0.89	S44600	446 SS	7.65	1.03
N05500	K-500 Alloy	8.44	0.93	S50100	5Cr-0.5Mo Steel	7.82	1.01
N06002	X Alloy	8.23	0.96	S50400	9Cr-1Mo Steel	7.67	1.02

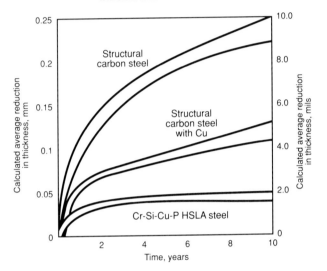

**ATMOSPHERIC CORROSION OF STEEL vs TIME IN AN
INDUSTRIAL ATMOSPHERE**

Corrosion of three types of steels in an industrial atmosphere.
Source: Metals Handbook, 9th ed., Volume 13, p. 1304, ASM 1987

CORROSION RATES OF CARBON STEEL
CALIBRATING SPECIMENS AT VARIOUS LOCATIONS

Location	Type of environment	Corrosion rate (a) μm/y	mpy
Norman Wells, NWT, Canada	Polar	0.76	0.03
Phoenix, AZ	Rural arid	4.6	0.18
Esquimalt, Vancouver Island BC, Canada	Rural marine	13	0.5
Detroit, MI	Industrial	14.5	0.57
Fort Amidor Pier, CZ	Marine	14.5	0.57
Morenci, MI	Urban	19.5	0.77
Potter County, PA	Rural	20	0.8
Waterbury, CT	Industrial	22.8	0.89
State College, PA	Rural	23	0.9
Montreal, Que. Canada	Urban	23	0.9
Durham, NH	Rural	28	1.1
Middletown, OH	Semi-industrial	28	1.1
Pittsburgh, PA	Industrial	30	1.2
Columbus, OH	Industrial	33	1.3
Trail, BC, Canada	Industrial	33	1.3
Cleveland, OH	Industrial	38	1.5
Bethlehem, PA	Industrial	38	1.5
London, Battersea, England	Industrial	46	1.8
Monroeville, PA	Semi-industrial	48	1.9
Newark, NJ	Industrial	51	2.0
Manila, Philippine Islands	Tropical marine	51	2.0
Limon Bay, Panama, CZ	Tropical marine	61	2.4
Bayonne, NJ	Industrial	79	3.1
East Chicago, IN	Industrial	84	3.3
Brazos River, TX	Industrial marine	94	3.7
Cape Kennedy, FL (60 ft elev., 60 yd from ocean)	Marine	132	5.2
Kure Beach, NC (800 ft from ocean)	Marine	147	5.8
Cape Kennedy, FL (30 ft elev., 60 yd from ocean)	Marine	165	6.5
Daytona Beach, FL	Marine	295	11.6
Cape Kennedy, FL (ground level, 60 yd from ocean)	Marine	442	17.4
Point Reyes, CA	Marine	500	19.7
Kure Beach, NC (80 ft from ocean)	Marine	533	21.0
Galeta Point Beach, Panama, CZ	Marine	686	27.0
Cape Kennedy, FL (beach)	Marine	1070	42.0

(a) Two-year average.

Source: Metals Handbook, 9th ed., Volume 1, p. 720, ASM 1978.

CORROSION OF STRUCTURAL STEEL
IN VARIOUS ENVIRONMENTS

| Type of Atmosphere | Time, Yr. | Average Reduction in Thickness, Mils[a] | | | | | |
		Structural Carbon Steel	Structural Copper Steel	UNS K11510[b]	UNS K11430[c]	UNS K11630[d]	UNS K11576[e]
Industrial	3.5	3.3	2.6	1.3	1.8	1.4	2.2
(Newark, NJ)	7.5	4.1	3.2	1.5	2.1	1.7	—
	15.5	5.3	4.0	1.8	—	2.1	—
Semi-industrial	1.5	2.2	1.7	1.1	1.4	1.2	1.6
(Monroeville, PA)	3.5	3.7	2.5	1.2	2.1	1.4	2.4
	7.5	5.1	3.2	1.4	2.4	1.7	—
	15.5	7.3	4.7	1.8	—	1.8	—
Semi-industrial	1.5	1.8	1.4	1.0	1.3	1.0	1.5
(South Bend, PA)	3.5	2.9	2.2	1.3	1.9	1.5	2.4
	7.5	4.6	3.2	1.8	2.7	1.9	—
	15.5	7.0	4.8	2.2	—	2.5	—
Rural	2.5	—	1.3	0.8	1.2	—	—
(Potter County, PA)	3.5	2.0	1.7	1.1	1.4	1.2	1.8
	7.5	3.0	2.5	1.3	1.5	1.5	—
	15.5	4.7	3.8	1.4	—	2.0	—
Moderate marine	0.5	0.9	0.8	0.6	0.8	0.7	1.0
(Kure Beach, NC,	1.5	2.3	1.9	1.1	1.7	1.2	1.7
800 ft from ocean)	3.5	4.9	3.3	1.8	2.5	1.9	2.2
	7.5	5.6	4.5	2.5	3.7	2.9	—
Severe marine	0.5	7.2	4.3	2.2	3.8	1.1	0.7
(Kure Beach, NC,	2.0	36.0	19.0	3.3	12.2	—	2.1
80 ft from ocean)	3.5	57.0	38.0	—	28.7	3.9	3.9
	5.0	f	f	19.4	38.8	5.0	—

a) To obtain equivalent values in μm, multiply listed value by 25. b) ASTM A242 (type 1). c) ASTM A588 (gradeA). d) ASTM A514 (type B) and A517 (grade B). e) ASTM A514 (type F) and A517 (grade F). f) Specimen corroded completely away.

Source: Metals Handbook, 9th ed., Volume 1, p. 723, ASM 1978.

EFFECT OF AMOUNT OF ZINC ON SERVICE LIFE OF GALVANIZED SHEET IN VARIOUS ENVIRONMENTS

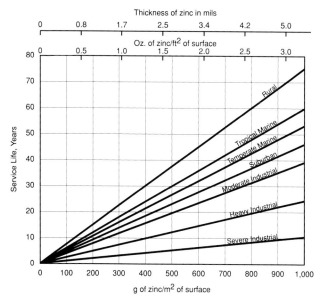

Surface life is measured in years to the appearance of first significant rusting.

Source: Metals Handbook, 9th Ed., Volume 1, p. 753, ASM 1978

DEVELOPMENT OF RUST ON ZINC AND CADMIUM-PLATED STEELS IN A MARINE ATMOSPHERE

25m lots at Kure Beach, NC
Coating thicknesses 0.0002 and 0.0005 inch.

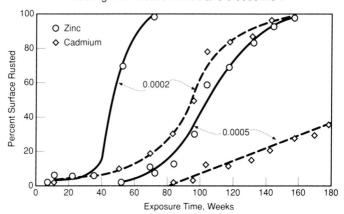

Source: F.W. Fink, et al, *Corrosion of Metals in Marine Environments,* Battelle Memorial Institute DMIC Report 254, NTIS AD-712 5B5-S, pp 7, 13, 1970

ATMOSPHERIC CORROSION OF ZINC IN VARIOUS LOCATIONS AS A FUNCTION OF TIME

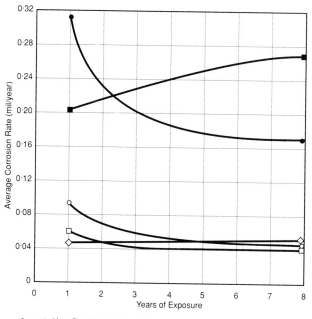

● ● Kure Beach, 80 ft site, marine
■ ■ New York, NY, industrial
□ □ State College, Pennsylvania, rural
○ ○ Kure Beach, 800 ft site, marine
◇ ◇ Middletown, Ohio, semi-industrial

Source: Zinc, Its Corrosion Resistance, 2nd Ed., p. 42,
International Lead Zinc Research Organization, 1983

ATMOSPHERIC CORROSION OF VARIOUS METALS AND ALLOYS

10-Year Exposure Times
Corrosion rates are given in mils/yr (1 mil/yr = 0.025 mm/yr).

	New York, NY (urban-industrial)	La Jolla, CA (marine)	State College, PA (rural)
Aluminum	0.032	0.028	0.025
Copper	0.047	0.052	0.050
Lead	0.017	0.016	0.021
Tin	0.047	0.091	0.112
Nickel	0.128	0.004	0.006
Ni-Cu Alloy 400	0.053	0.007	0.006
Zinc (99.9%)	0.202	0.063	0.069
Zinc (99.0%)	0.193	0.069	0.068
0.2% C Steel (a) (0.02% P, 0.05% S, 0.05% Cu, 0.02% Ni, 0.02% Cr)	0.48	—	—
Low-alloy steel(a) (0.1% C, 0.2% P, 0.04% S, 0.03% Ni, 1.1% Cr, 0.4% Cu)	0.09	—	—

Source: Metals Handbook, 9th ed., Volume 13, p. 82, ASM 1987.

CORROSION OF COPPER ALLOYS IN MARINE ATMOSPHERES

7 Year Exposure of Specimens at 25 Meter Lot at Kure Beach, NC (KB) or Point Reves, CA (PR). Corrosion Rates by Weight Loss

UNS	Common Name	Mils/Year		Appearance (KB)
		KB	PR	
C11000	ETP Copper	0.065	0.025	Brown film, smooth, slight patina near edges
C23000	Red Brass	0.033	0.026	Brown-maroon film, smooth
C26000	Cartridge Brass	0.030	0.017	Brown-maroon film, smooth, very slight patina
C42000	Tin Brass	0.024	—	Dark maroon, smooth
C50500	Phos. Bronze E (1.25% Sn)	0.069	0.017	"Mink brown", slight patina
C51000	Phos. Bronze, A (5% Sn)	0.099	—	Maroon film, heavy etch
—	Alum. Bronze (7.5% Al)	0.013	—	Light tan film, smooth
C70400	Copper-Nickel (5% Ni)	0.033	—	Dark brown, plus patina streaks on panel face
C70700	Copper-Nickel (10% Ni)	0.038	—	Uniform maroon with patina at edges
C71100	Copper-Nickel (22% Ni)	0.031	—	Greenish brown, green near edges, slight etch
C74500	Nickel Silver (10% Ni)	0.024	0.010	Brown with slight patina film in center, green near edges, smooth
C75200	Nickel Silver (18% Ni)	0.021	—	Brown film in center, green near edges, smooth

Source: F. W. Fink et al, Corrosion of Metals in Marine Environments, Battelle Memorial Institute DMIC Report 254, NTIS AD-712 585-S, pp 7, 13, 1970.

RELATIVE PERFORMANCE OF STAINLESS STEELS
EXPOSED IN A MARINE ATMOSPHERE

KURE BEACH – 26 YEARS

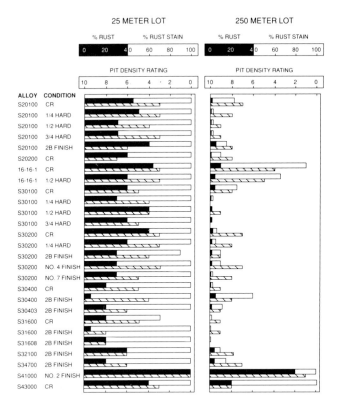

PIT DENSITY RATING

10 = No pits
9 = 0 to 0.1% coverage   4 = 2.5 to 5.0% coverage
8 = 0.1 to 0.25% coverage   3 = 5.0 to 10.0% coverage
7 = 0.25 to 0.5% coverage   2 = 10.0 to 25.0% coverage
6 = 0.5 to 1.0% coverage   1 = 25.0 to 50.0% coverage
5 = 1.0 to 2.5% coverage   0 = > 50% coverage

Source: Baker and Lee, ASTM Special Publication 965, p. 62 (1986)

CORROSION FACTORS FOR CARBON STEEL IN SEAWATER

Factor in Seawater	Effect on Iron and Steel
Chloride Ion	Highly corrosive to ferrous metals. Carbon steel and common ferrous metals cannot be passivated. (Sea salt is about 55% chloride.)
Electrical Conductivity	High conductivity makes it possible for anodes and cathodes to operate over long distances, thus corrosion possibilities are increased and the total attack may be much greater than that for the same structure in fresh water.
Oxygen	Steel corrosion, for the most part, is cathodically controlled. Oxygen, by depolarizing the cathode, facilitates the attack; thus a high oxygen content increases corrosivity.
Velocity	Corrosion rate is increased, especially in turbulent flow. Moving seawater may: (1) destroy rust barrier, and (2) provide more oxygen. Impingement attack tends to promote rapid penetration. Cavitation damage exposes the fresh steel surface to further corrosion.
Temperature	Increasing ambient temperature tends to accelerate attack. Heated seawater may deposit protective scale or lose its oxygen; either or both actions tend to reduce attack.
Biofouling	Hard-shell animal fouling tends to reduce attack by restricting access of oxygen. Bacteria can take part in the corrosion reaction in some cases.
Stress	Cyclic stress sometimes accelerates failure of a corroding steel member. Tensile stresses near yield also promote failure in special situations.
Pollution	Sulfides, which normally are present in polluted seawater, greatly accelerate attack on steel. However, the low oxygen content of polluted waters could favor reduced corrosion.
Silt and Suspended Sediment	Erosion of the steel surface by suspended matter in the flowing seawater greatly increases the tendency to corrode.
Film Formation	A coating of rust, or rust and mineral scale (calcium and magnesium salts), will interfere with the diffusion of oxygen to the cathode surface, thus slowing the attack.

Source: Fink, F. W., et al., The Corrosion of Metals in Marine Environment, Battelle Memorial Inst., DMIC Report 254, Distributed by NTIS, AD-712 585-S, pp. 7, 13, 1970.

ZONES OF CORROSION FOR STEEL PILING IN SEAWATER

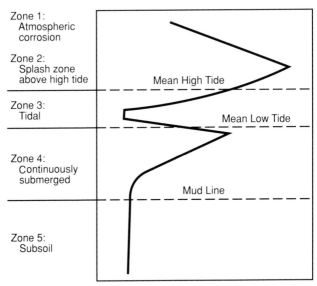

Zone 1:
Atmospheric
corrosion

Zone 2:
Splash zone
above high tide

Mean High Tide

Zone 3:
Tidal

Mean Low Tide

Zone 4:
Continuously
submerged

Mud Line

Zone 5:
Subsoil

Relative loss in metal thickness

Source: F.L. LaQue, *Marine Corrosion Causes and Prevention,* p. 116, John Wiley & Sons, 1975

RATES OF GENERAL WASTAGE OF METALS IN QUIET SEAWATER

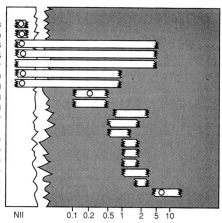

Typical Average Corrosion Rates, Mils/Year

○ Data from results of early tests at depths of 2300 to 5600 feet.

*Nickel chromium alloys designate a family of nickel base alloys with substantial chromium contents with or without alloying elements all of which, except those with high molybdenum contents, have related seawater corrosion characteristics.

Source: F.L. LaQue, *Marine Corrosion Causes and Prevention,* p. 146, John Wiley & Sons, 1975

SUGGESTED VELOCITY LIMITS
FOR CONDENSER TUBE ALLOYS IN SEAWATER

Alloy	Design Velocity That Should Not Be Exceeded (ft/s)	(m/s)
Copper	3[a]	0.9[a]
Silicon bronze	3[a]	0.9[a]
Admiralty brass	5[a]	1.5[a]
Aluminum brass	8[a]	2.4[a]
90-10 copper nickel	10[a]	3.0[a]
70-30 copper nickel	12[a]	3.7[a]
Ni-Cu alloy 400	No maximum velocity limit[b]	
Type 316 stainless steel	No maximum velocity limit[b]	
Ni-Cr-Fe-Mo alloys 825 and 20Cb3	No maximum velocity limit[b]	
Ni-Cr-Mo alloys 625 and C-276	No velocity limits	
Titanium	No velocity limits	

[a] In deaerated brines encountered in the heat recovery heat exchangers in desalination plants the critical velocities can be increased from 1 to 2 ft/sec. (0.3 to 0.6 m/s).
[b] Minimum velocity 5 ft/sec. (1.5 m/s).

Source: F. L. LaQue, Marine Corrosion Causes and Prevention, p. 147, John Wiley & Sons 1975.

GALVANIC SERIES IN SEAWATER
Flowing Seawater at 2.4 to 4.0 m/s for 5 to 15 days at 5 to 30°C

Volts vs Saturated Calomel Reference Electrode

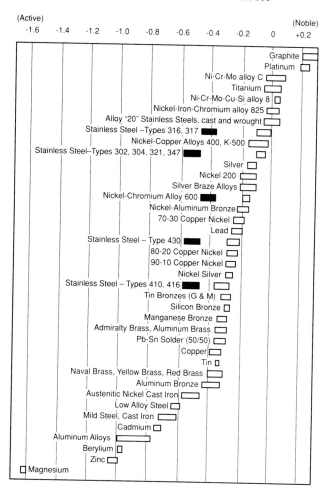

Note: Dark boxes indicate active behavior of active-passive alloys

Source: ASTM G-82

PRACTICAL GALVANIC SERIES

Open Circuit Potential Values Compared to Copper Alloy C11000
Test Medium: 5% NaCl at 25 C, 2.5-4 m/s

UNS	Name	Condition	Voltage	UNS	Name	Condition	Voltage
M11311	AZ31B Mg		−1.344	C11000	ETP Copper		0.000 (ref.)
M11912	AZ91B Mg		−1.314	S34700	347 SS	active	+0.006
Z33250	AG40A Zn		−0.786	—	Molybdenum		+0.006
—	Berylium		−0.780	C71500	Cu-Ni, 70-30		+0.012
A92014	2014 Al	T3	−0.639	S20200	202 SS	active (dull)	+0.014
A91160	1160 Al	H4	−0.609	—	Niobium		+0.018
A97075	7075 Al	T6	−0.604	C53400	Phosphor Bronze		+0.034
A97079	7079 Al		−0.584	S20200	202 SS	active (bright)	+0.051
M08990	Uranium		−0.556	N44000	400 Ni-Cu alloy		+0.051
A95052	5052 Al	H12	−0.545	S34700	347 SS	passive	+0.058
A95052	5052 Al	0	−0.534	N02200	Nickel 200		+0.064
A95083	5083 Al		−0.524	S20100	201 SS	active	+0.070
A96151	6151 Al	T6	−0.520	N08020	20Cb-3	active	+0.074
—	Cadmium		−0.519	S32100	321 SS	active	+0.077
A95456	5456 Al		−0.514	S31600	316 SS	active	+0.082
A95456	5456 Al	H343	−0.507	S30400	304 SS	passive	+0.098
A94043	4043 Al	H14	−0.507	S17700	17-7PH	passive	+0.098
A95052	5052 Al	H32	−0.502	S30900	309 SS	active	+0.108
A91100	1100 Al	0	−0.499	S31000	310 SS	passive	+0.109
A93003	3003 Al	H25	−0.496	S30100	301 SS	passive	+0.112
A96061	6061 Al	T6	−0.493	S32100	321 SS	passive	+0.116
A97071	7071 Al	T6	−0.484	S20100	201 SS	passive	+0.129
A13800	1380 Al	Cast	−0.444	S35500	AM 355	active	+0.167
A92014	2014 Al	0	−0.444	S66286	A286	active	+0.156
A92024	2024 Al	T4	−0.370	S31603	316L SS	passive	+0.156
A95056	5056 Al	H16	−0.369	S20200	202 SS	passive (dull)	+0.159
S43000	430 SS	active	−0.324	S35500	Am 355	active	+0.167
—	Lead		−0.316	S20200	202 SS	passive (bright)	+0.183
G10100	1010 steel		−0.297	N08020	20Cb-3	passive	+0.186
—	Tin		−0.281	S35500	Am 355	passive	+0.204
S41000	410 SS	active	−0.297	S66286	A286	passive	+0.311
—	Tantalum		−0.166	R54521	Ti, 5Al-2.5Sn		+0.423
S35000	AM 350	active	−0.149	R50810	Ti, 13V-11Cr-3Al	ANN (33.5 HRC)	+0.436
R50255	Ta-W, 90-10		−0.124	R56401	Ti, 6Al-4V	STA (41.5 HRC)	+0.455
S31000	310 SS	active	−0.124	—	Graphite		+0.473
S30100	301 SS	active	−0.120	R56401	Ti, 6Al-4V	ANN (36 HRC)	+0.481
S30400	304 SS	active	−0.106	R56080	Ti, 8Mn		+0.493
S43000	430 SS	passive	−0.094	R50810	Ti, 13V-11Cr-3Al	STA (45.5 HRC)	+0.498
S17700	17-7PH	active	−0.076	R50700	Ti, Gr.4 (75A)		+0.506
—	Tungsten		−0.047	S35000	AM 350	passive	+0.666
—	Niobium, 1% Zr		−0.044				
C26800	Yellow brass		−0.043				
—	Uranium, 8% Mo		−0.041				
C46400	Naval brass		−0.041				
C28000	Muntz metal		−0.034				
C75200	Nickel silver		−0.022				
S31603	316L SS	active	−0013				
C22000	Bronze, 90%		−0.012				
C65500	Si Bronze A		−0.007				

Source: C.M. Forman and E.A. Verchot, U.S. Army Missile Command Report No. RS-TR-67-11 (1967).

THE MAJOR CONSTITUENTS OF SEAWATER
Chlorinity = 19%

Anions	Parts Per Million	Milliequivalents Per Litre
Chloride	18980	535.3
Sulfate	2649	55.2
Bicarbonate	142	2.3
Bromide	65	0.8
Fluoride	1.4	0.07
Borate	24.9	0.58
		594.25
Cations		
Sodium	10561	459.4
Potassium	380	9.7
Magnesium	1272	104.4
Calcium	400	20.0
Strontium	13	0.3
		593.8

Notes:

1) The above composition shows slight difference between anions and cations, expressed as milliequivalents per litre because of the presence of traces of other components not listed in the above composition.

2) Chlorinity is the total amount of chlorine, bromine and iodine in grams contained in one kilogram of seawater assuming that the bromine and iodine have been expressed as chlorine.

3) Salinity is the total solid material in grams contained in one kilogram of sea water when all carbonate has been converted to oxide, the bromine and iodine replaced by chlorine, and all organic matter is completely oxidized. In the open sea the salinity varies between 32 and 36.

salinity = 1.807 x chlorinity

CHEMICAL COMPOSITION OF SUBSTITUTE SEAWATER[a]

Compound	Concentration (g/L)
NaCl	24.53
$MgCl_2$	5.20
Na_2SO_4	4.09
$CaCl_2$	1.16
KCl	0.695
$NaHCO_3$	0.201
KBr	0.101
H_3BO_3	0.027
$SrCl_2$	0.025
NaF	0.003
$Ba(NO_3)_2$	0.0000994
$Mn(NO_3)_2$	0.000034
$Cu(NO_3)_2$	0.0000308
$Zn(NO_3)_2$	0.0000096
$Pb(NO_3)_2$	0.0000066
$AgNO_3$	0.00000049

[a]Chlorinity = 19.38. Adjust pH to 8.2 with 0.1 N NaOH.

Source: ASTM D1141-86.

CALCULATION OF CALCIUM CARBONATE SATURATION INDEX
(LANGELIER INDEX)

A		C		D	
Total Solids mg/L	A	Calcium Hardness mg/L $CaCO_3$	C	M.O. Alkalinity mg/L $CaCO_3$	D
50-300	0.1	10-11	0.6	10-11	1.0
400-1000	0.2	12-13	0.7	12-13	1.1
		14-17	0.8	14-17	1.2
B		18-22	0.9	18-22	1.3
Temperature	B	23-27	1.0	23-27	1.4
°C °F		28-34	1.1	28-35	1.5
0-1 32-34	2.6	35-43	1.2	36-44	1.6
2-6 36-42	2.5	44-55	1.3	45-55	1.7
7-9 44-48	2.4	56-69	1.4	56-69	1.8
10-13 50-56	2.3	70-87	1.5	70-88	1.9
14-17 58-62	2.2	88-110	1.6	89-110	2.0
18-21 64-70	2.1	111-138	1.7	111-139	2.1
22-27 72-80	2.0	139-174	1.8	140-176	2.2
28-31 82-88	1.9	175-220	1.9	177-220	2.3
32-37 90-98	1.8	230-270	2.0	230-270	2.4
38-43 100-110	1.7	280-340	2.1	280-350	2.5
44-50 112-122	1.6	350-430	2.2	360-440	2.6
51-55 124-132	1.5	440-550	2.3	450-550	2.7
56-64 134-146	1.4	560-690	2.4	560-690	2.8
65-71 148-160	1.3	700-870	2.5	700-880	2.9
72-81 162-178	1.2	880-1000	2.6	890-1000	3.0

1. Obtain values of A, B, C and D from above table.
2. pHs = $(9.3 + A + B) - (C + D)$.
3. Saturation index = pH − pHs.
   If index is 0, water is in chemical balance.
   If index is a plus quantity, there is a tendency for calcium carbonate deposition.

   If the index is a minus quantity, calcium carbonate does not precipitate, and the probability of corrosion (if dissolved oxygen is present) will increase with an increase in the negative value of the index.

   To determine temperature at which scaling begins (i.e., pH = pHs), find the temperature equivalent to the following value of B:

   B = pH + $(C + D) - (9.3 + A)$

   Ryznar Stability Index = 2 (pHs) − pH
   With waters having a Stability Index of 6.0 or less, scaling increases and the tendency for corrosion decreases. When the Stability Index is above 7.0, a protective coating of calcium carbonate may not be developed.

WATER ANALYSIS CONVERSION FACTORS

	CaCO$_3$ ppm	CaCO$_3$ grains/US gal	English degree	French degree	German degree
CaCO$_3$, ppm	1.0	0.058	0.07	0.10	0.056
CaCO$_3$, grains/US gal	17.1	1.0	1.2	1.71	0.958
English degree	14.3	0.833	1.0	1.43	0.800
French degree	10.0	0.583	0.7	1.0	0.560
German degree	17.9	1.04	1.24	1.79	1.0

1 English degree = 1 grain CaCO$_3$/Imperial gallon.
1 French degree = 10 ppm CaCO$_3$.
1 German degree = 10 ppm CaO.

Courtesy H. P. Godard.

COMMON GROUPS OF ALGAE

Group	Temperature Range °C	°F	pH Range	Examples
Green Algae	30-35	86-95	5.5-8.9	Chlorella—common unicellular Ulothrix[1]—filamentous Spirogyra—filamentous
Blue-Green Algae (Contain Blue Pigment)	35-40	95-104	6.0-8.9	Anacystis—unicellular slime former Phormidium—filamentous[1] Oscillatoria[2] causes the most severe problems
Diatoms (Brown Pigment and Silica in Cell Wall)	18-85	64-186	5.5-8.9	Fragilaria Cyclotella Diatoms[1]

[1]These algae may occur in cooling water with as much as 120 ppm chromate present.
[2]Oscillatoria will grow at 186°F and at pH 9.5.

COMMON TYPES OF BACTERIA CAUSING SLIME PROBLEMS

Type	Example	pH Range	Problems Caused
Aerobic Capsulated	Aerobacter aerogenes Flavobacterium Proteus vulgaris Pseudomonas aeroginosa Serratia Alcaligenes	4.0-8.0 Optimum pH 7.5	Major slime forming bacteria. May produce green, yellow, and pink slimes in addition to usual grey or brown slime.
Aerobic Spore-Forming	Bacilus mycoides (in other Bacillus species)	5.0-8.0	Add to slime problem. Spores are more difficult to destroy.
Iron Bacteria	Crenothrix Leptothrix Gallionella	7.4-9.5	Precipitate ferric hydroxide in sheath-like coating around cell-forms bulking slime deposits.

NOTE: All of the above bacteria live in a temperature range of 68 to 104°F, with some species growing at 40 to 158°F.

Source: NACE, *Cooling Water Treatment Manual*, 1971.

MICROORGANISMS COMMONLY IMPLICATED
IN BIOLOGICAL CORROSION

Genus or Species	pH Range	Temperature Range °C	Oxygen Requirement	Metals Affected	Action
Bacteria					
Desulfovibrio					
Best known:					
D. desulfuricans	4-8	10-40	Anaerobic	Iron and steel, stainless steels, aluminum zinc, copper alloys	Utilize hydrogen in reducing SO_4^{2-} to S^{2-} and H_2S; promote formation of sulfide films
Desulfotomaculum					
Best known:					
D. nigrificans					
(also known as					
Clostridium)	6-8	10-40 (some 45-75)	Anaerobic	Iron and steel, stainless steels	Reduce SO_4^{2-} to S^{2-} and H_2S (spore formers)
Desulfomonas	...	10-40	Anaerobic	Iron and steel	Reduce SO_4^{2-} to S^{2-} and H_2S
Thiobacillus					
thiooxidans	0.5-8	10-40	Aerobic	Iron and steel, copper alloys, concrete	Oxidizes sulfur and sulfides to form H_2SO_4; damages protective coatings corrodes concrete in sewers
Thiobacillus					
ferrooxidans	1-7	10-40	Aerobic	Iron and steel	Oxidizes ferrous to ferric
Gallionella	7-10	20-40	Aerobic	Iron and steel, stainless steels	Oxidizes ferrous (and manganous) to ferric (and manganic); promotes tubercule formation
Sphaerotilus	7-10	20-40	Aerobic	Iron and steel, stainless steels	Oxidizes ferrous (and manganous) to ferric (sand manganic); promotes tubercule formation
S. natans	...	...	...	Aluminum alloys	
Pseudomonas	4-9	20-40	Aerobic	Iron and steel, stainless steels	Some strains can reduce ferric to ferrous
P. aeruginosa	4-8	20-40	Aerobic	Aluminum alloys	
Fungi					
Cladosporium					
resinae	3-7	10-45 (best at 30-35)	...	Aluminum alloys	Produces organic acids in metabolizing certain fuel constituents

Source: Metals Handbook, 9th ed., Volume 13, p. 118, ASM 1987.

MICROBIOCIDES USED IN COOLING WATER SYSTEMS

Microbiocide	Effectiveness[a] Bacteria	Fungi	Algae	Comments
Chlorine	E	S	E	Oxidizing: reacts with $-NH_2$ groups; effective at neutral pH; loses effectiveness at high pH. Use concentration: 0.1-0.2 mg/L continuous free residual; 0.5-1.0 mg/L intermittent free residual
Chlorine dioxide	E	G	G	Oxidizing; pH insensitive; can be used in presence of $-NH_2$ groups. Use concentration: 0.1-1 mg/L intermittent free residual
Bromine	E	S	E	Oxidizing; substitute for Cl_2 and ClO_2; effective over broad pH range. Use concentration: 0.05-0.1 mg/L continuous free residual; 0.2 to 0.4 mg/L intermittent free residual
Organo-bromide (DBNPA)	E	NA	S	Nonoxidizing; pH range 6-8.5. Use concentration: 0.5-24 mg/L intermittent feed
Methylene bisthiocyanate	E	S	S	Nonoxidizing; hydrolyzes above pH 8. Use concentration: 1.5-8 mg/L intermittent feed
Isothiazolinone	E	G	E	Nonoxidizing; pH insensitive; deactivated by HS^- and $-NH_2$ groups. Use concentration: 0.9-13 mg/L intermittent feed
Quaternary ammonium salts	E	G	E	Nonoxidizing; tendency to foam; surface active; ineffective in highly oil- or organic-fouled systems. Use concentration: 8-35 mg/L intermittent feed
Organo-tin/ quaternary ammonium salts	E	G	E	Nonoxidizing; tendency to foam; functions best in alkaline pH. Use concentration: 7-50 mg/L
Glutaraldehyde	E	E	E	Nonoxidizing; deactivated by $-NH_2$ groups; effective over broad pH range. Use concentration: 10-75 mg/L intermittent feed

[a]E, excellent; G, good; S, slight; NA, not applicable

Source: Metals Handbook, 9th ed., Volume 13, p. 493, ASM 1987.

APPROXIMATE CURRENT REQUIREMENTS FOR CATHODIC PROTECTION OF STEEL

Environmental Conditions	Current Density	
	mA/m^2	mA/ft^2
Immersed in Seawater[a]		
Stationary		
Well coated	1 to 2	0.1 to 0.2
Poor or old coating	2 to 20	0.2 to 2
Uncoated	20 to 30	2 to 3
Low velocity[b]		
Well coated	2 to 5	0.2 to 0.5
Poor coating	5 to 20	0.5 to 2
Uncoated	50 to 150	5 to 15
Medium velocity[c]		
Well coated	5 to 7	0.5 to 0.7
Poor coating	10 to 30	1 to 3
Uncoated	150 to 300	15 to 30
High velocity[d]		
Poor coating or uncoated	250 to 1000	25 to 100
Buried Underground[e]		
Soil resistivity		
0.5 to 5Ω·m	1 to 2	0.1 to 0.2
5 to 15Ω·m	0.5 to 1	0.05 to 0.1
15 to 40Ω·m	0.1 to 0.5	0.01 to 0.05

[a]Structures or vessels.
[b]0.3 to 1 m/s (1 to 3 ft/s).
[c]1 to 2 m/s (3 to 7 ft/s).
[d]Turbulent flow.
[e]Pipelines or structures, coated or wrapped.

Source: Metals Handbook, 9th ed., Volume 1, p. 758 ASM 1978.

DESIGN CRITERIA FOR OFFSHORE
CATHODIC PROTECTION SYSTEMS

Production Area	Water Resistivity[2] ohm-cm	Environmental Factors[1]			Lateral Water Flow	Typical Design Current Density[3]	
		Water Temp. °C	Turbulence Factor (Wave Action)			mA/ft²	mA/m²
Gulf of Mexico	20	22	Moderate		Moderate	5-6	54-65
U.S. West Coast	24	15	Moderate		Moderate	7-10	76-106
Cook Inlet	50	2	Low		High	35-40	380-430
North Sea[4]	26-33	0-12	High		Moderate	8-20	86-216
Persian Gulf	15	30	Moderate		Low	5-8	54-86
Indonesia	19	24	Moderate		Moderate	5-6	54-65

[1]Typical values and ratings based on average conditions, remote from river discharge.

[2]Water resistivities are a function of both chlorinity and temperature. In the Corrosion Handbook by H. Uhlig the following resitivities are given for chlorinities of 19 and 20 parts per thousand:

	Resistivities ohm-cm Temperature °C					
Chlorinity (ppt)	0	5	10	15	20	25
19	35.1	30.4	26.7	23.7	21.3	19.2
20	33.5	29.0	25.5	22.7	20.3	18.3

[3]In ordinary seawater, a current density less than the design value will suffice to hold the platform at protective potential once polarization has been accomplished and calcareous coatings are built up by the design current density. CAUTION: Depolarization can result from storm action.

[4]Conditions in the North Sea can vary greatly from the northern to the southern area, from winter to summer, and during storm periods.

Source: NACE Standard RP0176-83.

EFFECT OF APPLIED CATHODIC CURRENT ON CORROSION AND POTENTIAL OF STEEL IN FLOWING SEAWATER

15°C, 2.4 m/s, 14 days, 0.17 sq dm

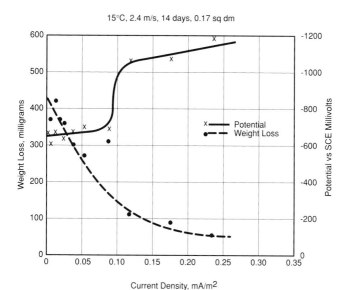

Source: F.L. LaQue, *Marine Corrosion Causes and Prevention,* p, 74, John Wiley and Sons, 1975

ENERGY CAPABILITIES AND CONSUMPTION RATES
OF GALVANIC ALLOYS IN SEAWATER

Galvanic Anode Alloy	Energy Capability[1]		Consumption Rate		Anode to Water[2] Closed Circuit Potentials, Negative Volts (Ag/AgCl electrode)
	amp-h/lb	amp-h/kg	lb/amp-y	kg/amp-y	
Aluminum-zinc-mercury	1250-1290	2750-2840	7.0-6.8	3.2-3.1	1.0-1.05
Aluminum-zinc-indium	1040-1180	2290-2600	8.4-7.4	3.8-3.4	1.05-1.1
Aluminum-zinc-tin	420-1180	925-2600	20.8-7.4	9.5-3.4	1.0-1.05
Zinc (MIL-A-18001 H)	354-370	780-815	24.8-23.7	11.2-10.7	1.0-1.05
Magnesium (H-1 Alloy)	500	1100	17.5	8.0	1.4-1.6

[1]The above data show ranges that are taken from field tests at Key West, Florida, by Naval Research Laboratory, Washington, DC and from manufacturers' long-term field tests. Modification to these numbers will be made only by recommendation from NACE Technical Unit Committee T-7L on Cathodic Protection in Natural Waters.

[2]Measured potentials can vary because of temperature and salinity differences.

Source: NACE Standard RP0176-83.

CONSUMPTION RATES OF IMPRESSED CURRENT ANODE MATERIALS

Impressed Current Anode Material	Typical Anode Current Density in Saltwater Service		Nominal Consumption Rate	
	amp/ft²	amp/m²	lb/amp-y	g/amp-y
Pb-6%Sb-1%Ag	15-20	160-220	0.03-0.2[1]	15-86
Pb-6%Sb-2%Ag	15-20	160-220	0.03-0.06[1]	13-25
Platinum (on Titanium, Niobium, or Tantalum substrate)	50-300	540-3200	0.008-0.016[2]	
Graphite	1-4	10-40	0.5-1.0	230-450
Fe.14.5% Si-4.5%Cr	1-4	10-40	0.5-1.0	230-450

[1]Very high consumption rates of Pb-Ag anodes have been experienced at depths in excess of 100 feet (30 m).

[2]This figure can increase when current density is extremely high and/or in waters of low salinity.

Source: NACE Standard RP0176-83.

PLATINUM CONSUMPTION RATES FOR CATHODIC
PROTECTION ANODES

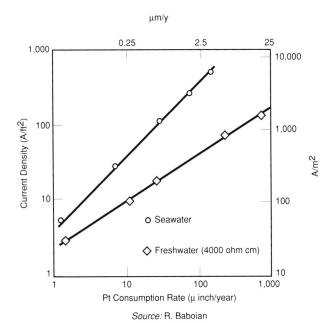

Source: R. Baboian

RESISTANCE OF GALVANIC ANODES—
DWIGHT'S EQUATION

One vertical ground rod:
length L, radius a

$$R = \frac{p}{2\pi L}\left(\log_n \frac{4L}{a} - 1\right)$$

Two vertical ground rods
separation s, s > L

$$R = \frac{p}{4\pi L}\left(\log_n \frac{4L}{a} - 1\right) +$$

$$\frac{p}{4\pi s}\left(1 - \frac{L^2}{3s^2} + \frac{2}{5}\cdot\frac{L^4}{s^4}\cdots\right)$$

s < L

$$R = \frac{p}{4\pi L}\left(\log_n \frac{4L}{a} + \log_n \frac{4L}{s} - 2 + \frac{s}{2L} - \frac{s^2}{16L^2} + \frac{s^4}{512L^4}\cdots\right)$$

Buried horizontal wire, length
2L, depth s/2

$$R = \frac{p}{4\pi L}\left(\log_n \frac{4L}{a} + \log_n \frac{4L}{s} - 2 + \frac{s}{2L} - \frac{s^2}{16L^2} + \frac{s^4}{512L^4}\cdots\right)$$

Right-angle turn of wire:
length of arm L, depth s/2

$$R = \frac{p}{4\pi L}\left(\log_n \frac{2L}{a} + \log_n \frac{2L}{s} - 0.24 + 0.2\frac{s}{L}\cdots\right)$$

Three-point star

$$R = \frac{p}{6\pi L}\left(\log_n \frac{2L}{a} + \log_n \frac{2L}{s} + 1.1 - 0.2\frac{s}{L}\cdots\right)$$

Four-point star

$$R = \frac{p}{8\pi L}\left(\log_n \frac{2L}{a} + \log_n \frac{2L}{s} + 3 - \frac{s}{L}\cdots\right)$$

Six-point star

$$R = \frac{p}{12\pi L}\left(\log_n \frac{2L}{a} + \log_n \frac{2L}{s} + 6.9 - \frac{3s}{L}\cdots\right)$$

Continued on Next Page

Eight-point star

$$R = \frac{p}{16\pi L}\left(\log_n \frac{2L}{a} + \log_n \frac{2L}{s} + 11 - 5.5 \frac{s}{L} \cdots \right)$$

Ring of wire. Diameter D of ring, diameter of wire, a depth s/2

$$R = \frac{p}{2\pi^2 D}\left(\log_n \frac{8D}{d} + \log_n \frac{4D}{s}\right)$$

Buried horizontal strip: length 2L, section a by b depth s/2, b < a/8

$$R = \frac{p}{4\pi L}\left(\log_n \frac{4L}{a} + \frac{a^2 - \pi ab}{2(a+b)^2} + \log_n \frac{4L}{s} - 1 + \frac{s}{2L} - \frac{s^2}{16L^2}\right)$$

Buried horizontal round plate radius a, depth s/2

$$R = \frac{p}{8a} + \frac{p}{4\pi s}\left(1 - \frac{7a^2}{12s^2} + \frac{33a^4}{40s^4} \cdots \right)$$

Buried vertical round plate

$$R = \frac{p}{8a} + \frac{p}{4\pi s}\left(1 - \frac{7a^2}{24s^2} + \frac{99a^4}{320s^4} \cdots \right)$$

Source: John H. Morgan, Cathodic Protection, 2nd ed., p. 104, NACE, 1987.

TYPICAL RESISTIVITIES OF SOME WATERS AND SOIL MATERIALS

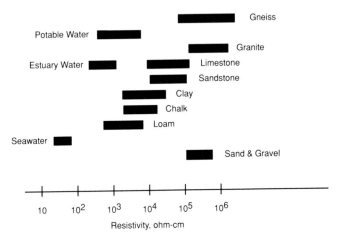

Source: John H. Morgan, *Cathodic Protection, 2nd ed.*, (NACE 1987)

PROPERTIES OF CONCENTRIC STRANDED
COPPER SINGLE CONDUCTORS

DIRECT BURIAL SERVICE, SUITABLY INSULATED

Size AWG	Overall Diameter Not Including Insulation (Inches)	Approx. Weight Not Including Insulation (lbs./M ft.)	Maximum Breaking Strength (lbs.)	Maximum D.C. Resistance @ 20°C Ohms/M ft.	Maximum Allowable D.C. Current Capacity (Amperes)
14	0.0726	12.68	130	2.5800	15
12	0.0915	20.16	207	1.6200	20
10	0.1160	32.06	329	1.0200	30
8	0.1460	50.97	525	0.6400	45
6	0.1840	81.05	832	0.4030	65
4	0.2320	128.90	1320	0.2540	85
3	0.2600	162.50	1670	0.2010	100
2	0.2920	204.90	2110	0.1590	115
1	0.3320	258.40	2660	0.1260	130
1/0	0.3730	325.80	3350	0.1000	150
2/0	0.4190	410.90	4230	0.0795	175
3/0	0.4700	518.10	5320	0.0631	200
4/0	0.5280	653.30	6453	0.0500	230
250 MCM	0.5750	771.90	7930	0.0423	255

Data Courtesy Rome Cable Division of ALCOA.
Source: W. T. Bryan, The Duriron Co., Inc.

TEMPERATURE CORRECTION FACTORS
FOR RESISTANCE OF COPPER

Temperature		Multiply Resistance at 25°C by:
°C	°F	
−10	14	0.862
−5	23	0.882
0	32	0.901
5	41	0.921
10	50	0.941
15	59	0.961
20	68	0.980
30	86	1.020
35	95	1.040
40	104	1.059

Source: A. W. Peabody, *Control of Pipeline Corrosion*, NACE, 1967.

STEEL PIPE RESISTANCE

Pipe Size, Inches	Outside Diameter, Inches	Wall Thickness, Inches	Weight Per Foot, Pounds	Resistance Micro ohms Per Foot	Resistance Micro ohms Per Metre
2	2.375	0.154	3.65	79.2	260.
4	4.5	0.237	10.8	26.8	87.9
6	6.625	0.280	19.0	15.2	49.9
8	8.625	0.322	28.6	10.1	33.1
10	10.75	0.365	40.5	7.13	23.4
12	12.75	0.375	49.6	5.82	19.1
14	14.00	0.375	54.6	5.29	17.4
16	16.00	0.375	62.6	4.61	15.1
18	18.00	0.375	70.6	4.09	13.4
20	20.00	0.375	78.6	3.68	12.1
22	22.00	0.375	86.6	3.34	11.0
24	24.00	0.375	94.6	3.06	10.0
26	26.00	0.375	102.6	2.82	9.25
28	28.00	0.375	110.6	2.62	8.60
30	30.00	0.375	118.7	2.44	8.0
32	32.00	0.375	126.6	2.28	7.48
34	34.00	0.375	134.6	2.15	7.05
36	36.00	0.375	142.6	2.03	6.66

Based on steel density of 489 pounds per cubic foot and steel resistivity of 18 microhm-cm.

Source: A. W. Peabody, *Control of Pipeline Corrosion*, NACE, 1967.

ALLOY PIPE RESISTANCE

Resistance of alloy piping can be estimated using the factor below, which is the ratio of alloy resistivity to that of a typical carbon steel (18 microhm-cm):

304 SS	4.0
316 SS	4.1
410 SS	3.2
400 Alloy	3.1
Al 3003	0.19
OF Copper	0.09

CORROSION OF GALVANIZED PIPE IN VARIOUS SOILS

Weight loss (oz/ft^2) and maximum pit depth (mil) after burial period 12.7 years

	oz/ft^2	mil
Inorganic oxidizing acid soils		
Cecil clay loam	0.6	<6
Hagerstown loam	0.6	<6
Susquehanna clay	0.8	<6
Inorganic oxidizing alkaline soils		
Chino silt loam	1.1	<6
Mohave fine gravelly loam	1.1	<6
Inorganic reducing acid soils		
Sharkey clay	1.1	6
Acadia clay	—	—
Inorganic reducing alkaline soils		
Docas clay	1.6	<6
Merced silt loam	1.3	8
Lake Charles clay	13.8	66
Organic reducing acid soils		
Carlisle muck	3.4	<6
Tidal marsh	4.8	52
Muck	10.7	76
Rifle peat	19.5	88
Cinders	11.9	48

Nominal weight of coating—3 oz/sq ft (915g/m^2) of exposed area.

Source: Zinc, Its Corrosion Resistance, 2nd Ed., p. 101, International Lead Zinc Research Organization 1983.

ESTIMATING SERVICE LIFE OF GALVANIZED STEEL IN SOILS

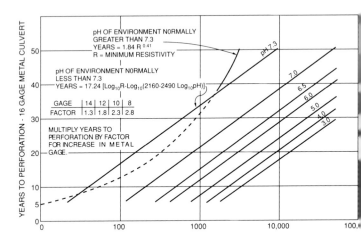

MINIMUM RESISTIVITY (R) - ohm cm

Method developed by the California Division of Highways for estimating service life of galvanized steel culverts, based on correlations involving pH and resistivity of soil. Base case is 16-gage galvanized steel pipe with a zinc coating thickness of 1.6 mm (0.064 in.).

Source: NACE-OSU Corrosion Course

CAUSTIC SODA SERVICE CHART

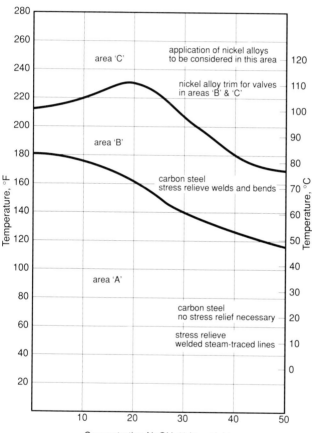

Concentration NaOH, % by weight

Source: Corrosion Data Survey, NACE, 1985

ALLOYS FOR SULFURIC ACID SERVICE
Isocorrosion Curves

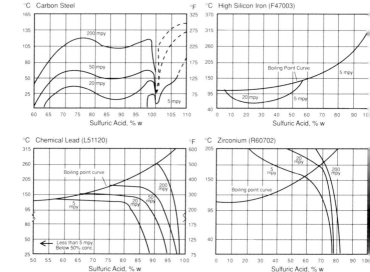

Note:
These graphs are based on laboratory tests with pure acid, and should be used for general guidance only. For detailed information see *Process Industries Corrosion - The Theory and Practice* (NACE, 1986)

Source:
Mars G. Fontana, Corrosion Engineering, McGraw-H (1986)

R.T. Webster and T.L. Yau, Materials Performance, Vol. 25, No. 2, p. 15 (1986) [Zirconium]

Continued on Next Pa

ALLOYS FOR SULFURIC ACID SERVICE (Cont.)

Isocorrosion Curves

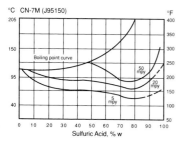

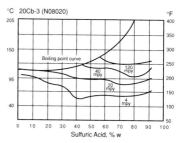

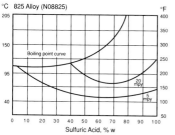

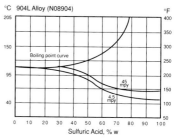

Note:
These graphs are based on laboratory tests with pure acid, and should be used for general guidance only. For detailed information see *Process Industries Corrosion - The Theory and Practice* (NACE, 1986)

Source:
Corrosion Engineering Bulletin 1, The International Nickel Company, 1983.

Continued on Next Page

ALLOYS FOR SULFURIC ACID SERVICE (Cont.)

Isocorrosion Curves

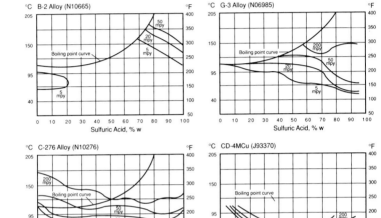

Note:
These graphs are based on laboratory tests with pure acid, and should be used for general guidance only. For detailed information see *Process Industries Corrosion - The Theory and Practice* (NACE, 1986)

Source:
Corrosion Engineering Bulletin 1, The International Nickel Company, 1983.

Hastelloy Alloy G-3 Booklet, Cabot Corp., 1983.

ALLOYS FOR NITRIC ACID SERVICE

Isocorrosion Curves

304 SS (S30400)

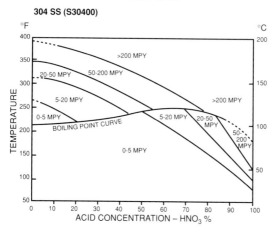

ALUMINUM 1100 (A91100)

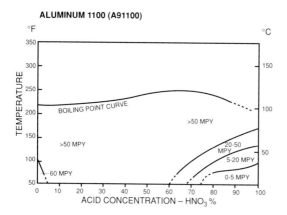

Source: Mars G. Fontana, Corrosion Engineering, McGraw-Hill (1986)

Note: These graphs are based on laboratory tests with pure acid, and should be used for general guidance only. For detailed information see Process Industries Corrosion – The Theory and Practice (NACE, 1986)

ALLOYS FOR HYDROCHLORIC ACID SERVICE

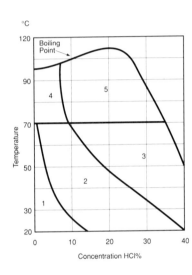

°C

Temperature

Concentration HCl%

Materials in numbered zones have reported corrosion rates of <20 mpy

Zone 1

CN-7M	(1) (3) (6)
400 Alloy	(2) (3) (6)
Copper	(2) (3) (6)
Nickel 200	(2) (3) (6)
Silicon Bronze	(2) (3) (6)
Silicon Cast Iron	(7)
Tungsten	
Titanium (Gr. 7)	
Titanium (Gr. 2)	(4)

Zone 2

Silicon Bronze	(2) (6)
Silicon Cast Iron	(7)

Zone 3

Silicon Cast Iron	(7)

Zone 4

400 Alloy	(2) (3) (8)
Tungsten	
Titanium (Gr.7)	(5)

All Zones (including 5)

Platinum	
Tantalum	
Silver	(3) (6)
Zirconium	(3) (6)
B-2 Alloy	(3) (6)
Molybdenum	(3) (6)

Note: This graph should be used for general guidance only. For more detailed information see *Process Industries Corrosion—The Theory and Practice* (NACE, 1986)

Source: Corrosion Data Survey, NACE 1985
Modified by T.F. Degnan

Notes:
1. <2% at 25°C
2. No Air
3. No FeCl$_3$ or CuCl$_2$
4. <10% at 25°C
5. <5% at B.P.
6. No Chlorine
7. Cr-Mo Alloy
8. <0.05% Concentration

ALLOYS FOR HYDROFLUORIC ACID SERVICE

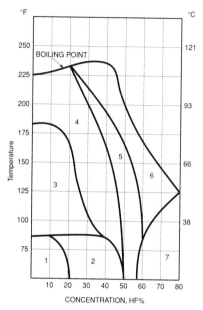

CONCENTRATION, HF%

Materials in numbered zones have reported corrosion rates of <20 mpy

Zone 1

CN-7M
310 SS
70-30 Cu-Ni*
400 Alloy*
C Alloy
Copper*
Gold
Lead*
Nickel*
Nickel cast iron
Platinum
Silver

Zone 2

CN-7M
70-30 Cu-Ni*
400 Alloy*
C Alloy
Copper*
Gold
Lead*
Nickel*
Platinum
Silver

Zone 3

CN-7M
70-30 Cu-Ni*
400 Alloy*
C Alloy
Copper*
Gold
Lead*
Platinum
Silver

Zone 4

70-30 Cu-Ni*
400 Alloy*
C Alloy
Copper*
Gold
Lead*
Platinum
Silver

Zone 5

70-30 Cu-Ni*
400 Alloy*
C Alloy
Gold
Lead*
Platinum
Silver

Zone 6

400 Alloy*
C Alloy
Gold
Platinum
Silver

Zone 7

400 Alloy*
C Alloy
Carbon Steel
Gold
Platinum
Silver

*No Air

Note: This graph should be used for general guidance only. For more detailed information see *Process Industries Corrosion—The Theory and Practice* (NACE, 1986)

Source: Corrosion Data Survey, NACE 1985

ESTIMATE OF SULFUR TRIOXIDE IN COMBUSTION GAS

Sulfur in Fuel (%)		0.5	1.0	2.0	3.0	4.0	5.0
Excess Air (%)	Oxygen in Gas (%)			Sulfur Trioxide Expected in Gas (ppm) *Oil Fired Units*			
5	1	2	3	3	4	5	6
11	2	6	7	8	10	12	14
17	3	10	13	15	19	22	25
25	4	12	15	18	22	26	30
				Coal Fired Units			
25	4	3-7	7-14	14-28	20-40	27-54	33-66

CALCULATED SULFURIC ACID DEWPOINT IN FLUE GAS

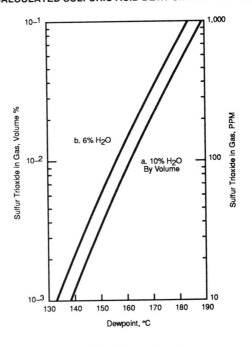

Calculated dewpoint vs sulfur trioxide concentration: (a) 10% H$_2$O from oil; and (b) 6% H$_2$O from coal.

Source: R.D. Tems and T.E. Mappes, *Materials Performance, Vol. 21, No. 12,* p. 26 (1982)

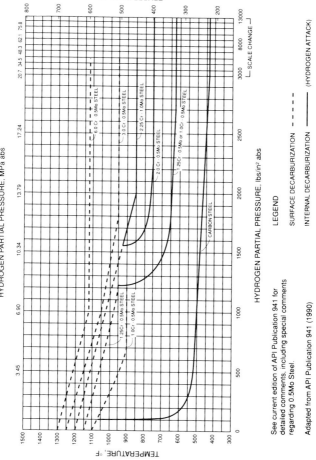

OPERATING LIMITS FOR STEELS IN HYDROGEN SERVICE TO AVOID DECARBURIZATION AND FISSURING

HYDROGEN PARTIAL PRESSURE, MPa abs

TEMPERATURE, °C

TEMPERATURE, °F

HYDROGEN PARTIAL PRESSURE, lbs/in² abs

6.0 Cr - 0.5Mo STEEL
3.0 Cr - 0.5Mo STEEL
2.25 Cr - 1.0Mo STEEL
2.0 Cr - 0.5Mo STEEL
1.25Cr - 0.5Mo or 1.0Cr - 0.5Mo STEEL
1.25Cr - 0.5Mo STEEL
1.0Cr - 0.5Mo STEEL
CARBON STEEL
SCALE CHANGE

LEGEND

SURFACE DECARBURIZATION – – – –

INTERNAL DECARBURIZATION ———— (HYDROGEN ATTACK)

See current edition of API Publication 941 for detailed comments, including special comments regarding 0.5Mo Steel.

Adapted from API Publication 941 (1990)

COMBINATIONS OF ALLOYS AND ENVIRONMENTS
SUBJECT TO DEALLOYING

Alloy	Environment	Element Removed
Brasses	Many waters, especially under stagnant conditions	Zinc (dezincification)
Gray iron	Soils, many waters	Iron (graphitic corrosion)
Aluminum bronzes	Hydrofluoric acid, acids containing chloride ions	Aluminum (dealuminification)
Silicon bronzes	High-temperature steam and acidic species	Silicon (desiliconification)
Tin bronzes	Hot brine or steam	Tin (destannification)
Copper-nickel alloys	High heat flux and low water velocity (in refinery condenser tubes)	Nickel (denickelification)
Copper-gold single crystals	Ferric chloride	Copper
Nickel-copper	Hydrofluoric and other acids	Copper in some acids, and nickel in others
Gold alloys with copper or silver	Sulfide solutions, human saliva	Copper, silver
High-nickel alloys	Molten salts	Chromium, iron, molybdenum, and tungsten
Medium- and high-carbon steels	Oxidizing atmospheres, hydrogen at high temperatures	Carbon (decarburization)
Iron-chromium alloys	High-temperature oxidizing atmospheres	Chromium, which forms a protective film
Nickel-molybdenum alloys	Oxygen at high temperature	Molybdenum

Source: Metals Handbook, 9th ed., Volume 13, p. 130, ASM 1987.

LIQUID METAL CRACKING

Couples identified below are those combinations of solid metal and liquid metal that have resulted in embrittlement in tests with either the pure element or an alloy. This compilation is a summary of data from the indicated source.

Solid	Liquid
Aluminum	Hg Ga Na In Sn Bi Cd Pb Zn
Bismuth	Hg
Cadmium	Cs Ga Sn
Copper	Hg Ga Na In Li Sn Bi Pb
Germanium	Ga In Sn Bi Tl Cd Pb Sb
Iron	Hg Ga In Li Sn Cd Pb Zn Te Sb Cu
Magnesium	Na Zn
Nickel	Hg Li Sn Pb
Palladium	Li
Silver	Hg Ga Li
Tin	Hg Ga
Titanium	Hg Cd
Zinc	Hg Ga In Sn Pb

Source: Metals Handbook, 9th ed., Volume 13, p. 182 ASM 1987.

STRESS CORROSION CRACKING SYSTEMS

Below is a partial listing of alloy/environment systems where stress corrosion cracking (anodic type) may occur. Whether cracking occurs in a specific system depends on temperature, environment composition, tensile stress level, alloy composition, and heat treatment.

Alloy	Environment
Aluminum alloys	Chloride solutions
Magnesium alloys	Chloride solutions
Copper alloys	Ammonia + oxygen + water Amines + oxygen + water Nitric acid vapor Steam
Carbon and low alloy steels	Nitrate solutions Caustic solutions Carbonate solutions Alkanolamines + carbon dioxide Carbon monoxide + carbon dioxide + water Anhydrous ammonia + air Hydrogen cyanide solutions
Austenitic stainless steels and some ferritic and duplex stainless steels	Chloride and bromide solutions Organic chlorides and bromides + water Caustic solutions H_2S solutions + chlorides or oxidants
Nickel alloys	Caustic solutions Fused caustic Hydrofluoric acid H_2S solutions + chlorides or oxidants
Titanium alloys	Aqueous salt systems Methanol plus halides Nitrogen tetroxide
Zirconium alloys	Aqueous salt systems Nitric acid
Sensitized austenitic stainless steels	Water + oxygen (high temperature) Chloride solutions Polythionic acid solutions Sulfurous acid

HYDROGEN STRESS CRACKING may occur with high strength alloys (carbon and low alloy steels, some stainless steels, and some nickel-base alloys) in a number of environments, including:

Hydrogen sulfide solutions   (SULFIDE STRESS CRACKING)

See following page for information on hydrogen degradation systems.
See graph on page 126 for H_2S concentration limit for SSC.

HYDROGEN DEGRADATION

| | Hydrogen Embrittlement | | | | | | | | | |
	Hydrogen Environment Embrittlement	Hydrogen Stress Cracking[a]	Loss in Tensile Ductility	Hydrogen Attack	Blistering	Shatter Cracks, Flakes, Fisheyes	Micro-perforation	Degradation in Flow Properties	Metal Hydride Formation
Typical materials	Steels, nickel-base alloys, metastable stainless steel, titanium alloys	Carbon and low-alloy steels	Steels, nickel-base alloys, Be-Cu bronze, aluminum alloys	Carbon and low-alloy steels	Steels, copper, aluminum	Steels (forgings and castings)	Steels (compressors)	Iron, steels, nickel-base alloys	V, Nb, Ta, Ti, Zr, V
Usual source of hydrogen (not exclusive)	Gaseous H_2	Thermal processing, electrolysis, corrosion	Gaseous hydrogen, internal hydrogen from electrochemical charging	Gaseous (above 200°C)	Hydrogen sulfide corrosion, electrolytic charging, gaseous	Water vapor reacting with molten steel	Gaseous hydrogen	Gaseous or internal hydrogen	Internal hydrogen from melting, corrosion, electrolyte charging, welding
Failure initiation	Surface or internal initiation; incubation period not observed	Internal crack initiation	Surface and/or internal effect	Surface (decarburization); internal carbide interfaces (methane bubble formation)	Internal defect	Internal defect	Unknown	. . .	Internal defect

[a]Includes sulfide stress cracking.

Source: J.P. Hirth and H.H. Johnson, Corrosion, Vol. 32, No. 1, p. 3 (1976).

POTENTIAL SULFIDE STRESS CRACKING REGION AS DEFINED BY THE 0.05 PSIA CRITERION

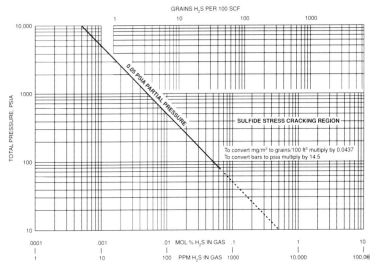

SOURCE: Adapted from MR0175-90, NACE.

MAXIMUM TEMPERATURE FOR CONTINUOUS SERVICE IN DRY HYDROGEN CHLORIDE AND DRY CHLORINE

Material	Hydrogen Chloride,		Chlorine	
	°C	°F	°C	°F
Platinum	1200	2200	260	500
Gold	870	1600	150	300
Nickel	510	950	540	1000
Ni-Cr alloy 600	480	900	540	1000
Ni-Mo alloy B	450	850	540	1000
Ni-Mo-Cr alloy C	450	850	540	950
Carbon steel	260	500	200	400
Ni-Cu alloy 400	230	450	430	800
Silver	230	450	65	150
Cast iron	200	400	180	350
304SS	400	750	310	600
316SS	400	750	340	650
Copper	93	200	200	400

Source: *Industrial Engineering Chemistry*, Vol. 39, p. 839 (1947).

MAXIMUM SERVICE TEMPERATURES IN AIR
FOR STAINLESS STEELS AND ALLOY STEELS

| UNS | Common Name | Maximum Service Temperature | | | |
| | | Intermittent Service | | Continuous Service | |
		°C	°F	°C	°F
S20100	201	815	1500	845	1550
S20200	202	815	1500	845	1550
S30100	301	840	1545	900	1650
S30200	302	870	1600	925	1700
S30400	304	870	1600	925	1700
S30800	308	925	1700	980	1795
S30900	309	980	1795	1095	2000
S31000	310	1035	1895	1150	2100
S31600	316	870	1600	925	1700
S31700	317	870	1600	925	1700
S32100	321	870	1600	925	1700
S33000	330	1035	1895	1150	2100
S34700	347	870	1600	925	1700
S40500	405	815	1500	705	1300
S43000	430	870	1600	815	1500
S44200	442	1035	1895	980	1795
S44600	446	1175	2145	1095	2000
S41000	410	815	1500	705	1300
S41600	416	760	1400	675	1250
S42000	420	735	1355	620	1150
S44000	440	815	1500	760	1400
	Carbon Steel	—	—	565	1050
K11522	½Mo Steel	—	—	565	1050
K11597	1Cr ½Mo Steel	—	—	595	1100
K21590	2-¼Cr 1Mo Steel	—	—	620	1150
K41545	5Cr ½Mo Steel	—	—	650	1200
S50400	9Cr 1Mo Steel	—	—	705	1300

Source: Adapted from Metals Handbook, 9th ed., Volume 13, p. 565, ASM 1987.

HIGH-TEMPERATURE SULFIDIC CORROSION OF STEELS AND STAINLESS STEELS

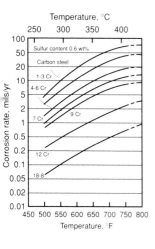

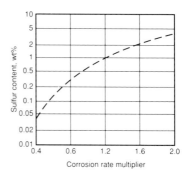

Modified McConomy curves showing the effect of temperature on high-temperature sulfidic corrosion of various steels and stainless steels.

Effect of sulfur content on corrosion rates predicted by modified McConomy curves in 290 to 400°C (550 to 750°F) temperature range.

Source: J. Gutzeit, *Process Industries Corrosion– The Theory and the Practice,* NACE 1986.

HIGH - TEMPERATURE H2S/H2 CORROSION OF 5Cr-0.5Mo STEEL

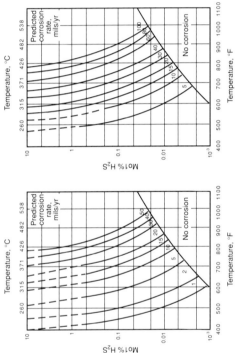

Effect of temperature and hydrogen sulfide content on high temperature H₂S/H₂ corrosion of 5Cr-0.5Mo steel (naphtha desulfurizers). 1 mil/yr = 0.025 mm/yr.

Effect of temperature and hydrogen sulfide content on high- temperature H₂S/H₂ corrosion of 5Cr-0.5Mo steel (gas oil desulfurizers). 1 mil/yr = 0.025 mm/yr.

Source: J. Gutzeit, *Process Industries Corrosion— The Theory and the Practice,* NACE 1986.

HIGH - TEMPERATURE H₂S/H₂ CORROSION OF STAINLESS STEELS

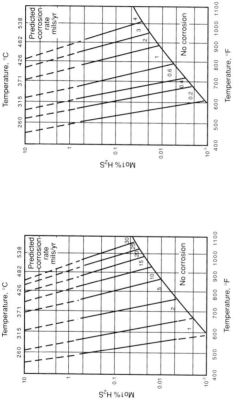

Effect of temperature and hydrogen sulfide content on high-temperature H₂S/H₂ corrosion of 12% Cr stainless steel. 1 mil/yr = 0.025 mm/yr.

Effect of temperature and hydrogen sulfide content on high-temperature H₂S/H₂ corrosion of 18Cr-8Ni austenitic stainless steel. 1 mil/yr = 0.025 mm/yr.

Source: J. Gutzeit, *Process Industries Corrosion— The Theory and the Practice,* NACE 1986.

ASH FUSION TEMPERATURES OF SLAG-FORMING COMPOUNDS

Chemical Compound	Chemical Formula	Ash Fusion Temperature	
		°C	°F
Vanadium pentoxide	V_2O_5	690	1274
Sodium sulfate	Na_2SO_4	890	1630
Nickel sulfate	$NiSO_4$	840	1545
Sodium metavanadate	$Na_2O \cdot V_2O_5$	630	1165
Sodium pyrovanadate	$2Na_2O \cdot V_2O_5$	655	1210
Sodium orthovanadate	$3Na_2O \cdot V_2O_5$	865	1590
Nickel orthovanadate	$3NiO \cdot V_2O_5$	900	1650
Sodium vanadyl vanadate	$Na_2O \cdot V_2O_4 \cdot 5V_2O_5$	625	1155
Sodium iron trisulfate	$2Na_3Fe[SO_4]_3$	620	1150

Source: Metals Handbook, 9th ed., Volume 13, p. 1273, ASM 1987.

DISTRIBUTION RATIO OF AMMONIA AND AMINES
IN STEAM AND STEAM CONDENSATE

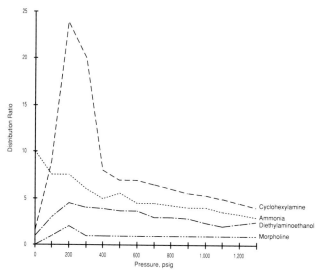

Distribution Ratio = $\dfrac{\text{Concentration of amine in steam}}{\text{Concentration of amine in liquid}}$

Source: Oil and Gas Journal, Nov. 9, 1987, p. 66

OIL FIELD CORROSION INHIBITORS – CATIONIC MOLECULAR STRUCTURES

The letter R denotes fatty acids derived from such oils as soya, coconut, tallow, and tall oil.

Imidazolines

$$R-C \begin{array}{c} \diagup N-CH_2 \\ \diagdown N-CH_2 \\ | \\ R^1 \end{array}$$

Primary Amine

$$R-NH_2$$

Diamines

$$R-NH-C_3H_6-NH_2$$

Amido-amines

$$\begin{array}{c} H \\ | \\ N-C_2H_4-N-C_2H_4-NH_2 \\ | \qquad\qquad | \\ C=O \qquad C=O \\ | \qquad\qquad | \\ R \qquad\qquad R \end{array}$$

Dimerized amido-amines

$$\begin{array}{c} \qquad H \qquad\qquad H \; O \qquad O \; H \qquad\qquad H \\ \qquad | \qquad\qquad | \; || \qquad || \; | \qquad\qquad | \\ H-N-C_2H_4-N-C_2H_4-N-C-R'-C-N-C_2H_4-N-C_2H_4-N-H \\ \qquad | \qquad\qquad\qquad\qquad\qquad\qquad\qquad\qquad\qquad | \\ \qquad C=O \qquad\qquad\qquad\qquad\qquad\qquad\qquad\qquad C=O \\ \qquad | \qquad\qquad\qquad\qquad\qquad\qquad\qquad\qquad\qquad | \\ \qquad R \qquad\qquad\qquad\qquad\qquad\qquad\qquad\qquad\qquad R \end{array}$$

Dimer acid amido-amines

$$NH_2-C_2H_4-N-C-R^1 \qquad R^2 \qquad\qquad\qquad O \\ \qquad\qquad | \quad || \qquad\qquad\qquad\qquad\qquad\qquad || \\ \qquad\qquad H \; O \qquad R^4 \qquad R^3 - C-N-C_2H_4- \\ \qquad\qquad\qquad\qquad\qquad\qquad\qquad\qquad\qquad | \\ \qquad\qquad\qquad\qquad\qquad\qquad\qquad\qquad\qquad H$$

Oxyethylated primary amines

$$R-N-[(C_2H_4O)n-H]_2$$

Alkyl pyridine

$$\begin{array}{c} R'' \\ | \\ C \\ \diagup \; \diagdown \\ C \qquad C \\ || \qquad\qquad || \\ R''-C \qquad C-R'' \\ \diagdown \; \diagup \\ N \end{array}$$

R'' may be CH_3 to C_3H_7 or higher

Quaternized amines

$$(CH_3)_3-N^+-Cl^- \\ \qquad\qquad | \\ \qquad\qquad R$$

Source: Metals Handbook, 9th ed., Vol. 13, p. 479, ASM, 1987

OIL FIELD CORROSION INHIBITORS –
ANIONIC MOLECULAR STRUCTURES

Dimer-trimer acids
(C$_{18}$ average)

Fatty acids

Naphthenic acids

Dodecyl benzene
sulfonic acid

R= C$_{12}$ branched or straight chained

Petroleum Oxidates

Petroleum-derived oxidates (from petrolatum, parrafin wax, lubricating oil, or fuel distillate) are complex mixtures composed primarily of organic acids and esters, but also contain alcohols, ketones, hydroxy and keto acids, estolides, lactides, and lactones.

Phosphate Esters of
Ethoxylated alcohol

R= C$_{10}$ to C$_{16}$ (average C$_{12}$)

Other Organic Acids

1. Acetic
2. Hydroxyacetic
3. Benzoic

Source: Metals Handbook, 9th ed., Vol. 13, p. 480, ASM, 1987

DESIGN DETAILS TO MINIMIZE CORROSION

On the opposite page are examples of how design and assembly can affect localized corrosion by creating crevices and traps where corrosive liquids can accumulate.

(a) Storage containers or vessels should allow complete drainage; otherwise, corrosive species can concentrate in bottom of vessel, and debris may accumulate if the vessel is open to the atmosphere.

(b) Structural members should be designed to avoid retention of liquids; L-shaped sections should be used with the open side down, and exposed seams should be avoided.

(c) Incorrect trimming or poor design of seals and gaskets can create crevice sites.

(d) Drain valves should be designed with sloping bottoms to avoid pitting of the base of the valve.

(e) Nonhorizontal tubing can leave pools of liquid at shutdown.

(f) to (i) Examples of poor assembly that can lead to premature corrosion problems.

(f) Nonvertical assembly of heat exchanger permits a dead space that may result in overheating if very hot gases are involved.

(g) Nonaligned assembly distorts the fastener, which creates a crevice and may result in a loose fitting that can contribute to vibration, fretting, and wear.

(h) Structural supports should allow good drainage; use of a slope at the bottom of the member allows liquid to run off, rather than impinging directly on the concrete support.

(i) Continuous welding is necessary for horizontal stiffeners to prevent the formation of traps and crevices.

Source: Metals Handbook, 9th ed., Volume 13, p. 339, ASM 1987.

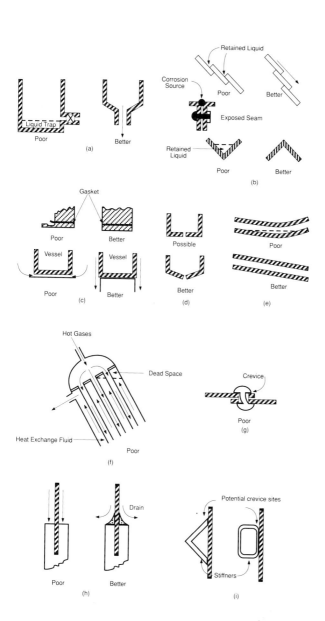

COMMON TYPES OF SCALE FORMING MINERALS

Scale	X-ray Analysis
Acmite - Sodium Iron Silicate	$NaFe (SiO_3)_2$
Barite - Barium Sulfate	$BaSO_4$
Anaicite - Sodium Aluminum Silicate	$NaAl\ Si_2O_5 \cdot H_2O$
Aragonite - (Rhombic Crystals)	$CaCO_3$
Calcium Carbonate - (Hexagonal Crystal)	$CaCO_3$
Calcium Sulfate - Anhydrite	$CaSO_4$
Hydromagnesite - Magnesium Carbonate and Hydroxide	$3MgCO_3 \cdot Mg(OH)_2 \cdot 3H_2O$
Hydroxyapatite - Calcium Phosphate	$Ca_{10}(OH)_2(PO_4)_6$
Iron Oxide	Alpha FeO (OH)
Iron Oxide - Magnetite	Fe_3O_4
Iron Oxide - Red	Fe_2O_3
Iron Chrome Spinels	$CrFe_2O_4$
Iron Sulfide -Troilite, Pyrrhotite	FeS
Magnesium Hydroxide - Bricite	$Mg(OH)_2$
Magnesium Oxide - Magnesia	MgO
Manganese Dioxide - Pyrolusite	MnO_2
Montmorillonite - Aluminum Silicate	$Al_2O_3 \cdot 4SiO_2 \cdot 4H_2O$
Noselite - Sodium Aluminum Silicate	$Na_8Al_6Si_6O_{24} \cdot SO_4$
Organic Deposits	
Pectolite - Calcium Sodium Silicate	$4CaO \cdot Na_2O \cdot 6SiO_2 \cdot H_2O$
Serpentine - Magnesium Silicate	$Mg_3Si_2O_7 \cdot 2H_2O$
Silica - Quartz	SiO_2
Sodalite - Sodium Aluminum Silicate	$Na_8Al_6Si_6O_{24} \cdot Cl_2$
Vermiculite - Magnesium Iron Aluminum Silicate	$(Mg \cdot Fe)_3(Si \cdot Al)_4O_{10}(OH)_2 \cdot 4H_2O$
Xonolite - Calcium Silicate	$5CaO \cdot 5SiO_2 \cdot H_2O$

The compounds listed below are usually found in industrial equipment which contains copper, brass, or bronze:

Copper Iron Sulfide	$CuFeS$
Copper Sulfide - Covellite, Chalcocite	CuS and Cu_2S
Basic Copper Chloride	$CuCl_2 \cdot 3Cu(OH)_2$
Copper Oxide - Cuprite	Cu_2O
Chalcopyrite	$CuFeS_2$
Beta Zinc Sufide - Sphalerite	ZnS
Green Basic Carbonate - Malachite	$CuCO_3Cu(OH)_2$

Source: Corrosion Testing of Chemical Cleaning Solvents Publication 3M182, NACE, 1982.

CHEMICAL CLEANING SOLUTIONS FOR SPECIFIC SCALES

Scale Component	Solvent*	Testing Conditions
Iron Oxide Fe_3O_4 magnetite or mill scale Fe_2O_3 red iron oxide or red rust	5 to 15% HCl 2% Hydroxyacetic/1 Formic Monoammoniated Citric Acid Ammonium EDTA EDTA organic acid mixtures	150-175 F (66-80 C) 180-220 F (82-104 C) circulating 180-220 F (82-104 C) circulating 170-300 F (77-149 C) circulating 100-150 F (38-66 C) circulating
Copper Copper Oxides	Copper complexor in HCl Ammoniacal Bromate Monoammoniated Citric Acid Ammonium Persulfate Ammonium EDTA	150 F (66 C) 120-180 F (50-82 C) 140-180 F (60-82), pH 9 to 11 below 100 F (38 C) 150-180 F (66-82 C), pH 9 to 11
Calcium Carbonate $CaCO_3$	5 to 15% HCl 7 to 10% Sulfamic Acid Sodium EDTA	Preferable not above 150 F (66 C) Do not exceed 140 F (60 C) Circulate at 150-300 F (66-149 C)
Calcium Sulfate $CaSO_4$	Sodium EDTA 1% NaOH·5% HCl EDTA organic acid mixtures	Circulate at 150-300 F (66-149 C) Circulate at 120-150 F (49-66 C) 100-150 F (38-66 C) circulating
Hydroxyapatite or Phosphate Compounds $Ca_{10}(OH)_2 (PO_4)_6$	5 to 10% HCl Sodium EDTA Sulfamic Acid 7 to 10%	Preferable not above 150 F (66 C) Undesirable to add fluoric Circulate 150-300 F (66-149 C) Do not exceed 140 F (60 C)
Silicate Compounds Ex: Acmite, $NaFe(SiO_3)_2$ Analcite, $NaAlSi_2O_5·H_2O$	Prolonged treatment with 0.5 to 1% soda ash at 50 psi (345 kPa), follow with HCl containing fluoride	Alkaline preboil at 50 to 100 psi (345 to 690 kPa) for 12 to 16 hours
Pectolite, $4Ca·Na_2O·$ $6SiO_2·H_2O$ Serpentine, $Mg_3Si_2O_2·2H_2O$	HCl containing Ammonium Bi-fluoride	150-175 F (66-80 C)
Sulfides, ferrous Trolite, FeS Pyrrhotite, FeS	HCl, inhibited	Heat slowly to avoid sudden release of toxic H_2S gas
Disulfides FeS_2, marcasite FeS_2 pyrite	Chromic Acid, CrO_3 followed by HCl	Boiling 7 to 10% chromic acid followed by HCl inhibited
Organic residues Organo · lignins Algae Some polymeric residues	Potassium Permanganate ($KMnO_4$) followed by HCl containing Oxalic Acid	Circulate not 212 F (100 C) a 1 to 2% $KMnO_4$ · solution. Oxalic acid added to HCl controls release of toxic chlorine gas

*The chemicals are to be considered possible solvents only. There are many alternative solvents for each deposit listed.

Source: Corrosion Testing of Chemical Cleaning Solvents Publication 3M182, NACE, 1982.

COMPONENTS OF BOILER DEPOSITS

Mineral	Formula	Nature of Deposit	Usual Location and Form
Acmite	$Na_2O \cdot Fe_2O_3 \cdot 4SiO_2$	Hard, adherent	Tube scale under hydroxyapatite or serpentine
Alpha quartz	SiO_2	Hard, adherent	Turbine blades, mud drum, tube scale
Amphibole	$MgO \cdot SiO_2$	Adherent binder	Tube scale and sludge
Analcite	$Na_2O \cdot Al_2O_3 \cdot 4SiO_2 \cdot 2H_2O$	Hard, adherent	Tube scale under hydroxyabatite or serpentine
Anhydrite	$CaSO_4$	Hard, adherent	Tube scale, generating tubes
Aragonite	$CaCO_3$	Hard, adherent	Tube scale, feed lines, sludge
Brucite	$Mg(OH)_2$	Flocculent	Sludge in mud drum and water wall headers
Copper	Cu	Electroplated layer	Boiler tubes and turbine blades
Cuprite	Cu_2O	Adherent layer	Turbine blades, boiler deposits
Gypsum	$CaSO_4 \cdot 2H_2O$	Hard, adherent	Tube scale, generating tubes
Hematite	Fe_2O_3	Binder	Throughout boiler
Hydroxyapatite	$Ca_{10}(PO_4)_6(OH)_2$	Flocculent	Mud drum, water walls, sludge
Magnesium phosphate	$Mg_3(PO_4)_2$	Adherent binder	Tubes, mud drum, water walls
Magnetite	Fe_3O_4	Protective film	All internal surfaces
Noselite	$4Na_2O \cdot 3Al_2O_3 \cdot 6SiO_2 \cdot SO_4$	Hard, adherent	Tube scale
Pectolite	$Na_2O \cdot 4CaO \cdot 6SiO_2 \cdot H_2O$	Hard, adherent	Tube scale
Serpentine	$3MgO \cdot 2SiO_2 \cdot H_2O$	Flocculent	Sludge
Sodalite	$3Na_2O \cdot 3Al_2O_2 \cdot 6SiO_2 \cdot 2NaCl$	Hard, adherent	Tube scale
Xonotlite	$5CaO \cdot 5SiO_2 \cdot H_2O$	Hard, adherent	Tube scale

Source: J.W. McCoy, Industrial Chemical Cleaning, Chemical Publishing Co., New York, 1984.

NONDESTRUCTIVE METHODS FOR EVALUATING MATERIALS

Method	Measures or Defects	Applications	Advantages	Limitations
oustic ission	Crack initiation and growth rate Internal cracking in welds during cooling Boiling or cavitation Friction or wear Plastic deformation Phase transformations	Pressure vessels Stressed structures Turbine or gear boxes Fracture mechanics research Weldments Sonic signature analysis	Remote and continuous surveillance Permanent record Dynamic (rather than static) detection of cracks Portable Triangulation techniques to locate flaws	Transducers must be placed on part surface Highly ductile materials yield low amplitude emissions Part must be stressed or operating Test system noise needs to be filtered out
ustic- act oping	Debonded areas or delaminations in metal or nonmetal composites or laminates Cracks in turbine wheels or turbine blades Loose rivets or fasteners Crushed core	Brazed or adhesive-bonded structures Bolted or riveted assemblies Turbine blades Turbine wheels Composite structures Honeycomb assemblies	Portable Easy to operate May be automated Permanent record or positive meter readout No couplant required	Part geometry and mass influences test results Impactor and probe must be repositioned to fit geometry of part Reference standards required Pulser impact rate is critical for repeatability
rkhausen se lysis	Residual stresses in ferromagnetic steels	Jet engine components such as compressor blades, discs, diffuser cases	Nondestructive stress analysis Permanent record Fully automatic	Expensive Requires reference standard Need trained operator Not yet a production tool
ly current Hz to kHz)	Subsurface cracks around fastener holes in aircraft structure	Aluminum and titanium structure	Detect subsurface cracks not detectable by radiography	Part geometry Will not detect short cracks
ly current kHz to Hz)	Surface and subsurface cracks and seams Alloy content Heat treatment variations Wall thickness, coating thickness Crack depth Conductivity Permeability	Tubing Wire Ball bearings "Spot checks" on all types of surfaces Proximity gage Metal detector Metal sorting Measure conductivity in % IACS	No special operator skills required High speed, low cost Automation possible for symmetrical parts Permanent record capability for symmetrical parts No couplant or probe contact required	Conductive materials Shallow depth or penetration (thin walls only) Masked or false indications caused by sensitivity to variations, such as part geometry, lift-off Reference standards required Permeability variations
y-sonic	Debonded areas in metal-core or metal-faced honeycomb structures Delaminations in metal laminates or composites Crushed core	Metal-core honeycomb Metal-faced honeycomb Conductive laminates such as boron or graphite fiber composites Bonded metal panels	Portable Simple to operate No couplant required May be automated	Specimen or part must contain conductive materials to establish eddy-current field Reference standards required Part geometry

Continued on Next Page

Method	Measures or Defects	Applications	Advantages	Limitations
Electric current (direct current conduction method)	Cracks Crack depth Resistivity Wall thickness Corrosion-induced wall-thinning	Metallic materials Electrically conductive materials Train rails Nuclear fuel elements Bars, plates, other shapes	Access to only one surface required Battery or dc source Portable	Edge effect Surface contaminat Good surface conta required Difficult to automate Electrode spacing Reference standard required
Electrified particle	Surface defects in nonconducting material Through-to-metal pinholes on metal-backed material Tension, compression, cyclic cracks Brittle-coating stress cracks	Glass Porcelain enamel Nonhomogeneous materials such as plastic or asphalt coatings Glass-to-metal seals	Portable Useful on materials not practical for penetrant inspection	Poor resolution on thin coatings False indications from moisture streaks or lint Atmospheric condit High voltage discha
Exo-electron emission	Fatigue in metals	Metals	Access to only one surface required Permanent record Quantitative	No surface films or contamination Geometry limitation Skilled technician required
Filtered particle	Cracks Porosity Differential absorption	Porous materials such as clay, carbon, powedered metals, concrete Grinding wheels High-tension insulators Sanitary ware	Colored or fluorescent particles Leaves no residue after baking part over 400 F Quickly and easily applied Portable	Size and shape of particles must be selected before use Penetrating power suspension medium is critical Particle concentrati must be controlled Skin irritation
Fluoroscopy (Cine-fluoro-graphy (kine-fluoro-graphy)	Level of fill in containers Foreign objects Internal components Density variations Voids, thickness Spacing or position	Particles in liquid flow Presence of cavitation Operation of valves and switches Burning in small solid-propellant rocket motors	High-brightness images Real-time viewing Image magnification Permanent record Moving subject can be observed	Costly equipment Geometric unsharp thick specimens Speed of event to be studied Viewing area
Holography (acoustical-liquid surface levitation)	Lack of bond Delaminations Voids Porosity Resin-rich or resin-starved areas Inclusions Density variations	Metals Plastics Composites Laminates Honeycomb structures Ceramics Biological specimens	No hologram film development required Real-time imaging provided Liquid-surface responds rapidly to ultrasonic energy	Through-transmission techniques only Object and reference beams must superi pose on special liquid surface Immersion test only Laser required

Continued on Next Pa

ethod	Measures or Defects	Applications	Advantages	Limitations
graphy rfero- y)	Strain Plastic deformation Cracks Debonded areas Voids and inclusions Vibration	Bonded and composite structures Automotive or aircraft tires Three-dimensional imaging	Surface of test object can be uneven No special surface preparations or coatings required No physical contact with test specimen	Vibrationfree environment is required Heavy base to dampen vibrations Difficult to identify type of flaw detected
day ctor voltage k)	Integrity of coatings or linings	Defects holidays in coatings of thickness >15 mils	Portable Easy to operate	Possible damage if dielectric strength exceeded
day ctor voltage	Integrity of coatings	Detects holidays in coatings of thickness <20 mils	Portable Easy to operate	Requires contact with substrate
red ome-	Lack of bond Hot spots Heat transfer Isotherms Temperature ranges	Brazed joints Adhesive-bonded joints Metallic platings or coatings; debonded areas or thickness Electrical assemblies Temperature monitoring	Sensitive to 1.5 F temperature variation Permanent record or thermal picture Quantitative Remote sensing; need not contact part Portable	Emissivity Liquid-nitrogen-cooled detector Critical time-temperature relationship Poor resolution for thick specimens Reference standards required

k ng	Leaks
	Helium
	Ammonia
	Smoke
	Water
	Air bubbles
	Radioactive gas
	Halogens
		Joints:
		  Welded
		  Brazed
		  Adhesive-bonded
		Sealed assemblies
		Pressure or vacuum chambers
		Fuel or gas tanks
			High sensitivity to extremely small, tight separations not detectable by other NDT methods
			Sensitivity related to method selected
				Accessibility to both surfaces of part required
				Smeared metal or contaminants may prevent detection
				Cost related to sensitivity

netic	Cracks Wall thickness Hardness Coercive force Magnetic anisotropy Magnetic field Nonmagnetic coating thickness on steel	Ferromagnetic materials Ship degaussing Liquid level control Treasure hunting Wall thickness of nonmetallic materials Material sorting	Measurement of magnetic material properties May be automated Easily detect magnetic objects in nonmagnetic material Portable	Permeability Reference standards required Edge-effect Probe lift-off
netic icle	Surface and slightly subsurface defects; cracks, seams, porosity, inclusions Permeability variations Extremely sensitive for locating small tight cracks	Ferromagnetic materials; bar, forgings, weldments, extrusions, etc.	Advantage over penetrant in that it indicates subsurface defects, particularly inclusions Relatively fast and low cost May be portable	Alignment of magnetic field is critical Demagnetization of parts required after tests Parts must be cleaned before and after inspection Masking by surface coatings

Continued on Next Page

Method	Measures or Defects	Applications	Advantages	Limitations
Magnetic perturbation	Cracks Crack depth Broken strands in steel cables Permeability effects Nonmetallic inclusions Grinding burns and cracks under chromium plating	Ferromagnetic metals Broken steel cables in reinforced concrete	May be automated Easily detects magnetic objects in nonmagnetic materials Detects subsurface defects	Requires reference standard Need trained oper Part geometry Expensive equipm
Microwave (300 MHz- 300 GHz)	Cracks, holes, debonded areas, etc. in non- metallic parts Changes in composition, degree of cure, moisture content Thickness measurement Dielectric constant Loss tangent	Reinforced plastics Chemical products Ceramics Resins Rubber Liquids Polyurethane foam Radomes	Between radio waves and infrared in the electromagnetic spectrum Portable Contact with part surface not nor- mally required Can be automated	Will not penetrate metals Reference standar required Horn to part spacing critical Part geometry Wave interference Vibration
Mossbauer effect	Nuclear magnetic resonance in materials, most common being iron-57 Polarization of magnetic domains in steel	Detect and identify iron in specimen or sample Detect iron films on stainless steel Measure retained austenite (2 to 35%) in steels Determine nitrided surfaces on steel Interaction of domains with dislocation in ferromagnetic materials	Provide unique infor- mation about the surroundings of the iron-57 nuclei	Radiation hazard Trained engineers physicists required Nonportable Precision equipme for vibrating source and spectrum anal
Neutron activation analysis (Reactor, accelerator, or radio- isotope)	Radiation emission resulting from neutron activation Oxygen in steel Nitrogen in food products Silicon in metals and ores	Metallurgical Prospecting Well logging Oceanography On-line process control of liquid or solid materials	Automatic systems Accurate (ppm range) Fast No contact with sample Sample preparation minimal	Radiation hazard Fast decay time Reference standar required Sensitivity varies with irradiation tim

Continued on Next Page

Method	Measures or Defects	Applications	Advantages	Limitations
netrants e or rescent)	Defects open to sur-face of parts: cracks, porosity, seams, laps, etc. Through-wall leaks	All parts with non-absorbing surfaces (forgings, weldments, castings, etc.) Note: Bleed-out from porous surfaces can mask indications of defects	Low cost Portable Indications may be further examined visually Results easily interpreted	Surface films, such as coatings, scale, and smeared metal may prevent detection of defects Parts must be cleaned before and after inspection Defect must be open to surface
diography rmal utrons n reactor, elerator, fornium)	Hydrogen contamination of titanium or zirconium alloys Defective or improperly loaded pyrotechnic devices Improper assembly of metal, nonmetal parts Corrosion products	Pyrotechnic devices Metallic, nonmetallic assemblies Biological specimens Nuclear reactor fuel elements and control rods Adhesive bonded structures	High neutron absorp-tion by hydrogen, boron, lithium, cadmium, uranium, plutonium Low neutron absorption by most metals Complement to X-ray or gamma-ray radiography	Very costly equipment Nuclear reactor or accelerator required Trained physicists required Radiation hazard Nonportable Indium or gadolinium screens required
iography nma s) alt-60 um-192	Internal defects and variations: porosity, inclusions, cracks, lack of fusion, geometry variations, corrosion thinning Density variations Thickness, gap and position	Usually where X-ray machines are not suitable because source cannot be placed in part with small openings and/or power source not available Panoramic imaging	Low initial cost Permanent records; film Small sources can be placed in parts with small openings Portable Low contrast	One energy level per source Source decay Radiation hazard Trained operators needed Lower image resolution Cost related to source size
iography ays—	Internal defects and variations: porosity; inclusions; cracks; lack of fusion; geometry variations; corrosion thinning Density variations Thickness, gap and position Misassembly Misalignment	Castings Electrical assemblies Weldments Small, thin, complex wrought products Nonmetallics Solid propellant rocket motors Composites	Permanent records; film Ajustable energy levels (5 kv·25 mev) High sensitivity to density changes No couplant required Geometry variations do not effect direction of X-ray beam	High initial costs Orientation of linear defects in part may not be favorable Radiation hazard Depth of defect not indicated Sensitivity decreases with increase in scattered radiation
ometry ay, ma-ray, -ray) ns-ion or scatter)	Wall thickness Plating thickness Variations in density or composition Fill level in cans or containers Inclusions or voids	Sheet, plate, foil, strip, tubing Nuclear reactor fuel rods Cans or containers Plated parts Composites	Fully automatic Fast Extremely accurate In-line process control Portable	Radiation hazard Beta-ray useful for ultrathin coatings only Source decay Reference standards required

Continued on Next Page

Method	Measures or Defects	Applications	Advantages	Limitations
Sonic (Less than 0.1 MHz)	Debonded areas or delaminations in metal or nonmetal composites or laminates Cohesive bond strength under controlled conditions Crushed or fractured core Bond integrity of metal insert fasteners	Metal or nonmetal composite or laminates brazed or adhesive-bonded Plywood Rocket motor nozzles Honeycomb	Portable Easy to operate Locates far-side debonded areas May be automated Access to only one surface required	Surface geometry influences test resu Reference standard required Adhesive or core thickness variations influence results
Thermal (thermochromic paint, liquid crystals)	Lack of bond Hot spots Heat transfer Isotherms Temperature ranges Blockage in coolant passages	Brazed joints Adhesive-bonded joints Metallic platings or coatings Electrical assemblies Temperature monitoring	Very low initial cost Can be readily applied to surfaces which may be difficult to inspect by other methods No special operator skills	Thin-walled surface only Critical time-tempe ture relationship Image retentivity effected by humidi Reference standard required
Thermoelectric probe	Thermoelectric potential Coating thickness Physical properties Thompson effect P-N junctions in semiconductors	Metal sorting Ceramic coating thickness on metals Semiconductors	Portable Simple to operate Access to only one surface required	Hot probe Difficult to automate Reference standard required Surface contaminan Conductive coatings
Tomography	Boundaries Surface reconstruction Crack size, location, and orientation	Metals research Medicine	Pinpoint defect location Image display is computer controlled	Very expensive Need highly trained operator
Ultrasonic (0.1·25 MHz)	Internal defects and variations; cracks, lack of fusion, porosity, inclusions, delaminations, lack of bond, texturing Thickness or velocity Poisson's ratio, elastic modulus	Wrought metals Welds Brazed joints Adhesive-bonded joints Nonmetallics In-service parts	Most sensitive to cracks Test results known immediately Automating and permanent record capability Portable High penetration capability	Couplant required Small, thin, complex parts may be difficu to check Reference standard required Trained operators fc manual inspection Special probes
Ultrasonic angle reflectivity	Elastic properties acoustic attenuation in solids Near-surface metallic property gradients, e.g. carburization in steel Metallic grain structure and size	Metals Nonmetals	Access to only one surface required Permanent record Quantitative No physical contact of sample required Sample preparation minimal	Test parts must be immersed Geometry limitations test part must have flat, smooth area Goniometer device required Skilled technician required

Source: *Metals Progress Databook*, Mid-June 1985, pp. 127-131 with additions.

DIMENSIONS OF SEAMLESS AND WELDED WROUGHT STEEL PIPE
(All dimensions are in inches)

NPS Desig-nator	Outside Diameter	Sche-dule 5S	Sche-dule 10S	Sche-dule 10	Sche-dule 20	Sche-dule 30	Sche-dule 40S	Stan-dard Wall	Sche-dule 40	Sche-dule 60	Sche-dule 80S	Extra Strong	Sche-dule 80	Sche-dule 100	Sche-dule 120	Sche-dule 140	Sche-dule 160	Double Extra Strong
								Nominal Wall Thickness										
1/8	0.405	—	0.049	—	—	—	0.068	0.068	0.068	—	0.095	0.095	0.095	—	—	—	—	—
1/4	0.540	—	0.065	—	—	—	0.088	0.088	0.088	—	0.119	0.119	0.119	—	—	—	—	—
3/8	0.675	—	0.065	—	—	—	0.091	0.091	0.091	—	0.126	0.126	0.126	—	—	—	—	—
1/2	0.840	0.065	0.083	—	—	—	0.109	0.109	0.109	—	0.147	0.147	0.147	—	—	—	0.188	0.294†
3/4	1.060	0.065	0.083	—	—	—	0.113	0.113	0.113	—	0.154	0.154	0.154	—	—	—	0.219	0.308†
1	1.315	0.065	0.109	—	—	—	0.133	0.133	0.133	—	0.179	0.179	0.179	—	—	—	0.250	0.358†
1 1/4	1.660	0.065	0.109	—	—	—	0.140	0.140	0.140	—	0.191	0.191	0.191	—	—	—	0.250	0.382†
1 1/2	1.900	0.065	0.109	—	—	—	0.145	0.145	0.145	—	0.200	0.200	0.200	—	—	—	0.281	0.400†
2	2.375	0.065	0.109	—	—	—	0.154	0.154	0.154	—	0.218	0.218	0.218	—	—	—	0.344	0.436†
2 1/2	2.875	0.083	0.120	—	—	—	0.203	0.203	0.203	—	0.276	0.276	0.276	—	—	—	0.375	0.522†
3	3.500	0.083	0.120	—	—	—	0.216	0.216	0.216	—	0.300	0.300	0.300	—	—	—	0.438	0.600†
3 1/2	4.000	0.083	0.120	—	—	—	0.226	0.226	0.226	—	0.318	0.318	0.318	—	—	—	—	—
4	4.500	0.083	0.120	—	—	—	0.237	0.237	0.237	—	0.337	0.337	0.337	—	0.438	—	0.531	0.674†
5	5.563	0.109	0.134	—	—	—	0.258	0.258	0.258	—	0.375	0.375	0.375	—	0.500	—	0.625	0.750†
6	6.625	0.109	0.134	—	—	—	0.280	0.280	0.280	—	0.432	0.432	0.432	—	0.562	—	0.719	0.864†

Continued on Next Page

DIMENSIONS OF SEAMLESS AND WELDED WROUGHT STEEL PIPE (Cont)
(All dimensions are in inches)

NPS Desig-nator	Outside Diameter	Nominal Wall Thickness																
		Sche-dule 5S	Sche-dule 10S	Sche-dule 10	Sche-dule 20	Sche-dule 30	Sche-dule 40S	Stan-dard Wall	Sche-dule 40	Sche-dule 60	Sche-dule 80S	Extra Strong	Sche-dule 80	Sche-dule 100	Sche-dule 120	Sche-dule 140	Sche-dule 160	Double Extra Strong
8	8.625	0.109	0.148	—	0.250	0.277	0.322	0.322	0.322	0.406	0.500	0.500	0.500	0.594	0.719	0.812	0.906	0.875†
10	10.750	0.134	0.165	—	0.250	0.307	0.365	0.365	0.365	0.500	0.500	0.500	0.594	0.719	0.844	1.000	1.125	1.000
12	12.750	0.156	0.180	—	0.250	0.330	0.375	0.375†	0.406	0.562	0.500	0.500†	0.688	0.844	1.000	1.125	1.312	1.000
14	14.000	0.156*	0.188*	0.250	0.312	0.375		0.375	0.438	0.594		0.500†	0.750	0.938	1.094	1.250	1.406	—
16	16.000	0.165*	0.188*	0.250	0.312	0.375		0.375	0.500	0.656		0.500	0.844	1.031	1.219	1.438	1.594	—
18	18.000	0.165*	0.188*	0.250	0.312	0.438		0.375†	0.562	0.750		0.500†	0.938	1.156	1.375	1.562	1.781	—
20	20.000	0.188*	0.218*	0.250	0.375	0.500		0.375	0.594	0.812		0.500	1.031	1.281	1.500	1.750	1.969	—
22	22.000	0.188*	0.218*	0.250	0.375	0.500		0.375	0.625	0.875		0.500	1.125	1.375	1.625	1.875	2.125	—
24	24.000	0.218*	0.250*	0.250	0.375	0.562		0.375	0.688	0.969		0.500†	1.219	1.531	1.812	2.062	2.344	—
26	26.000	—	—	0.312	0.500	—		0.375†				0.500						—
28	28.000	—	—	0.312	0.500	0.625		0.375†				0.500						—
30	30.000	—	—	0.312	0.500	0.625		0.375†				0.500						—
32	32.000	—	—	0.312	0.500	0.625		0.375†	0.688			0.500						—
34	34.000	—	—	0.312	0.500	0.625		0.375†	0.688			0.500						—
36	36.000	—	—	0.312	0.500	0.625		0.375†	0.750			0.500						—

For tolerances on outside diameter and wall thickness see appropriate specifications.

Schedule 5S, 10S, 40S and 80S apply to austenitic chromium-nickel steel pipe only.

Except when marked † Standard. Extra Strong and Double Extra Strong wall thicknesses have pipe of corresponding wall thickness listed under one of the schedule numbers.

Dimensions in this table are abstracted from ASA Standard B36.10 and ASA Standard B36.19, except that wall thicknesses marked* do not appear in ASA Standard B36.19.

METRIC DIMENSIONS OF SEAMLESS AND WELDED WROUGHT STEEL PIPE
(Dimensions in mm)

NPS Designator	Outside Diameter	Schedule 5S	Schedule 10S	Schedule 10	Schedule 20	Schedule 30	Schedule 40S	Standard Wall	Schedule 40	Schedule 60	Schedule 80S	Extra Strong	Schedule 80	Schedule 100	Schedule 120	Schedule 140	Schedule 160	Double Extra Strong
1/8	10.3	—	1.24	—	—	—	1.73	1.73	1.73	—	2.41	2.41	2.41	—	—	—	—	—
1/4	13.7	—	1.65	—	—	—	2.24	2.24	2.24	—	3.02	3.02	3.02	—	—	—	—	—
3/8	17.1	—	1.65	—	—	—	2.31	2.31	2.31	—	3.2	3.2	3.2	—	—	—	—	—
1/2	21.3	1.65	2.11	—	—	—	2.77	2.77	2.77	—	3.73	3.73	3.73	—	—	—	4.78	7.47†
3/4	26.7	1.65	2.11	—	—	—	2.87	2.87	2.87	—	3.91	3.91	—	—	—	—	5.56	7.82†
1	33.4	1.65	2.77	—	—	—	3.38	3.38	3.38	—	4.55	4.55	4.55	—	—	—	6.35	9.09†
1 1/4	42.2	1.65	2.77	—	—	—	3.56	3.56	3.56	—	4.85	4.85	4.85	—	—	—	6.35	9.70†
1 1/2	48.3	1.65	2.77	—	—	—	3.68	3.68	3.68	—	5.08	5.08	5.08	—	—	—	7.14	10.16†
2	60.3	1.65	2.77	—	—	—	3.91	3.91	3.91	—	5.54	5.54	5.54	—	—	—	8.74	11.07†
2 1/2	73.0	2.11	3.05	—	—	—	5.16	5.16	5.16	—	7.01	7.01	7.01	—	—	—	9.52	14.02†
3	88.9	2.11	3.05	—	—	—	5.49	5.49	5.49	—	7.62	7.62	7.62	—	—	—	11.13	15.24†
3 1/2	101.6	2.11	3.05	—	—	—	5.74	5.74	5.74	—	8.08	8.08	8.08	—	—	—	—	—
4	114.3	2.11	3.05	—	—	—	6.02	6.02	6.02	—	8.56	8.56	8.56	—	11.13	—	13.49	17.12†
5	141.3	2.77	3.40	—	—	—	6.55	6.55	6.55	—	9.52	9.52	9.52	—	12.7	—	15.88	19.05†
6	168.3	2.77	3.40	—	—	—	7.11	7.11	7.11	—	10.97	10.97	10.97	—	14.27	—	18.26	21.95†
8	219.1	2.77	3.76	—	6.35	7.04	8.18	8.18	8.18	10.31	12.7	12.7	12.7	15.09	18.26	20.62	23.01	22.22†
10	273	3.40	4.19	—	6.35	7.8	9.27	9.27	9.27	12.7	12.7	12.7	15.09	18.26	21.44	25.40	28.58	25.40†
12	323.9	3.96*	4.57*	—	6.35	8.38	9.52	9.52†	10.31	14.27	12.7	12.7†	17.47	21.44	25.40	28.58	33.34	25.40†

Continued on Next Page

METRIC DIMENSIONS OF SEAMLESS AND WELDED WROUGHT STEEL PIPE (Cont)
(Dimensions in mm)

NPS Desig-nator	Outside Diameter	Sche-dule 5S	Sche-dule 10S	Sche-dule 10	Sche-dule 20	Sche-dule 30	Sche-dule 40S	Stan-dard Wall	Sche-dule 40	Sche-dule 60	Sche-dule 80S	Extra Strong	Sche-dule 80	Sche-dule 100	Sche-dule 120	Sche-dule 140	Sche-dule 160	Double Extra Strong
14	355.6	3.96*	4.78*	6.35	7.92	9.52	—	9.52†	11.13	15.09	—	12.7*	19.05	23.82	27.79	31.75	35.71	—
16	406.4	4.19*	4.78*	6.35	7.92	9.52	—	9.52†	12.7	16.64	—	12.7	21.44	26.19	30.96	36.52	40.49	—
18	457.2	4.19*	4.78*	6.35	7.92	11.13	—	9.52†	14.27	19.05	—	12.7†	23.82	29.36	34.92	39.69	45.24	—
20	508	4.78*	5.54*	6.35	9.52	12.7	—	9.52	15.9	20.62	—	12.7*	26.19	32.54	38.10	44.45	50.01	—
22	558.8	4.78*	5.54*	6.35	9.52	12.7	—	9.52	15.88	22.22	—	12.7	28.58	34.92	41.28	47.62	53.98	—
24	609.6	5.54*	6.35*	6.35	9.52	14.27	—	9.52	17.48	24.61	—	12.7*	30.96	38.89	46.02	52.39	59.54	—
26	660.4	—	—	7.92	12.7	15.88	—	9.52†	—	—	—	12.7	—	—	—	—	—	—
28	711.2	—	—	7.92	12.7	15.88	—	9.52†	—	—	—	12.7	—	—	—	—	—	—
30	762	—	—	7.92	12.7	15.88	—	9.52†	—	—	—	12.7	—	—	—	—	—	—
32	812.8	—	—	7.92	12.7	15.88	—	9.52†	—	—	—	12.7	—	—	—	—	—	—
34	863.6	—	—	7.92	12.7	15.88	—	9.52†	—	—	—	12.7	—	—	—	—	—	—
36	914.4	—	—	7.92	12.7	15.88	—	9.52†	—	—	—	12.7	—	—	—	—	—	—

For tolerance on outside diameter and wall thickness see appropriate specifications

Schedule 5S, 10S, and 40S and 80S apply to austenitic chromium-nickel steel pipe only

Except when marked † Standard, Extra Strong and Double Extra Strong wall thicknesses have pipe of corresponding wall thickness listed under one of the schedule numbers

Dimensions in this table are based on ASA Standard B36 10 and ASA Standard B36 19, except that the thicknesses marked* do not appear in ASA Standard B36 19

STANDARD WALL STEEL PIPE DIMENSIONS, CAPACITIES AND WEIGHTS

NPS Desig- nator	Outside Diameter in.	Wall Thickness in.	Inside Diameter in.	Cross Sectional Area of Metal in.^{2}	Circumference, ft or Surface Area ft^2/ft of length Outside	Inside	Capacity at 1 ft/s Velocity U.S. Gal. per min.	Weig of Pi lb/f
1/8	0.405	0.068	0.269	0.072	0.106	0.0705	0.179	0.25
1/4	0.540	0.088	0.364	0.125	0.141	0.0954	0.323	0.43
3/8	0.675	0.091	0.493	0.167	0.177	0.1293	0.596	0.57
1/2	0.840	0.109	0.622	0.250	0.220	0.1630	0.945	0.85
3/4	1.050	0.113	0.824	0.333	0.275	0.2158	1.665	1.13
1	1.315	0.133	1.049	0.494	0.344	0.2745	2.690	1.68
1 1/4	1.660	0.140	1.380	0.669	0.435	0.362	4.57	2.28
1 1/2	1.900	0.145	1.610	0.799	0.498	0.422	6.34	2.72
2	2.375	0.154	2.067	1.075	0.622	0.542	10.45	3.66
2 1/2	2.875	0.203	2.469	1.704	0.753	0.647	14.92	5.80
3	3.500	0.216	3.068	2.228	0.917	0.804	23.00	7.58
3 1/2	4.000	0.226	3.548	2.680	1.047	0.930	30.80	9.11
4	4.500	0.237	4.026	3.173	1.178	1.055	39.6	10.8
5	5.563	0.258	5.047	4.304	1.456	1.322	62.3	14.7
6	6.625	0.280	6.065	5.584	1.734	1.590	90.0	19.0
8	8.625	0.322	7.981	8.396	2.258	2.090	155.7	28.6
10	10.750	0.365	10.020	11.90	2.814	2.620	246.0	40.5
12	12.750	0.375	12.000	14.58	3.338	3.142	352.5	49.6
14	14.000	0.375	13.250	16.05	3.665	3.469	430.0	54.6
16	16.000	0.375	15.250	18.41	4.189	3.992	568.0	62.6
18	18.000	0.375	17.250	20.76	4.712	4.516	728.4	70.6
20	20.000	0.375	19.250	23.12	5.236	5.040	902.0	78.6
22	22.000	0.375	21.250	25.48	5.760	5.563	1105.	86.6
24	24.000	0.375	23.250	27.83	6.283	6.087	1325.	94.6
26	26.000	0.375	25.250	30.19	6.807	6.610	1561.	102.6
28	28.000	0.375	27.250	32.54	7.330	7.134	1818.	110.6
30	30.000	0.375	29.250	34.90	7.854	7.658	2094.	118.7
32	32.000	0.375	31.250	37.26	8.378	8.181	2391.	126.7
34	34.000	0.375	33.250	39.61	8.901	8.705	2706.	134.7
36	36.000	0.375	35.250	41.97	9.424	9.228	3042.	142.7

METRIC CONVERSION FACTORS

—	×25.4 = mm	×25.4 = mm	×25.4 = mm	×645. = mm^2	×0.305 = m^2/m	×0.305 = m^2/m	×0.01242 = m^3/min. at 1 m/s	×1.4 = kg

UNIFIED NUMBERING SYSTEM FOR METALS AND ALLOYS

UNS Series	Metal
Nonferrous metals and alloys	
A00001-A99999	Aluminum and aluminum alloys
C00001-C99999	Copper and copper alloys
E00001-E99999	Rare earth and rare earth-like metals and alloys
L00001-L99999	Low melting metals and alloys
M00001-M99999	Miscellaneous nonferrous metals and alloys
N00001-N99999	Nickel and nickel alloys
P00001-P00999	Gold
P01001-P01999	Iridium
P02001-P02999	Osmium
P03001-P03999	Palladium
P04001-P04999	Platinum
P05001-P05999	Rhodium
P06001-P06999	Ruthenium
P07001-P07999	Silver
R01011-R01999	Boron
R02001-R02999	Hafnium
R03001-R03999	Molybdenum
R04001-R04999	Niobium (Columbium)
R05001-R05999	Tantalum
R06001-R06999	Thorium
R07001-R07999	Tungsten
R08001-R08999	Vanadium
R10001-R19999	Beryllium
R20001-R29999	Chromium
R30001-R39999	Cobalt
R40001-R49999	Rhenium
R50001-R59999	Titanium
R60001-R69999	Zirconium
Z00001-Z99999	Zinc and zinc alloys
Ferrous metals and alloys	
D00001-D99999	Specified mechanical properties steels
F00001-F99999	Cast irons
G00001-G99999	AISI and SAE carbon and alloys steels (except tool steels)
H00001-H99999	AISI H-steels
J00001-J99999	Cast steels (except tool steels)
K00001-K99999	Miscellaneous steels and ferrous alloys
S00001-S99999	Heat and corrosion resistant (stainless) steels
T00001-T99999	Tool steels
Welding filler metals, classified by weld deposit composition	
W00001-W09999	Carbon steel with no significant alloying elements
W10000-W19999	Manganese-molybdenum low alloy steels
W20000-W29999	Nickel low alloy steels
W30000-W39999	Austenitic stainless steels
W40000-W49999	Ferritic stainless steels
W50000-W59999	Chromium low alloy steels
W60000-W69999	Copper base alloys
W70000-W79999	Surfacing alloys
W80000-W89999	Nickel base alloys

The following tables on composition and mechanical properties of metallic materials were compiled from data given in the indicated API, ASTM, ASME, and UNS standards.

COMMON NAMES OF UNS ALLOYS - Nonferrous

A02420 Al 242.0	A02950 Al 295.0	A03560 Al 356.0
A05140 Al 514.0	A05200 Al 520.0	A24430 Al B443.0
A91060 Al 1060	A91100 Al 1100	A92014 Al 2014
A92024 Al 2024	A93003 Al 3003	A95052 Al 5052
A95083 Al 5083	A95086 Al 5086	A95154 Al 5154
A96061 Al 6061	A96063 Al 6063	A97075 Al 7075

C10200 OF Copper	C11000 ETP Copper	C12200 DHP Copper
C14200 DPA Copper	C22000 Commercial Bronze	C23000 Red Brass
C26000 Cartridge Brass	C27000 Yellow Brass	C28000 Muntz Metal
C44300 Admiralty Brass, As	C44400 Admiralty Brass, Sb	C44500 Admiralty Brass, P
C46500 Naval Brass, As	C51000 Phosphor Bronze A	C52400 Phosphor Bronze D
C60800 Aluminum Bronze, 6%	C61300 Aluminum Bronze, 7%	C61400 Aluminum Bronze D
C63000 Nickel Aluminum Bronze	C65500 High-Silicon Bronze	C67500 Manganese Bronze A
C68700 Aluminum Brass, As	C70600 90-10 Copper-Nickel	C71500 70-30 Copper-Nickel
C75200 Nickel Silver	C83600 Ounce Metal	C86500 Manganese Bronze
C90500 Gun Metal	C92200 M Bronze	C95700 Cast Mn-Ni-Al Bronze
C95800 Cast Ni-Al Bronze	C96400 Cast 70-30 Cu-Ni	

L50045 Common Lead	L51120 Chemical Lead	L55030 50/50 Solder

M11311 Mg AZ31B	M11914 Mg AZ91C	M12330 Mg EZ33A
M13310 Mg HK31A		

N02200 Nickel 200	N02201 Nickel 201	N02230 Nickel 230
N04400 400 Alloy	N04405 R-405 Alloy	N05500 K-500 Alloy
N05502 502 Alloy	N06002 X Alloy	N06007 G Alloy
N06022 C-22 Alloy	N06030 G-30 Alloy	N06110 Allcor
N06333 RA333 Alloy	N06455 C-4 Alloy	N06600 600 Alloy
N06601 601 Alloy	N06617 617 Alloy	N06625 625 Alloy
N06690 690 Alloy	N06975 2550 Alloy	N06985 G-3 Alloy
N07001 Waspaloy	N07031 31 Alloy	N07041 Rene 41
N07090 90 Alloy	N07716 625 Plus	N07718 718 Alloy
N07750 X-750 Alloy	N08020 20Cb-3	N08024 20Mo-4
N08026 20Mo-6	N08028 Sanicro 28	N08320 20 Mod
N08330 RA-330	N08366 AL-6X	N08367 AL-6XN
N08700 JS700	N08800 800 Alloy	N08801 801 Alloy
N08810 800H Alloy	N08811 800HT Alloy	N08825 825 Alloy
N08904 904L Alloy	N08925 25-6Mo	N09706 706 Alloy
N09925 925 Alloy	N10001 B Alloy	N10002 C Alloy
N10003 N Alloy	N10004 W Alloy	N10276 C-276 Alloy
N10665 B-2 Alloy		

R03600 Molybdenum	R03630 Molybdenum Alloy	R03650 Molybdenum, low C
R04210 Niobium (Columbium)	R05200 Tantalum	R07005 Tungsten
R30003 Elgiloy	R30004 Havar	R30006 Stellite 6
R30031 Stellite 31	R30035 MP35N	R30155 N-155
R30188 HS-188 Alloy	R30260 Duratherm 2602	R30556 HS-556
R30605 L-605 Alloy	R50250 Titanium, Gr 1	R50400 Titanium, Gr 2
R50550 Titanium, Gr 3	R50700 Titanium, Gr 4	R52250 Titanium, Gr 11
R52400 Titanium, Gr 7	R53400 Titanium, Gr 12	R54520 Titanium, Gr 6
R56260 Ti6Al6Mo2Sn4Zr	R56320 Titanium, Gr 9	R56400 Titanium, Gr 5
R58640 Beta-C	R60702 Zr 702	R60704 Zr 704
R60705 Zr 705		

Z13000 Zinc Anode Type II	Z32120 Zinc Anode Type I	Z32121 Zinc Anode Type III

COMMON NAMES OF UNS ALLOYS - Ferrous

F10006 Gray Cast Iron	F20000 Malleable Cast Iron	F32800 Ductile Iron 60-40-18
F41000 Ni-Resist Type 1	F41002 Ni-Resist Type 2	F41006 Ni-Resist Type 5
F43000 Ductile Ni-Resist D2	F43006 Ductile Ni-Resist D5	F47003 Duriron
G10200 1020 Carbon Steel	G41300 4130 Steel	G43400 4340 Steel
J91150 CA-15	J91151 CA-15M	J91153 CA-40
J91540 CA-6NM	J91803 CB-30	J92500 CF-3
J92600 CF-8	J92602 CF-20	J92603 HF
J92605 HC	J92615 CC-50	J92701 CF-16F
J92710 CF-8C	J92800 CF-3M	J92900 CF-8M
J93000 CG-8M	J93001 CG-12	J93005 HD
J93370 CD-4MCu	J93402 CH-20	J93423 CE-30
J93503 HH	J94003 HI	J94202 CK-20
J94203 HK-30	J94204 HK-40	J94213 HN
J94224 HK	N08604 HL	N08705 HP
N08007 CN-7M	N08004 HU	
K01800 A516-55	K02100 A516-60	K02403 A516-65
K02700 A516-70	K02801 A285-C	K03005 A53-B
K03006 A106-B	K11510 0.2Cu Steel	K11522 C-0.5Mo
K11576 HSLA Steel	K11597 1.25Cr-0.5Mo	K21590 2.25Cr-1Mo
K41545 5Cr-0.5Mo	K81340 9Ni Steel	K90941 9Cr-1Mo
K94610 KOVAR		
S13800 PH 13-8 Mo	S15500 15-5 PH	S15700 PH 15-7 Mo
S17400 17-4 PH	S17600 Stainless W	S17700 17-7 PH
S20100 201 SS	S20200 202 SS	S20910 22-13-5
S21400 Tenelon	S21600 216 SS	S21800 Nitronic 60
S21900 21-6-9	S24000 18-3 Mn	S28200 18-18 Plus
S30200 302 SS	S30300 303 SS	S30400 304 SS
S30403 304L SS	S30409 304H SS	S30451 304N SS
S30453 304LN SS	S30500 305 SS	S30800 308 SS
S30815 253MA	S30900 309 SS	S30908 309S SS
S31000 310 SS	S31008 310S SS	S31200 44LN
S31254 254 SMO	S31260 DP-3	S31400 314 SS
S31500 3RE60	S31600 316 SS	S31603 316L SS
S31609 316H SS	S31635 316Ti SS	S31640 316Cb SS
S31651 316N SS	S31653 316LN SS	S31700 317 SS
S31703 317L SS	S31725 317LM SS	S31726 317L4 SS
S31803 2205 Alloy	S32100 321 SS	S32109 321H SS
S32304 SAF 2304	S32404 Uranus 50	S32550 Ferralium 255
S32900 329 SS	S32950 7-Mo Plus	S34700 347 SS
S34709 347H SS	S34800 348 SS	S35000 AM 350
S35500 AM 355	S36200 Almar 362	S38100 18-18-2
S40300 403 SS	S40500 405 SS	S40900 409 SS
S41000 410 SS	S41400 414 SS	S41600 416 SS
S41800 Greek Ascoloy	S42000 420 SS	S42200 422 SS
S42400 F6NM	S42900 429 SS	S43000 430 SS
S43100 431 SS	S43400 434 SS	S43600 436 SS
S44002 440A SS	S44003 440B SS	S44004 440C SS
S44200 442 SS	S44400 18-2	S44600 446 SS
S44625 26-1	S44626 26-1 Ti	S44627 26-1 Cb
S44635 26-4-4	S44660 SC-1	S44700 29-4
S44735 29-4C	S44800 29-4-2	S45000 Custom 450
S45500 Custom 455	S50100 5Cr-0.5Mo	S50200 5Cr-0.5Mo
S50300 7Cr-0.5Mo	S50400 9Cr-1Mo	S66286 A286

COMPARABLE ALLOY DESIGNATIONS

UNS	Name	Germany DIN	U.K. BS	France AFNOR	UNS	Name	Germany DIN	U.K. BS	France AFNOR
C22000	Com. Bronze	2.0230	—	—	N08800	800 Alloy	1.4876	NA15	Z5NC35-20
C44300	Adm. Brass. As	2.0470	CZ111	—	N08825	825 Alloy	2.4858	NA16	NFe32C20DU
C61400	Al Bronze D	2.0932	CA106	—	N08904	904L Alloy	1.4539	—	—
C63000	Ni Al Bronze	2.0966	CA105	—	N08925	25-6Mo	1.4529	—	—
C68700	Al Brass. As	2.0460	CZ110	—	N10276	C-276 Alloy	2.4819	—	—
C70600	90-10 Cu-Ni	2.0872	CN102	—	N10665	B-2 Alloy	2.4617	—	—
C71500	70-30 Cu-Ni	2.0882	CN107	UZ30	R50250	Titanium Gr 1	3.7025	—	—
N02200	Nickel 200	2.4066	NA11	—	R50400	Titanium Gr 2	3.7035	—	—
N02201	Nickel 201	2.4068	NA12	—	R50550	Titanium Gr 3	3.7055	—	—
N04400	400 Alloy	2.4360	NA13	NU30	S15700	PH 15-7 Mo	1.4532	—	—
N05500	K-500 Alloy	2.4375	NA18	—	S17700	17-7 PH	1.4568	—	—
N06002	X Alloy	2.4665	HR204	NC22FeD	S30300	303 SS	1.4305	303S21	Z10CNF1809
N06022	C-22 Alloy	2.4602	—	—	S30400	304 SS	1.4301	304S18	Z6CN18-9
N06030	G-30 Alloy	2.4603	—	—	S30403	304L SS	1.4306	304S14	Z2CN18-10
N06333	RA333 Alloy	2.4608	—	—	S30409	304H SS	1.4948	—	—
N06455	C-4 Alloy	2.4610	—	—	S30453	304LN SS	1.4311	—	—
N06600	600 Alloy	2.4816	NA14	NC15Fe	S30500	305 SS	1.4303	—	—
N06601	601 Alloy	2.4851	—	—	S30800	308 SS	1.4303	—	—
N06625	625 Alloy	2.4856	NA21	NC22DNb	S30900	309 SS	1.4828	—	—
N06985	G-3 Alloy	2.4619	—	—	S30908	309S SS	1.4833	—	Z15CN2413
N07718	718 Alloy	2.4668	—	NC19FeNb	S31000	310 SS	1.4841	310S24	Z12CNS2520
N07750	X-750 Alloy	2.4669	HR505	NC15Fe (Nb)	S31008	310S SS	1.4845	—	—
N08020	20Cb-3	2.4660	—	—	S31200	44LN	1.4460	—	—
N08028	Sanicro 28	1.4563	—	—	S31400	314 SS	1.4841	310S24	Z12CNS2520

Continued on Next Page

COMPARABLE ALLOY DESIGNATIONS (Cont)

UNS	Name	Germany DIN	U.K. BS	France AFNOR
S31500	3RE60	1.4417	—	—
S31600	316 SS	1.4401	316S16	Z6CND17-11
S31603	316L SS	1.4404	316S14	Z2CND17-12
S31609	316H SS	14919	316S59	—
S31651	316N SS	1.4429	—	—
S31653	316LN SS	1.4406	—	—
S31700	317 SS	1.4449	—	—
S31703	317L SS	1.4438	317S12	Z2CND1915
S31803	2205 Alloy	1.4462	—	—
S32100	321 SS	1.4541	321S12	Z6CNT18-10
S32109	321H SS	1.4541	321S59	—
S34700	347 SS	1.4550	347S17	Z6CNNb18-10

UNS	Name	Germany DIN	U.K. BS	France AFNOR
S34709	347H SS	1.4550	347S59	—
S34800	348 SS	1.4546	347S17	—
S40500	405 SS	1.4002	405S17	Z6CA13
S41000	410 SS	1.4006	—	—
S43000	430 SS	1.4016	430S15	Z8C17
S43100	431 SS	1.4057	431S29	—
S43400	434 SS	1.4113	434S19	Z8CD1701
S44003	440B SS	1.4112	—	—
S44004	440C SS	1.4125	—	—
S44400	18-2 SS	1.4521	—	—
S66286	A286	1.4980	—	—

ASTM	UNS	Name
615	S41800	Greek Ascoloy
630	S17400	17-4 PH
631	S17700	17-7 PH
632	S15700	PH 15-7 Mo
634	S35500	AM 355
635	S17600	Stainless W
660	S66286	A286 alloy

ASTM	UNS	Name
P1, T1	K11522	C-0.5Mo
P5, T5	K41545	5Cr-0.5Mo
P7, T7	S50300	7Cr-0.5Mo
P9, T9	S50400	9Cr-1Mo
P11, T11	K11597	1.25Cr-0.5Mo
P22, T22	K21590	2.25Cr-1Mo

ASTM	UNS	Name
XM-9	S36200	Almar 362
XM-10	S21900	21-6-9
XM-12	S15500	15-5 PH
XM-13	S13800	PH 13-8 Mo
XM-15	S38100	18-18-2
XM-16	S45500	Custom 455
XM-17	S21600	216 SS
XM-19	S20910	22-13-5
XM-25	S45000	Custom 450
XM-27	S44625	26-1
XM-29	S24000	18-3 Mn
XM-31	S21400	Tenelon
XM-33	S44626	26-1 Ti

ALUMINUM ALLOYS - Composition, %

UNS	Common Name	Cr	Cu	Mg	Mn	Si	Zn	Other	UNS
A02420	Al 242.0	0.25 max	3.5-4.5	1.2-1.8	0.35 max	0.7 max	0.35 max	Ni 1.7-2.3	A02420
A02950	Al 295.0	—	4.5-5.0	0.03 max	0.35 max	0.7-1.5	0.35 max	—	A02950
A03560	Al 356.0	—	0.25 max	0.20-0.45	0.35 max	6.5-7.5	0.35 max	—	A03560
A05140	Al 514.0	—	0.15 max	3.5-4.5	0.35 max	0.35 max	0.15 max	—	A05140
A05200	Al 520.0	—	0.25 max	9.5-10.6	0.15 max	0.25 max	0.15 max	—	A05200
A24430	Al B443.0	—	0.15 max	0.05 max	0.35 max	4.5-6.0	0.35 max	—	A24430
A91060	Al 1060	—	0.05 max	0.03 max	0.03 max	0.25 max	0.05 max	Al 99.60 min	A91060
A91100	Al 1100	—	0.05-0.20	—	0.05 max	—	0.10 max	Al 99.00 min	A91100
A92014	Al 2014	0.10 max	3.9-5.0	0.20-0.8	0.40-1.2	0.50-1.2	0.25 max	—	A92014
A92024	Al 2024	0.10 max	3.8-4.9	1.2-1.8	0.30-0.9	0.50 max	0.25 max	—	A92024
A93003	Al 3003	—	0.05-0.20	—	1.0-1.5	0.6 max	0.10 max	—	A93003
A95052	Al 5052	0.15-0.35	0.10 max	2.2-2.8	0.10 max	—	0.10 max	—	A95052
A95083	Al 5083	0.05-0.25	0.10 max	4.0-4.9	0.40-1.0	0.40 max	0.25 max	—	A95083
A95086	Al 5086	0.05-0.25	0.10 max	3.5-4.5	0.20-0.7	0.40 max	0.25 max	—	A95086
A95154	Al 5154	0.15-0.35	0.10 max	3.1-3.9	0.10 max	0.25 max	0.20 max	—	A95154
A96061	Al 6061	0.04-0.35	0.15-0.40	0.8-1.2	0.15 max	0.40-0.8	0.25 max	—	A96061
A96063	Al 6063	0.10 max	0.10 max	0.45-0.9	0.10 max	0.20-0.6	0.10 max	—	A96063
A97075	Al 7075	0.18-0.28	1.2-2.0	2.1-2.9	0.30 max	0.40 max	5.1-6.1	—	A97075

ALUMINUM ALLOYS - Mechanical Properties

UNS	Common Name	ASTM	Form	Temper	YS-ksi	TS-ksi	YS-MPa	TS-MPa	UNS
A02420	Al 242.0	B26	Casting	O	—	23 min	—	159 min	A02420
A02950	Al 295.0	B26	Casting	T6	20 min	32 min	138 min	221 min	A02950
A03560	Al 356.0	B26	Casting	T6	20 min	30 min	138 min	207 min	A03560
A05140	Al 514.0	B26	Casting	F	9.0 min	22 min	62 min	152 min	A05140
A05200	Al 520.0	B26	Casting	T4	22 min	42 min	152 min	290 min	A05200
A24430	Al B443.0	B26	Casting	F	6.0 min	17 min	41 min	117 min	A24430
A91060	Al 1060	B210	Tube	O	2.5 min	8.5-13.5	17 min	59-93	A91060
A91100	Al 1100	B210	Tube	O	3.5 min	11-15.5	24 min	76-107	A91100
A92014	Al 2014	B210	Tube	T6	55 min	65 min	380 min	450 min	A92014
A92024	Al 2024	B210	Tube	T3	42 min	64 min	290 min	440 min	A92024
A93003	Al 3003	B210	Tube	O	5 min	14-19	35 min	95-130	A93003
A95052	Al 5052	B210	Tube	O	10 min	25-35	70 min	170-240	A95052
A95083	Al 5083	B210	Tube	O	16 min	39-51	110 min	270-350	A95083
A95086	Al 5086	B210	Tube	O	14 min	35-46	95 min	240-315	A95086
A95154	Al 5154	B210	Tube	O	11 min	30-41	75 min	205-280	A95154
A96061	Al 6061	B210	Tube	T6	35 min	42 min	240 min	290 min	A96061
A96063	Al 6063	B210	Tube	T6	28 min	33 min	195 min	225 min	A96063
A97075	Al 7075	B210	Tube	T6	66 min	77 min	455 min	530 min	A97075

COPPER ALLOYS - Composition, %

UNS	Common Name	Cu*	Al	Fe	Mn	Ni	Si	Sn	Zn	P	Pb	Other	UNS
C10200	OF Copper	99.95 min											C10200
C11000	ETP Copper	99.90 min											C11000
C12200	DHP Copper	99.9 min								0.015-0.040			C12200
C14200	DPA Copper	99.4 min								0.015-0.040		As 0.15-0.50	C14200
C22000	Com. Bronze	89.0-91.0							bal		0.05 max		C22000
C23000	Red Brass	84.0-86.0		0.05 max					bal		0.06 max		C23000
C26000	Cartridge Brass	68.5-71.5		0.05 max					bal		0.07 max		C26000
C27000	Yellow Brass	63.0-68.5		0.07 max					bal		0.10 max		C27000
C28000	Muntz Metal	59.0-63.0		0.07 max					bal		0.30 max		C28000
C44300	Adm. Brass, As	70.0-73.0		0.06 max				0.8-1.2**	bal		0.07 max	As 0.02-0.06	C44300
C44400	Adm. Brass, Sb	70.0-73.0		0.06 max				0.8-1.2**	bal		0.07 max	Sb 0.02-0.10	C44400
C44500	Adm. Brass, P	70.0-73.0		0.06 max				0.8-1.2**	bal	0.02-0.10	0.07 max		C44500
C46500	Naval Brass, As	59.0-62.0		0.10 max				0.5-1.0	bal		0.20 max	As 0.02-0.06	C46500
C51000	Phos. Bronze A	99.5 min**		0.10 max				4.2-5.8	0.30 max	0.03-0.35	0.05 max		C51000
C52400	Phos. Bronze D	99.5 min**		0.10 max				9.0-11.0	0.20 max	0.03-0.35	0.05 max		C52400
C60800	Al Bronze 6%	bal	6.0-7.5	0.10 max							0.10 max		C60800
C61300	Al Bronze, 7%	bal	6.0-8.0	2.0-3.0				0.20-0.50		0.015-0.040	0.01 max		C61300
C61400	Al Bronze D	bal	9.0-11.0	1.5-3.5	1.0 max		0.10 max			0.015 max	0.01 max		C61400
C63000	Ni-Al Bronze	bal	9.0-11.0	2.0-4.0	1.5 max	4.0-5.5	0.25 max	0.20 max		0.015 max			C63000
C65500	Silicon Bronze	bal		0.8 max		0.6 max	2.8-3.8		1.5 max		0.05 max		C65500
C67500	Mn Bronze A	57.0-60.0	0.25 max	0.8-2.0	0.05-0.50			0.50-1.5	bal		0.20 max		C67500
C68700	Al Brass, As	76.0-79.0	1.8-2.5	0.06 max					bal		0.07 max	As 0.02-0.06	C68700
C70600	90-10 Cu-Ni	86.5 min		1.0-1.8	1.0 max	9.0-11.0			1.0 max		0.05 max		C70600
C71500	70-30 Cu-Ni	bal		0.4-1.0	1.0 max	29.0-33.0			1.0 max		0.05 max		C71500
C75200	Nickel Silver	63.5-66.5		0.25 max	0.50 max	16.5-19.5			bal		0.05 max		C75200
C83600	Ounce Metal	84.0-86.0	0.005 max	0.30 max		1.0 max	0.005 max	4.0-6.0	4.0-6.0	0.05 max	4.0-6.0	Sb 0.25 max	C83600
C86500	Mn Bronze	55.0-60.0	0.50-1.5	0.40-2.0	0.10-1.5	1.00 max		1.0 max	36-42		0.40 max		C86500
C90500	Gun Metal	86.0-89.0	0.005 max	0.20 max		1.0 max	0.005 max	9.0-11.0	1.0-3.0	0.05 max	0.30 max	Sb 0.20 max	C90500
C92200	M Bronze	86.0-89.0	0.005 max	0.25 max	0.10-1.5	1.0 max	0.005 max	5.5-6.5	3.0-5.0	0.05 max	1.00-2.00	Sb 0.25 max	C92200
C95700	Mn-Ni-Al Bronze	71.0 min	7.0-8.5	2.0-4.0	11.0-14.0	1.5-3.0	0.10 max				0.03 max		C95700
C95800	Ni-Al Bronze	79.0 min	8.5-9.5	3.5-4.5	0.8-1.5	4.5-5.0	0.10 max				0.03 max		C95800
C96400	Cast 70-30 Cu-Ni	65.0-69.0		0.25-1.5	1.5 max	28.0-32.0	0.50 max			0.02 max	0.01 max		C96400

*Ag included in Cu
**Includes P + Sn

COPPER ALLOYS - Mechanical Properties

UNS	Common Name	ASTM	Form	Temper	YS-ksi	TS-ksi	YS-MPa	TS-MPa	Hardness	UNS
C11000	ETP Copper	B152	Plate, Sheet	O25	10 min	30 min	70 min	205 min	65 HRF max	C11000
C22000	Commercial Bronze	B135	Tube	O60	–	–	–	–	70 HRF max	C22000
C23000	Red Brass	B111	Tube	O61	12 min	40 min	85 min	275 min	–	C23000
C26000	Cartridge Brass	B135	Tube	O60	–	–	–	–	70 HRF max	C26000
C27000	Yellow Brass	B135	Tube	O60	–	–	–	–	80 HRF max	C27000
C28000	Muntz Metal	B111	Tube	O61	20 min	50 min	140 min	345 min	–	C28000
C44300	Admiralty Brass, As	B111	Tube	O61	15 min	45 min	105 min	310 min	–	C44300
C44400	Admiralty Brass, Sb	B111	Tube	O61	15 min	45 min	105 min	310 min	–	C44400
C44500	Admiralty Brass, P	B111	Tube	O61	15 min	45 min	105 min	310 min	–	C44500
C46500	Naval Brass, As	B171	Plate	–	20 min	50 min	140 min	345 min	–	C46500
C51000	Phosphor Bronze A	B139	Bar	O60	–	40-58	–	380 min	–	C51000
C52400	Phosphor Bronze D	B139	Bar	O60	–	60-75	–	480 min	–	C52400
C61300	Aluminum Bronze, 7%	B171	Plate	–	36 min	75 min	250 min	515 min	–	C61300
C61400	Aluminum Bronze D	B315	Pipe, Tube	O61	28 min	65 min	195 min	450 min	–	C61400
C63000	Nickel Aluminum Bronze	B171	Plate	–	34 min	90 min	235 min	620 min	–	C63000
C65500	High-Silicon Bronze	B315	Pipe, Tube	O61	15-29	50 min	103-200	345 min	–	C65500
C67500	Manganese Bronze A	B138	Bar	O61	22 min	55 min	150 min	380 min	–	C67500
C68700	Aluminum Brass, As	B111	Tube	O61	18 min	50 min	125 min	345 min	–	C68700
C70600	90-10 Copper-Nickel	B111	Tube	O61	15 min	40 min	105 min	275 min	–	C70600
C71500	70-30 Copper-Nickel	B111	Tube	O61	18 min	52 min	125 min	360 min	–	C71500
C75200	Nickel Silver	B122	Plate, Sheet	H04	–	78-91	–	540-625	80-90 HRB	C75200
C83600	Ounce Metal	B62	Casting	M01	14 min	30 min	095 min	205 min	–	C83600
C86500	Manganese Bronze	B584	Casting	M01	25 min	65 min	172 min	448 min	–	C86500
C90500	Gun Metal	B584	Casting	M01	18 min	40 min	124 min	275 min	–	C90500
C92200	M Bronze	B61	Casting	M01	16 min	34 min	110 min	235 min	–	C92200
C95700	Cast Mn-Ni-Al Bronze	B148	Casting	M01	40 min	90 min	275 min	620 min	–	C95700
C95800	Cast Ni-Al Bronze	B148	Casting	M01	35 min	85 min	240 min	585 min	–	C95800
C96400	Cast 70-30 Copper-Nickel	B369	Casting	M01	32 min	60 min	220 min	415 min	–	C96400

CARBON AND LOW ALLOY STEELS - Composition %

UNS	Common Name	C	Cr	Cu	Mn	Mo	Ni	P	S	Si	UNS
G10200	1020 Carbon Steel	0.17-0.23	—	—	0.30-0.60	—	—	0.040 max	0.050 max	—	G10200
G41300	4130 Steel	0.28-0.33	0.80-1.10	—	0.40-0.60	0.15-0.25	—	0.035 max	0.040 max	0.15-0.35	G41300
G43400	4340 Steel	0.38-0.43	0.70-0.90	—	0.60-0.80	0.20-0.30	1.65-2.00	0.035 max	0.040 max	0.15-0.30	G43400
K02001	A515-55 Steel	0.20 max	—	—	0.90 max	—	—	0.04 max	0.05 max	0.15-0.30	K02001
K02401	A515-60 Steel	0.24 max	—	—	0.90 max	—	—	0.04 max	0.05 max	0.15-0.30	K02401
K02800	A515-65 Steel	0.28 max	—	—	0.90 max	—	—	0.035 max	0.04 max	0.13-0.45	K02800
K03101	A515-70 Steel	0.31 max	—	—	0.90 max	—	—	0.035 max	0.04 max	0.13-0.33	K03101
K01800	A516-55 Steel	0.18 max	—	—	0.55-0.98	—	—	0.035 max	0.04 max	0.13-0.45	K01800
K02100	A516-60 Steel	0.21 max	—	—	0.55-0.98	—	—	0.035 max	0.04 max	0.13-0.45	K02100
K02403	A516-65 Steel	0.24 max	—	—	0.79-1.30	—	—	0.035 max	0.04 max	0.13-0.45	K02403
K02700	A516-70 Steel	0.27 max	—	—	0.79-1.30	—	—	0.035 max	0.04 max	0.13-0.45	K02700
K02801	A285-C Steel	0.28 max	—	—	0.90 max	—	—	0.035 max	0.040 max	—	K02801
K03005	A53-B Steel	0.30 max	—	—	1.20 max	—	—	0.05 max	0.06 max	—	K03005
K03006	A106-B Steel	0.30 max	—	—	0.29-1.06	—	—	0.048 max	0.058 max	0.10 min	K03006
K11510	U.2Cu Steel	0.15 max	—	0.20 min	1.00 max	—	—	0.15 max	0.05 max	—	K11510
K11576	HSLA Steel	0.10-0.20	0.40-0.65	0.15-0.50	0.60-1.00	0.40-0.60	0.70-1.00	0.035 max	0.040 max	0.15-0.35	K11576
K11522	C-0.5Mo Steel	0.10-0.20	—	—	0.30-0.80	0.44-0.65	—	0.045 max	0.045 max	0.10-0.50	K11522
K11597	1.25Cr-0.5Mo Steel	0.15 max	1.00-1.50	—	0.30-0.60	0.44-0.65	—	0.030 max	0.030 max	0.50-1.00	K11597
K21590	2.25Cr-1Mo Steel	0.15 max	2.00-2.50	—	0.30-0.60	0.90-1.10	—	0.030 max	0.030 max	0.50 max	K21590
K41545	5Cr-0.5Mo Steel	0.15 max	4.00-6.00	—	0.30-0.60	0.45-0.65	—	0.030 max	0.030 max	0.50 max	K41545
K81340	9Ni Steel	0.13 max	—	—	0.90 max	—	8.4-9.6	0.045 max	0.045 max	0.13-0.32	K81340
K90941	9Cr-1Mo Steel	0.15 max	8.0-10.0	—	0.30-0.60	0.90-1.10	—	0.030 max	0.030 max	0.50-1.00	K90941
S50300	7Cr-0.5Mo Steel	0.15 max	6.0-8.0	—	1.00 max	0.45-0.65	—	0.040 max	0.040 max	0.50-1.00	S50300
S50400	9Cr-1Mo Steel	0.15 max	8.0-10.0	—	1.00 max	0.90-1.10	—	0.040 max	0.040 max	1.00 max	S50400
K94610	Kovar	0.04 max	0.20 max	0.20 max	0.50 max	0.20 max	29. nom	—	—	0.20 max	K94610

CARBON AND LOW ALLOY STEELS - Mechanical Properties

UNS	Common Name	ASTM	Form	Heat Tr.	YS-ksi	TS-ksi	YS-MPa	TS-MPa	UNS
G10200	1020 Carbon Steel	A519 Gr 1020	Tube	ANN	28 typ	48 typ	193 typ	331 typ	G10200
K02001	A515-55 Steel	A515 Gr 55	Plate	—	30 min	55-75	205 min	380-515	K02001
K02401	A515-60 Steel	A515 Gr 60	Plate	—	32 min	60-80	220 min	414-550	K02401
K02800	A515-65 Steel	A515 Gr 65	Plate	—	35 min	65-85	240 min	450-585	K02800
K03101	A515-70 Steel	A515 Gr 70	Plate	—	38 min	70-90	260 min	485-620	K03101
K01800	A516-55 Steel	A516 Gr 55	Plate	—	30 min	55-75	205 min	380-515	K01800
K02100	A516-60 Steel	A516 Gr 60	Plate	—	32 min	60-80	220 min	415-550	K02100
K02403	A516-65 Steel	A516 Gr 65	Plate	—	35 min	65-85	240 min	450-585	K02403
K02700	A516-70 Steel	A516 Gr 70	Plate	—	38 min	70-90	260 min	485-620	K02700
K02801	A285-C Steel	A285 Gr C	Plate	—	30 min	55-75	205 min	380-515	K02801
K03005	A53-B Steel	A53 Gr B	Pipe	—	35 min	60 min	240 min	414 min	K03005
K03006	A106-B Steel	A106 Gr B	Pipe	—	35 min	60 min	240 min	414 min	K03006
K11510	0.2Cu Steel	A242	Plate	—	50 min	70 min	345 min	480 min	K11510
K11576	HSLA Steel	A517 Gr F	Plate	QT	100 min	115-135	690 min	795-930	K11576
K11522	C-0.5Mo Steel	A335 Gr P1	Pipe	ANN. NT	30 min	55 min	207 min	379 min	K11522
K11597	1.25Cr-0.5Mo Steel	A335 Gr P11	Pipe	ANN. NT	30 min	60 min	207 min	414 min	K11597
K21590	2.25Cr-1Mo Steel	A335 Gr P22	Pipe	ANN. NT	30 min	60 min	207 min	414 min	K21590
K41545	5Cr-0.5Mo Steel	A335 Gr P5	Pipe	ANN. NT	30 min	60 min	207 min	414 min	K41545
K81340	9Ni Steel	A333 Gr 8	Pipe	QT. DNT	75 min	100 min	517 min	689 min	K81340
S50300	7Cr-0.5Mo Steel	A335 Gr P7	Pipe	ANN. NT	30 min	60 min	207 min	414 min	S50300
S50400	9Cr-1Mo Steel	A335 Gr P9	Pipe	ANN. NT	30 min	60 min	207 min	414 min	S50400

CAST IRONS - Composition, %

UNS	Common Name	C	Cr	Cu	Mn	Mo	Ni	Si	UNS
F10006	Gray Cast Iron	3.10-3.40	—	—	0.60-0.90	—	—	2.30-1.90	F10006
F20000	Malleable Cast Iron	2.20-2.90	—	—	0.15-1.25	—	—	0.90-1.90	F20000
F32800	Ductile Iron 60-40-18	—	—	—	—	—	—	—	F32800
F41000	Ni-Resist Type 1	3.00 max	1.5-2.5	5.50-7.50	0.5-1.5	—	13.5-17.5	1.00-2.80	F41000
F41002	Ni-Resist Type 2	3.00 max	1.5-2.5	0.50 max	0.5-1.5	—	18.0-22.0	1.00-2.80	F41002
F41006	Ni-Resist Type 5	2.40 max	0.10 max	0.50 max	0.5-1.5	—	34.0-36.0	1.00-2.00	F41006
F43000	Ductile Ni-Resist Type D2	3.00 max	1.75-2.75	—	0.70-1.25	—	18.0-22.0	1.50-3.00	F43000
F43006	Ductile Ni-Resist Type D5	2.40 max	0.10 max	—	1.00 max	—	34.0-36.0	1.00-2.80	F43006
F47003	Duriron	0.70-1.10	0.50 max	0.50 max	1.50 max	0.50 max	—	14.20-14.75	F47003

CAST IRONS - Mechanical Properties

UNS	Common Name	ASTM	Heat Tr.	YS-ksi	TS-ksi	YS-MPa	TS-MPa	Hardness	UNS
F10006	Gray Cast Iron, G3000	A159	As cast	—	30 min	—	210 min	187-241 HB	F10006
F20000	Malleable Cast Iron, M3210	A602	ANN	32 min	50 min	221 min	345 min	156 HB max	F20000
F32800	Ductile Iron 60-40-18	A395	—	40 min	60 min	276 min	414 min	143-187 HB	F32800
F41000	Ni-Resist Type 1	A436	As cast	—	25 min	—	172 min	131-183 HB	F41000
F41002	Ni-Resist Type 2	A436	As cast	—	25 min	—	172 min	118-174 HB	F41002
F41006	Ni-Resist Type 5	A436	As cast	—	20 min	—	138 min	99-124 HB	F41006
F43000	Ductile Ni-Resist Type D2	A439	As cast	30 min	58 min	207 min	400 min	139-202 HB	F43000
F43006	Ductile Ni-Resist Type D5	A439	As cast	30 min	55 min	207 min	379 min	131-185 HB	F43006

CAST HEAT RESISTANT STAINLESS STEELS - Composition, %

UNS	ACI	C	Cr	Ni	Mo	Si	Mn	UNS
J92603	HF	0.20-0.40	18.0-23.0	8.0-12.0	0.50 max	2.00 max	2.00 max	J92603
J92605	HC	0.50 max	26.0-30.0	4.00 max	0.50 max	2.00 max	1.00 max	J92605
J93005	HD	0.50 max	26.0-30.0	4.0-7.0	0.50 max	2.00 max	1.50 max	J93005
J93503	HH	0.20-0.50	24.0-28.0	11.0-14.0	0.50 max	2.00 max	2.00 max	J93503
J94003	HI	0.20-0.50	26.0-30.0	14.0-18.0	0.50 max	2.00 max	2.00 max	J94003
J94203	HK-30	0.25-0.35	23.0-27.0	19.0-22.0	—	1.75 max	1.50 max	J94203
J94204	HK-40	0.35-0.45	23.0-27.0	19.0-22.0	—	1.75 max	1.50 max	J94204
J94213	HN	0.20-0.50	19.0-23.0	23.0-27.0	0.50 max	2.00 max	2.00 max	J94213
J94224	HK	0.20-0.60	24.0-28.0	18.0-22.0	0.50 max	2.00 max	2.00 max	J94224
N08604	HL	0.20-0.60	28.0-32.0	18.0-22.0	0.50 max	2.00 max	2.00 max	N08604
N08002	HT	0.35-0.75	15.0-19.0	33.0-37.0	—	2.50 max	2.00 max	N08002
N08004	HU	0.35-0.75	17.0-21.0	37.0-41.0	0.50 max	2.50 max	2.00 max	N08004
N08705	HP	0.35-0.75	24.0-28.0	35.0-37.0	0.50 max	2.50 max	2.00 max	N08705

CAST HEAT RESISTANT STAINLESS STEELS - Mechanical Properties

UNS	ACI	ASTM	Heat Tr.	YS-ksi	TS-ksi	YS-MPa	TS-MPa	UNS
J92603	HF	A297	As cast	35 min	70 min	240 min	485 min	J92603
J92605	HC	A297	As cast	—	55 min	—	380 min	J92605
J93005	HD	A297	As cast	35 min	75 min	240 min	515 min	J93005
J93503	HH	A297	As cast	35 min	75 min	240 min	515 min	J93503
J94003	HI	A297	As cast	35 min	70 min	240 min	485 min	J94003
J94203	HK-30	A608	As cast	—	26 min*	—	179 min*	J94203
J94204	HK-40	A608	As cast	—	29 min*	—	200 min*	J94204
J94213	HN	A297	As cast	—	63 min	—	435 min	J94213
J94224	HK	A297	As cast	35 min	65 min	240 min	450 min	J94224
N08604	HL	A297	As cast	35 min	65 min	240 min	450 min	N08604
N08002	HT	A297	As cast	—	65 min	—	450 min	N08002
N08004	HU	A297	As cast	—	65 min	—	450 min	N08004
N08705	HP	A297	As cast	34 min	62.5 min	235 min	430 min	N08705

*At 1400 F (760 C)

CAST CORROSION RESISTANT STAINLESS STEELS - Composition, %

UNS	ACI	(AISI)	C	Cr	Ni	Mo	Cu	Si	Mn	UNS
J91150	CA-15	410	0.15 max	11.5-14.0	1.00 max	0.50 max	—	1.50 max	1.00 max	J91150
J91151	CA-15M	—	0.15 max	11.5-14.0	1.00 max	0.15-1.0	—	0.65 max	1.00 max	J91151
J91153	CA-40	420	0.20-0.40	11.5-14.0	1.00 max	0.5 max	—	1.50 max	1.00 max	J91153
J91540	CA-6NM	—	0.06 max	11.5-14.0	3.5-4.5	0.40-1.0	—	1.00 max	1.00 max	J91540
J91803	CB-30	431	0.30 max	18.0-21.0	2.00 max	—	—	1.50 max	1.00 max	J91803
J92500	CF-3	304L	0.03 max	17.0-21.0	8.0-12.0	—	—	2.00 max	1.50 max	J92500
J92600	CF-8	304	0.08 max	18.0-21.0	8.0-11.0	—	—	2.00 max	1.50 max	J92600
J92602	CF-20	302	0.20 max	18.0-21.0	8.0-11.0	—	—	2.00 max	1.50 max	J92602
J92615	CC-50	446	0.50 max	26.0-30.0	4.00 max	—	—	1.50 max	1.00 max	J92615
J92701	CF-16F	303	0.16 max	18.0-21.0	9.0-12.0	—	—	2.00 max	1.50 max	J92701
J92710	CF-8C	347	0.08 max	18.0-21.0	9.0-12.0	—	—	2.00 max	1.50 max	J92710
J92800	CF-3M, 3F	316L	0.03 max	17.0-21.0	9.0-13.0	2.0-3.0	—	1.50 max	1.50 max	J92800
J92900	CF-8M	316	0.08 max	18.0-21.0	9.0-12.0	2.0-3.0	—	2.00 max	1.50 max	J92900
J93000	CG-8M	317	0.08 max	18.0-21.0	9.0-13.0	3.0-4.0	—	1.50 max	1.50 max	J93000
J93001	CG-12	—	0.12 max	20.0-23.0	10.0-13.0	—	—	2.00 max	1.50 max	J93001
J93370	CD-4MCu	—	0.04 max	24.5-26.5	4.75-6.00	1.75-2.25	2.75-3.25	1.00 max	1.00 max	J93370
J93402	CH-20	309	0.20 max	22.0-26.0	12.0-15.0	—	—	2.00 max	1.50 max	J93402
J93423	CE-30	312	0.30 max	26.0-30.0	8.0-11.0	—	—	2.00 max	1.50 max	J93423
J94202	CK-20	310	0.20 max	23.0-27.0	19.0-22.0	—	—	2.00 max	2.00 max	J94202
N08007	CN-7M	—	0.07 max	19.0-22.0	27.5-30.5	2.0-3.0	3.0-4.0	1.50 max	1.50 max	N08007

CAST CORROSION RESISTANT STAINLESS STEELS - Mechanical Properties

UNS	ACI	ASTM	Heat Tr.	YS-ksi	TS-ksi	YS-MPa	TS-MPa	UNS
J91150	CA-15	A743	ACT	65 min	90 min	450 min	620 min	J91150
J91151	CA-15M	A743	ACT	65 min	90 min	450 min	620 min	J91151
J91153	CA-40	A743	ACT	70 min	100 min	485 min	690 min	J91153
J91540	CA-6NM	A487	QT	80 min	110-135	550 min	760-930	J91540
J91803	CB-30	A743	ANN	30 min	65 min	205 min	450 min	J91803
J92500	CF-3	A743	STQ	30 min	70 min	205 min	485 min	J92500
J92600	CF-8	A743	STQ	30 min	70 min	205 min	485 min	J92600
J92602	CF-20	A743	STQ	30 min	70 min	205 min	485 min	J92602
J92615	CC-50	A743	ANN	—	55 min	—	380 min	J92615
J92701	CF-16F	A743	STQ	30 min	70 min	205 min	485 min	J92701
J92710	CF-8C	A743	STQ	30 min	70 min	205 min	485 min	J92710
J92800	CF-3F	A743	STQ	30 min	70 min	205 min	485 min	J92800
J92900	CF-8M	A743	STQ	30 min	70 min	205 min	485 min	J92900
J93000	CG-8M	A743	STQ	35 min	75 min	240 min	520 min	J93000
J93001	CG-12	A743	STQ	28 min	70 min	195 min	485 min	J93001
J93370	CD-4MCu	A743	STQ	70 min	100 min	485 min	690 min	J93370
J93402	CH-20	A743	STQ	30 min	70 min	205 min	485 min	J93402
J93423	CE-30	A743	STQ	40 min	80 min	275 min	550 min	J93423
J94202	CK-20	A743	STQ	28 min	65 min	195 min	450 min	J94202
N08007	CN-7M	A743	STQ	25 min	62 min	170 min	425 min	N08007

AUSTENITIC STAINLESS STEELS - Composition, %

UNS	Common Name	C	Cr	Mn	Mo	N	Ni	P	S	Si	Other	UNS
S30200	302 SS	0.15 max	17.0-19.0	2.00 max	—	—	8.0-10.0	0.045 max	0.030 max	1.00 max	—	S30200
S30300	303 SS	0.15 max	17.0-19.0	2.00 max	—	—	8.0-10.0	0.20 max	0.15-0.35	1.00 max	—	S30300
S30400	304 SS	0.08 max	18.0-20.0	2.00 max	—	—	8.0-10.5	0.045 max	0.030 max	1.00 max	—	S30400
S30403	304L SS	0.03 max	18.0-20.0	2.00 max	—	—	8.0-12.0	0.045 max	0.030 max	1.00 max	—	S30403
S30409	304H SS	0.04-0.10	18.0-20.0	2.00 max	—	—	8.0-11.0	0.040 max	0.030 max	1.00 max	—	S30409
S30451	304N SS	0.08 max	18.0-20.0	2.00 max	—	0.10-0.16	8.0-10.5	0.045 max	0.030 max	1.00 max	—	S30451
S30453	304LN SS	0.030 max	18.0-20.0	2.00 max	—	0.10-0.16	8.0-12.0	0.045 max	0.030 max	1.00 max	—	S30453
S30500	305 SS	0.12 max	17.0-19.0	2.00 max	—	—	10.0-13.0	0.045 max	0.030 max	1.00 max	—	S30500
S30800	308 SS	0.08 max	19.0-21.0	2.00 max	—	—	10.0-12.0	0.045 max	0.030 max	1.00 max	—	S30800
S30815	253MA	0.10 max	20.0-22.0	0.80 max	—	0.14-0.20	10.0-12.0	0.040 max	0.030 max	1.40-2.00	Ce 0.03-0.08	S30815
S30900	309 SS	0.20 max	22.0-24.0	2.00 max	—	—	12.0-15.0	0.045 max	0.030 max	1.00 max	—	S30900
S30908	309S SS	0.08 max	22.0-24.0	2.00 max	—	—	12.0-15.0	0.045 max	0.030 max	1.00 max	—	S30908
S31000	310 SS	0.25 max	24.0-26.0	2.00 max	—	—	19.0-22.0	0.045 max	0.030 max	1.50 max	—	S31000
S31008	310S SS	0.08 max	24.0-26.0	2.00 max	—	—	19.0-22.0	0.045 max	0.030 max	1.50 max	—	S31008
S31254	254 SMO	0.020 max	19.5-20.5	1.00 max	6.0-6.5	0.18-0.22	17.5-18.5	0.030 max	0.010 max	0.80 max	Cu 0.50-1.00	S31254
S31400	314 SS	0.25 max	23.0-26.0	2.00 max	—	—	19.0-22.0	0.045 max	0.030 max	1.50-3.00	—	S31400
S31600	316 SS	0.08 max	16.0-18.0	2.00 max	2.0-3.0	—	10.0-14.0	0.045 max	0.030 max	1.00 max	—	S31600
S31603	316L SS	0.03 max	16.0-18.0	2.00 max	2.0-3.0	—	10.0-14.0	0.045 max	0.030 max	1.00 max	—	S31603
S31609	316H SS	0.04-0.10	16.0-18.0	2.00 max	2.0-3.0	—	10.0-14.0	0.040 max	0.030 max	1.00 max	—	S31609
S31635	316Ti SS	0.08 max	16.0-18.0	2.00 max	2.0-3.0	0.10 max	10.0-14.0	0.045 max	0.030 max	1.00 max	Ti 5XC - N 0.70	S31635
S31640	316Cb SS	0.08 max	16.0-18.0	2.00 max	2.0-3.0	0.10 max	10.0-14.0	0.045 max	0.030 max	1.50 max	—	S31640
S31651	316N SS	0.08 max	16.0-18.0	2.00 max	2.0-3.0	0.10-0.16	10.0-14.0	0.045 max	0.030 max	1.00 max	—	S31651
S31653	316LN SS	0.03 max	16.0-18.0	2.00 max	2.0-3.0	0.10-0.16	10.0-14.0	0.045 max	0.030 max	1.00 max	—	S31653
S31700	317 SS	0.08 max	18.0-20.0	2.00 max	3.0-4.0	—	11.0-15.0	0.045 max	0.030 max	1.00 max	—	S31700
S31703	317L SS	0.030 max	18.0-20.0	2.00 max	3.0-4.0	—	11.0-15.0	0.045 max	0.030 max	1.00 max	—	S31703
S31725	317LM SS	0.03 max	18.0-20.0	2.00 max	4.0-5.0	0.10 max	13.0-17.0	0.045 max	0.030 max	0.75 max	Cu 0.75 max	S31725
S31726	317L4 SS	0.03 max	17.0-20.0	2.00 max	4.0-5.0	0.10-0.20	13.5-17.5	0.045 max	0.030 max	0.75 max	Cu 0.75 max	S31726
S32100	321 SS	0.08 max	17.0-19.0	2.00 max	—	—	9.0-12.0	0.045 max	0.030 max	1.00 max	Ti 5XC min	S32100
S32109	321H SS	0.04-0.10	17.0-20.0	2.00 max	—	—	9.0-12.0	0.040 max	0.030 max	1.00 max	Ti 4XC-0.60	S32109
S34700	347 SS	0.08 max	17.0-19.0	2.00 max	—	—	9.0-13.0	0.045 max	0.030 max	1.00 max	Cb 10XC min	S34700
S34709	347H SS	0.04-0.10	17.0-20.0	2.00 max	—	—	9.0-13.0	0.040 max	0.030 max	1.00 max	Cb 8XC-1.00	S34709
S34800	348 SS	0.08 max	17.0-19.0	2.00 max	—	—	9.0-13.0	0.045 max	0.030 max	1.00 max	Cb 10XC min Co 0.20 max Ta 0.10 max	S34800
S38100	18-18-2	0.08 max	17.0-19.0	2.00 max	—	—	17.5-18.5	0.030 max	0.030 max	1.50-2.50	—	S38100

AUSTENITIC STAINLESS STEELS - Mechanical Properties

UNS	Common Name	ASTM	Form	Heat Tr.	YS-ksi	TS-ksi	YS-MPa	TS-MPa	UNS
S30200	302 SS	A240	Plate	STQ	30 min	75 min	205 min	515 min	S30200
S30300	303 SS	A320	Bolting	STQ	30 min	75 min	205 min	515 min	S30300
S30400	304 SS	A312	Pipe	STQ	30 min	75 min	205 min	515 min	S30400
S30403	304L SS	A312	Pipe	STQ	25 min	70 min	170 min	485 min	S30403
S30409	304H SS	A312	Pipe	STQ	30 min	75 min	205 min	515 min	S30409
S30451	304N SS	A312	Pipe	STQ	35 min	80 min	240 min	550 min	S30451
S30453	304LN SS	A312	Pipe	STQ	30 min	75 min	205 min	515 min	S30453
S30500	305 SS	A249	Tube	STQ	30 min	75 min	205 min	515 min	S30500
S30800	308 SS	A473	Forging	STQ	30 min	75 min	205 min	515 min	S30800
S30815	253NA	A312	Pipe	STQ	45 min	87 min	310 min	600 min	S30815
S30900	309 SS	A312	Pipe	STQ	30 min	75 min	205 min	515 min	S30900
S30908	309S SS	A312	Pipe	STQ	30 min	75 min	205 min	515 min	S30908
S31000	310 SS	A312	Pipe	STQ	30 min	75 min	205 min	515 min	S31000
S31008	310S SS	A312	Pipe	STQ	30 min	75 min	205 min	515 min	S31008
S31254	254 SMO	A312	Pipe	STQ	44 min	94 min	300 min	650 min	S31254
S31400	314 SS	A473	Forging	STQ	30 min	75 min	205 min	515 min	S31400
S31600	316 SS	A312	Pipe	STQ	30 min	75 min	205 min	515 min	S31600
S31603	316L SS	A312	Pipe	STQ	25 min	70 min	170 min	485 min	S31603
S31609	316H SS	A312	Pipe	STQ	30 min	75 min	205 min	515 min	S31609
S31635	316Ti SS	A240	Plate	STQ	30 min	75 min	205 min	515 min	S31635
S31651	316N SS	A312	Pipe	STQ	35 min	80 min	240 min	550 min	S31651
S31653	316LN SS	A312	Pipe	STQ	30 min	75 min	205 min	515 min	S31653
S31700	317 SS	A312	Pipe	STQ	30 min	75 min	205 min	515 min	S31700
S31703	317L SS	A312	Pipe	STQ	30 min	75 min	205 min	515 min	S31703
S31725	317LM SS	A312	Tube	STQ	30 min	75 min	205 min	515 min	S31725
S31726	317L4 SS	A312	Pipe	STQ	35 min	80 min	240 min	550 min	S31726
S32100	321 SS	A312	Pipe	STQ	30 min	75 min	205 min	515 min	S32100
S32109	321H SS	A312	Pipe	STQ	30 min	75 min	205 min	515 min	S32109
S34700	347 SS	A312	Pipe	STQ	30 min	75 min	205 min	515 min	S34700
S34709	347H SS	A312	Pipe	STQ	30 min	75 min	205 min	515 min	S34709
S34800	348 SS	A312	Pipe	STQ	30 min	75 min	205 min	515 min	S34800
S38100	18-18-2	A312	Pipe	STQ	30 min	75 min	205 min	515 min	S38100

AUSTENITIC STAINLESS STEELS (High Mn) - Composition, %

UNS	Common Name	C	Cr	Mn	Mo	N	Ni	P	S	Si	Other	UNS
S20100	201 SS	0.15 max	16.0-18.0	5.5-7.5	—	0.25 max	3.5-5.5	0.060 max	0.030 max	1.00 max	—	S20100
S20200	202 SS	0.15 max	17.0-19.0	7.5-10.0	—	0.25 max	4.0-6.0	0.060 max	0.030 max	1.00 max	—	S20200
S20910	22-13-5 SS	0.06 max	20.5-23.5	4.0-6.0	1.50-3.0	0.20-0.40	11.5-13.5	0.040 max	0.030 max	1.00 max	Cb 0.10-0.30 V 0.10-0.30	S20910
S21400	Tenelon	0.12 max	17.0-18.5	14.5-16.0	—	0.35 max	0.75 max	0.045 max	0.030 max	0.30-1.00	—	S21400
S21600	216 SS	0.08 max	17.5-22.0	7.5-9.0	2.0-3.0	0.25-0.50	5.0-7.0	0.045 max	0.030 max	1.00 max	—	S21600
S21800	Nitronic 60	0.10 max	16.0-18.0	7.0-9.0	—	0.08-0.18	8.0-9.0	0.040 max	0.030 max	3.5-4.5	—	S21800
S21900	21-6-9 SS	0.08 max	19.0-21.5	8.0-10.0	—	0.15-0.40	5.5-7.5	0.060 max	0.030 max	1.00 max	—	S21900
S24000	18-3 Mn	0.08 max	17.0-19.0	11.5-14.5	—	0.20-0.40	2.50-3.75	0.060 max	0.030 max	1.00 max	—	S24000
S28200	18-8 Plus	0.15 max	17.0-19.0	17.0-19.0	0.50-1.50	0.40-0.60	—	0.045 max	0.030 max	1.00 max	Cu 0.50-1.50	S28200

AUSTENITIC STAINLESS STEELS (High Mn) - Mechanical Properties

UNS	Common Name	ASTM	Form	Heat Tr.	YS-ksi	TS-ksi	YS-MPa	TS-MPa	UNS
S20100	201 SS	A412	Plate	STQ	38 min	95 min	260 min	655 min	S20100
S20200	202 SS	A473	Forging	STQ	45 min	90 min	310 min	620 min	S20200
S20910	22-13-5 SS	A312	Pipe	STQ	55 min	100 min	380 min	690 min	S20910
S21400	Tenelon	A240	Plate	STQ	70 min	125 min	485 min	860 min	S21400
S21600	216 SS	A479	Bar	STQ	50 min	90 min	345 min	620 min	S21600
S21800	Nitronic 60	A479	Bar	STQ	50 min	95 min	345 min	655 min	S21800
S21900	21-6-9 SS	A473	Forging	STQ	50 min	90 min	345 min	620 min	S21900
S24000	18-3 Mn	A312	Pipe	STQ	55 min	100 min	380 min	690 min	S24000
S28200	18-18 Plus	A473	Forging	STQ	60 min	110 min	410 min	760 min	S28200

UNS	Common Name	C	Cr	Mn	Mo	Ni	P	S	Si	V	W	UNS
S40300	403 SS	0.15 max	11.5-13.0	1.00 max	—	—	0.040 max	0.030 max	0.50 max	—	—	S40300
S41000	410 SS	0.15 max	11.5-13.5	1.00 max	—	—	0.040 max	0.030 max	1.00 max	—	—	S41000
S41400	414 SS	0.15 max	11.5-13.5	1.00 max	—	1.25-2.50	0.040 max	0.030 max	1.00 max	—	—	S41400
S41600	416 SS	0.15 max	12.0-14.0	1.25 max	—	—	0.060 max	0.15 min	1.00 max	—	—	S41600
S41800	Greek Ascoloy	0.15-0.20	12.0-14.0	0.50 max	—	1.80-2.20	0.040 max	0.030 max	0.50 max	—	2.50-3.50	S41800
S42000	420 SS	0.15 min	12.0-14.0	1.00 max	—	—	0.040 max	0.030 max	1.00 max	—	—	S42000
S42200	422 SS	0.20-0.25	11.5-13.5	1.00 max	0.75-1.25	0.50-1.00	0.040 max	0.030 max	0.75 max	0.15-0.03	0.75-1.25	S42200
S42400	F6NM	0.06 max	12.0-14.0	0.50-1.00	0.30-0.70	3.50-4.50	0.03 max	0.03 max	0.30-0.60	—	—	S42400
S43100	431 SS	0.20 max	15.0-17.0	1.00 max	—	1.25-2.50	0.040 max	0.030 max	1.00 max	—	—	S43100
S44002	440A SS	0.60-0.75	16.0-18.0	1.00 max	0.75 max	—	0.040 max	0.030 max	1.00 max	—	—	S44002
S44003	440B SS	0.75-0.95	16.0-18.0	1.00 max	0.75 max	—	0.040 max	0.030 max	1.00 max	—	—	S44003
S44004	440C SS	0.95-1.20	16.0-18.0	1.00 max	0.75 max	—	0.040 max	0.030 max	1.00 max	—	—	S44004

MARTENSITIC STAINLESS STEELS - Mechanical Properties

UNS	Common Name	ASTM	Form	Heat Tr.	YS-ksi	TS-ksi	YS-MPa	TS-MPa	Hardness	UNS
S40300	403 SS	A176	Plate	—	30 min	70 min	205 min	485 min	88 HRB max	S40300
S41000	410 SS	A268	Tube	ANN	30 min	60 min	205 min	415 min	95 HRB max	S41000
S41400	414 SS	A473	Forging	OQT	90 min	115 min	620 min	795 min	321 HB max	S41400
S41600	416 SS	A473	Forging	ANN	40 min	70 min	275 min	485 min	223 HB max	S41600
S41800	Greek Ascoloy	A565	Forging	Q2T	110 min	140 min	760 min	965 min	302-352 HB	S41800
S42000	420 SS	A473	Forging	ANN	—	—	—	—	223 HB max	S42000
S42200	422 SS	A565	Forging	QT	110 min	140 min	760 min	965 min	302-352 HB	S42200
S42400	F6NM	A182	Forging	NT	90 min	110 min	620 min	760 min	235-285 HB	S42400
S43100	431 SS	A473	Forging	OQT	90 min	115 min	620 min	795 min	321 HB max	S43100
S44002	440A SS	A473	Forging	ANN	—	—	—	—	269 HB max	S44002
S44003	440B SS	A473	Forging	ANN	—	—	—	—	269 HB max	S44003
S44004	440C SS	A473	Forging	ANN	—	—	—	—	269 HB max	S44004

FERRITIC STAINLESS STEELS - Composition, %

UNS	Common Name	C	Cr	Mn	Mo	N	Ni	P	S	Si	Other
S40500	405 SS	0.08 max	11.5-14.5	1.00 max	—	—	—	0.040 max	0.030 max	1.00 max	Al 0.10-0.30
S40900	409 SS	0.08 max	10.50-11.75	1.00 max	—	—	0.50 max	0.045 max	0.045 max	1.00 max	Ti 6XC-0.75
S42900	429 SS	0.12 max	14.0-16.0	1.00 max	—	—	—	0.040 max	0.030 max	1.00 max	—
S43000	430 SS	0.12 max	16.0-18.0	1.00 max	—	—	—	0.040 max	0.030 max	1.00 max	—
S43400	434 SS	0.12 max	16.0-18.0	1.00 max	0.75-1.25	—	—	0.040 max	0.030 max	1.00 max	—
S43600	436 SS	0.12 max	16.0-18.0	1.00 max	0.75-1.25	—	—	0.040 max	0.030 max	1.00 max	Cb + Ta 5XC-0.70
S44200	442 SS	0.20 max	18.0-23.0	1.00 max	—	—	—	0.040 max	0.030 max	1.00 max	—
S44400	18-2	0.025 max	17.5-19.5	1.00 max	1.75-2.50	0.025 max	1.00 max	0.040 max	0.030 max	1.00 max	Cb + Ti 0.20 + 4 (C + N)-0.80
S44600	446 SS	0.20 max	23.0-27.0	1.50 max	—	0.25 max	—	0.040 max	0.030 max	1.00 max	—
S44625	26-1	0.01 max	25.0-27.5	0.40 max	0.75-1.50	0.015 max	0.50 max	0.020 max	0.020 max	0.40 max	Ni + Cu 0.50 max, Cu 0.20 max
S44626	26-1 Ti	0.06 max	25.0-27.0	0.75 max	0.75-1.50	0.04 max	0.50 max	0.040 max	0.020 max	0.75 max	Ti 0.20-1.00, 7X (C + N) min
S44627	26-1 Cb	0.010 max	25.0-27.0	0.40 max	0.75-1.50	0.015 max	0.50 max	0.020 max	0.020 max	0.40 max	Cb 0.05-0.20
S44635	26-4-4	0.025 max	24.5-26.0	1.00 max	3.50-4.50	0.035 max	3.50-4.50	0.040 max	0.030 max	0.75 max	Cb + Ti 0.20 + 4 (C + N)-0.80
S44660	SC-1	0.025 max	25.0-27.0	1.00 max	2.50-3.50	0.035 max	1.50-3.50	0.040 max	0.030 max	1.00 max	Cb + Ti 0.20 + 4 (C + N)-0.80
S44700	29-4	0.010 max	28.0-30.0	0.30 max	3.5-4.2	0.020 max	0.15 max	0.025 max	0.020 max	0.20 max	C + N 0.025 max
S44735	29-4C	0.030 max	28.0-30.0	1.00 max	3.60-4.20	0.045 max	1.00 max	0.040 max	0.030 max	1.00 max	Cb + Ti 6(C + N) 0.20-1.00
S44800	29-4-2	0.010 max	28.0-30.0	0.30 max	3.5-4.2	0.020 max	2.0-2.5	0.025 max	0.020 max	0.20 max	C + N 0.025 max, Cu 0.15 max

FERRITIC STAINLESS STEELS - Mechanical Properties

UNS	Common Name	ASTM	Form	Heat Tr.	YS-ksi	TS-ksi	YS-MPa	TS-MPa	Hardness	UNS
S40500	405 SS	A268	Tube	ANN	30 min	60 min	205 min	415 min	95 HRB max	S40500
S40900	409 SS	A176	Plate	—	30 min	55 min	205 min	380 min	80 HRB max	S40900
S42900	429 SS	A268	Tube	ANN	35 min	60 min	240 min	415 min	90 HRB max	S42900
S43000	430 SS	A268	Tube	ANN	35 min	60 min	240 min	415 min	90 HRB max	S43000
S44200	442 SS	A176	Plate	—	40 min	65 min	275 min	515 min	95 HRB max	S44200
S44400	18-2	A268	Tube	ANN	40 min	60 min	275 min	415 min	95 HRB max	S44400
S44600	446 SS	A268	Tube	ANN	40 min	70 min	275 min	485 min	95 HRB max	S44600
S44626	26-1 Ti	A268	Tube	ANN	45 min	68 min	310 min	470 min	100 HRB max	S44626
S44627	26-1 Cb	A268	Tube	ANN	40 min	65 min	275 min	450 min	90 HRB max	S44627
S44635	26-4-4	A268	Tube	ANN	75 min	90 min	515 min	620 min	27 HRC max	S44635
S44660	SC-1	A268	Tube	ANN	65 min	85 min	450 min	585 min	25 HRC max	S44660
S44700	29-4	A268	Tube	ANN	60 min	80 min	415 min	550 min	100 HRB max	S44700
S44735	29-4C	A268	Tube	ANN	60 min	75 min	415 min	515 min	100 HRB max	S44735
S44800	29-4-2	A268	Tube	ANN	60 min	80 min	415 min	550 min	100 HRB min	S44800

DUPLEX STAINLESS STEELS - Composition, %

UNS	Common Name	C	Cr	Cu	Mn	Mo	N	Ni	P	S	Si	W
S31200	44LN	0.030 max	24.0-26.0	—	2.00 max	1.2-2.0	0.14-0.20	5.5-6.5	0.045 max	0.030 max	1.00 max	
S31260	DP-3	0.030 max	24.0-26.0	0.20-0.80	1.00 max	2.5-3.5	0.10-0.30	5.5-7.5	0.030 max	0.030 max	0.75 max	0.10-0.50
S31500	3RE60	0.030 max	18.0-19.0	—	1.20-2.00	2.5-3.0	—	4.25-5.25	0.030 max	0.030 max	1.40-2.00	
S31803	2205 Alloy	0.030 max	21.0-23.0	—	2.0 max	2.5-3.5	0.08-0.20	4.5-6.5	0.030 max	0.020 max	1.00 max	
S32304	SAF 2304	0.030 max	21.5-24.5	0.05-0.60	2.50 max	0.05-0.60	0.05-0.20	3.0-5.5	0.040 max	0.040 max	1.0 max	
S32404	Uranus 50	0.04 max	20.5-22.5	1.0-2.0	2.0 max	2.0-3.0	0.20 max	5.5-8.5	0.030 max	0.010 max	1.0 max	
S32550	Ferralium 255	0.04 max	24.0-27.0	1.5-2.5	1.50 max	2.0-4.0	0.10-0.25	4.5-6.5	0.04 max	0.03 max	1.00 max	
S32900	329 SS	0.20 max	23.0-28.0	—	1.00 max	1.0-2.0	—	2.5-5.0	0.040 max	0.030 max	0.75 max	
S32950	7-Mo Plus	0.030 max	26.0-29.0	—	2.00 max	1.0-2.5	0.15-0.35	3.5-5.2	0.035 max	0.010 max	0.60 max	

DUPLEX STAINLESS STEELS - Mechanical Properties

UNS	Common Name	ASTM	Form	Heat Tr.	YS-ksi	TS-ksi	YS-MPa	TS-MPa	Hardness	UNS
S31200	44LN	A790	Pipe	STQ	65 min	100 min	450 min	690 min	280 HB max	S31200
S31260	DP-3	A790	Pipe	STQ	64 min	92 min	440 min	630 min	30.5 HRC max	S31260
S31500	3RE60	A790	Pipe	STQ	64 min	92 min	440 min	630 min	30.5 HRC max	S31500
S31803	2205 Alloy	A790	Pipe	STQ	65 min	90 min	450 min	620 min	30.5 HRC max	S31803
S32304	SAF 2304	A790	Pipe	STQ	58 min	87 min	400 min	600 min	30.5 HRC max	S32304
S32550	Ferralium 255	A790	Pipe	STQ	80 min	110 min	550 min	760 min	31.5 HRC max	S32550
S32900	329 SS	A789	Tube	STQ	70 min	90 min	485 min	620 min	28 HRC max	S32900
S32950	7-Mo Plus	A790	Pipe	STQ	70 min	90 min	480 min	620 min	30.5 HRC max	S32950

PRECIPITATION-HARDENABLE STAINLESS STEELS - Composition, %

UNS	Common Name	C	Al	Cr	Cu	Mn	Mo	Ni	Si	P	S	Other
S13800	PH 13-8 Mo	0.05 max	0.90-1.35	12.25-13.25	—	0.20 max	2.00-2.50	7.50-8.50	0.10 max	0.01 max	0.008 max	N 0.01 max
S15500	15-5 PH	0.07 max	—	14.0-15.5	2.50-4.50	1.00 max	—	3.50-5.50	1.0 max	0.040 max	0.030 max	Cb 0.15-0.45
S15700	PH 15-7 Mo	0.09 max	0.75-1.50	14.0-16.0	—	1.00 max	2.00-3.00	6.50-7.75	1.00 max	0.04 max	0.03 max	—
S17400	17-4 PH	0.07 max	—	15.5-17.5	3.00-5.00	1.00 max	—	3.00-5.00	1.00 max	0.040 max	0.030 max	Cb 0.15-0.45
S17600	Stainless W	0.08 max	0.40 max	16.0-18.0	—	1.00 max	—	6.00-7.50	1.00 max	0.040 max	0.030 max	Ti 0.40-1.20
S17700	17-7 PH	0.09 max	0.75-1.50	16.0-18.0	—	1.00 max	—	6.50-7.75	1.00 max	0.040 max	0.040 max	—
S35000	AM 350	0.07-0.11	—	16.0-17.0	—	0.50-1.25	2.50-3.25	4.00-5.00	0.50 max	0.040 max	0.030 max	N 0.07-0.13
S35500	AM 355	0.10-0.15	—	15.0-16.0	—	0.50-1.25	2.50-3.25	4.00-5.00	0.50 max	0.040 max	0.030 max	N 0.07-0.13
S36200	Almar 362	0.05 max	0.10 max	14.0-15.0	—	0.50 max	—	6.00-7.00	0.30 max	0.030 max	0.030 max	Ti 0.55-0.90
S45000	Custom 450	0.05 max	—	14.0-16.0	1.25-1.75	1.00 max	0.50-1.00	5.00-7.00	1.00 max	0.030 max	0.030 max	Cb 8xC min
S45500	Custom 455	0.05 max	—	11.0-12.5	1.50-2.50	0.50 max	0.50 max	7.50-9.50	0.50 max	0.040 max	0.030 max	Cb 0.10-0.50 Ti 0.80-1.40
S66286	A286	0.08 max	0.35 max	13.5-16.0	—	2.00 max	1.00-1.50	24.0-27.0	1.00 max	0.040 max	0.030 max	Ti 1.90-2.35 V 0.10-0.50 B 0.001-0.01

PRECIPITATION-HARDENABLE STAINLESS STEELS - Mechanical Properties

UNS	Common Name	ASTM	Form	Heat Tr.	YS-ksi	TS-ksi	YS-MPa	TS-MPa	Hardness	UNS
S13800	PH 13-8 Mo	A705	Forging	H1150	90 min	135 min	620 min	931 min	30 HRC min	S13800
S15500	15-5 PH	A705	Forging	H1150	105 min	135 min	725 min	930 min	28 HRC min	S15500
S15700	PH 15-7 Mo	A705	Forging	TH1050	160 min	180 min	1100 min	1240 min	375 HB min	S15700
S17400	17-4 PH	A705	Forging	H1150	105 min	135 min	725 min	930 min	28 HRC min	S17400
S17600	Stainless W	A705	Forging	H1050	150 min	170 min	1030 min	1170 min	35 HRC min	S17600
S17700	17-7 PH	A705	Forging	TH1050	140 min	170 min	965 min	1170 min	38 HRC min	S17700
S35500	AM 355	A705	Forging	H1000	155 min	170 min	1070 min	1170 min	37 HRC min	S35500
S36200	Almar 362	A705	Forging	IS	145 min	155 min	1000 min	1070 min	33 HRC min	S36200
S45000	Custom 450	A705	Forging	H1150	75 min	125 min	515 min	860 min	26 HRC min	S45000
S45500	Custom 455	A705	Forging	H1000	185 min	205 min	1275 min	1410 min	40 HRC min	S45500
S66286	A286	A638	Forging	STA	85 min	130 min	585 min	895 min	248 HB min	S66286

NICKEL ALLOYS - Composition, %

UNS	Common Name	Ni	Co	Cr	Cu	Fe	Mo	Al	C	Mn	UNS
N02200	Nickel 200	99.0 min	—	—	0.25 max	0.40 max	—	—	0.15 max	0.35 max	N02200
N02201	Nickel 201	99.0 min	—	—	0.25 max	0.40 max	—	—	0.02 max	0.35 max	N02201
N02230	Nickel 230	99.0 min	—	—	0.10 max	0.10 max	—	—	0.15 max	0.15 max	N02230
N04400	400 Alloy	63.0-70.0	—	—	bal	2.50 max	—	—	0.3 max	2.00 max	N04400
N04405	R-405 Alloy	63.0-70.0	—	—	bal	2.5 max	—	—	0.30 max	2.0 max	N04405
N05500	K-500 Alloy	63.0-70.0	—	—	bal	2.00 max	—	2.30-3.15	0.25 max	1.50 max	N05500
N05502	502 Alloy	63.0-70.0	—	—	bal	2.00 max	—	2.50-3.50	0.10 max	1.50 max	N05502
N06600	600 Alloy	72.0 min	—	14.0-17.0	0.50 max	6.0-10.0	—	—	0.15 max	1.00 max	N06600
N06601	601 Alloy	58.0-63.0	—	21.0-25.0	1.0 max	bal	—	1.0-1.7	0.1 max	1.0 max	N06601
N06690	690 Alloy	58.0 min	—	27.0-31.0	0.50 max	7.0-11.0	—	—	0.05 max	0.50 max	N06690
N07090	90 Alloy	bal	15.0-21.0	18.0-21.0	—	3.0 max	—	0.8-2.0	0.13 max	1.0 max	N07090
N07750	X-750 Alloy	70.0 min	—	14.0-17.0	0.5 max	5.0-9.0	—	0.40-1.0	0.08 max	1.0 max	N07750
N08330	RA-330	34.0-37.0	—	17.0-20.0	1.00 max	bal	—	—	0.08 max	2.00 max	N08330
N08800	800 Alloy	30.0-35.0	—	19.0-23.0	0.75 max	bal	—	0.15-0.60	0.10 max	1.5 max	N08800
N08801	801 Alloy	30.0-34.0	—	19.0-22.0	0.5 max	bal	—	—	0.10 max	1.5 max	N08801
N08810	800H Alloy	30.0-35.0	—	19.0-23.0	0.75 max	bal	—	0.15-0.60	0.05-0.10	1.5 max	N08810
N08811	800HT Alloy	30.0-35.0	—	19.0-23.0	0.75 max	39.5 min	—	0.15-0.60	0.06-0.10	1.5 max	N08811
N09706	706 Alloy	39.0-44.0	—	14.5-17.5	0.30 max	bal	—	0.40 max	0.06 max	0.35 max	N09706
N10001	B Alloy	bal	2.50 max	1.00 max	—	6.00 max	26.0-33.0	—	0.12 max	1.00 max	N10001
N10665	B-2 Alloy	bal	1.00 max	1.00 max	—	2.00 max	26.0-30.0	—	0.020 max	1.00 max	N10665

Continued on Next Page

NICKEL ALLOYS - Composition, % (Cont)

	Si	P	S	Ti	Other	
Nickel 200	0.35 max	—	0.010 max	—	—	N02200
Nickel 201	0.35 max	—	0.010 max	—	—	N02201
Nickel 230	0.010-0.035	—	0.008 max	0.005 max	Mg 0.04-0.08	N02230
400 Alloy	0.50 max	—	0.024 max	—	—	N04400
R-405 Alloy	0.50 max	—	0.025-0.06	—	—	N04405
K-500 Alloy	0.50 max	—	0.01 max	0.35-0.85	—	N05500
502 Alloy	0.5 max	—	0.010 max	0.50 max	—	N05502
600 Alloy	0.50 max	—	0.015 max	—	—	N06600
601 Alloy	0.50 max	—	0.015 max	—	—	N06601
690 Alloy	0.50 max	—	0.015 max	—	—	N06690
90 Alloy	1.5 max	—	—	1.8-3.0	—	N07090
X-750 Alloy	0.50 max	—	0.01 max	2.25-2.75	—	N07750
RA-330	0.75-1.50	0.03 max	0.03 max	—	Pb 0.005 max,Sn 0.025 max	N08330
800 Alloy	1.0 max	—	0.015 max	0.15-0.60	—	N08800
801 Alloy	1.0 max	—	0.015 max	0.75-1.5	—	N08801
800H Alloy	1.0 max	—	0.015 max	0.15-0.60	—	N08810
800HT Alloy	1.0 max	—	0.015 max	0.15-0.60	—	N08811
706 Alloy	0.35 max	0.020 max	0.015 max	1.5-2.0	B 0.006 max	N09706
B Alloy	1.00 max	0.040 max	0.030 max	—	V 0.60 max	N10001
B-2 Alloy	0.10 max	0.04 max	0.03 max	—	—	N10665

See also CrMo NICKEL ALLOYS, page 180.

CrMo NICKEL ALLOYS - Composition, %

UNS	Common Name	Ni	Co	Cr	Cu	Fe	Mo	Al	C	Mn	UNS
N06002	X Alloy	bal	0.5-2.5	20.5-23.0	—	17.0-20.0	8.0-10.0	—	0.05-0.15	1.00 max	N06002
N06007	G Alloy	bal	2.5 max	21.0-23.5	1.5-2.5	18.0-21.0	5.5-7.5	—	0.05 max	1.0-2.0	N06007
N06022	C-22 Alloy	bal	2.50 max	20.0-22.5	—	2.0-6.0	12.5-14.5	—	0.015 max	0.50 max	N06022
N06030	G-30 Alloy	bal	5.0 max	28.0-31.5	1.0-2.4	13.0-17.0	4.0-6.0	—	0.03 max	1.5 max	N06030
N06110	Allcor	bal	12.0 max	27.0-33.0	—	—	8.0-12.0	1.50 max	0.15 max	—	N06110
N06333	RA333 Alloy	44.0-47.0	2.50-4.00	24.0-27.0	0.50 max	bal	2.5-4.0	—	0.08 max	2.00 max	N06333
N06455	C-4 Alloy	bal	2.0 max	14.0-18.0	—	3.0 max	14.0-17.0	—	0.015 max	1.00 max	N06455
N06617	617 Alloy	44.5 min	10.0-15.0	20.0-24.0	0.50 max	3.0 max	8.0-10.0	0.80-1.50	0.05-0.15	1.00 max	N06617
N06625	625 Alloy	bal	—	20.0-23.0	—	5.0 max	8.0-10.0	0.40 max	0.10 max	0.50 max	N06625
N06975	2550 Alloy	47.0-52.0	—	23.0-26.0	0.70-1.20	bal	5.0-7.0	—	0.03 max	1.00 max	N06975
N06985	G-3 Alloy	bal	5.0 max	21.0-23.5	1.5-2.5	18.0-21.0	6.0-8.0	—	0.015 max	1.00 max	N06985
N07001	Waspaloy	bal	12.0-15.0	18.0-21.0	0.50 max	2.00 max	3.5-5.0	1.20-1.60	0.03-0.10	1.00 max	N07001
N07031	Alloy 31	55.0-58.0	—	22.0-23.0	0.60-1.20	bal	1.70-2.30	1.00-1.70	0.03-0.06	0.20 max	N07031
N07041	Rene 41	bal	10.0-12.0	18.0-20.0	—	5.00 max	9.0-10.5	1.40-1.80	0.12 max	0.10 max	N07041
N07716	625 Plus	59.0-63.0	—	19.0-22.0	—	bal	7.0-9.5	0.35 max	0.03 max	0.20 max	N07716
N07718	718 Alloy	50.0-55.0	1.00 max	17.0-21.0	0.30 max	bal	2.80-3.30	0.20-0.80	0.08 max	0.35 max	N07718
N08020	20Cb-3	32.0-38.0	—	19.0-21.0	3.00-4.00	bal	2.00-3.00	—	0.07 max	2.00 max	N08020
N08024	20Mo-4	35.0-40.0	—	22.5-25.0	0.50-1.50	bal	3.50-5.00	—	0.03 max	1.00 max	N08024
N08026	20Mo-6	33.0-37.2	—	22.0-26.0	2.00-4.00	bal	5.00-6.70	—	0.030 max	1.00 max	N08026
N08028	Sanicro 28	29.5-32.5	—	26.0-28.0	0.6-1.4	bal	3.0-4.0	—	0.030 max	2.50 max	N08028
N08320	20 Mod	25.0-27.0	—	21.0-23.0	—	bal	4.0-6.0	—	0.05 max	2.5 max	N08320
N08366	AL-6X	23.5-25.5	—	20.0-22.0	—	bal	6.0-7.0	—	0.035 max	2.00 max	N08366
N08367	AL-6XN	23.5-25.5	—	20.0-22.0	—	bal	6.00-7.00	—	0.030 max	2.00 max	N08367
N08700	JS700	24.0-26.0	—	19.0-23.0	0.50 max	bal	4.3-5.0	—	0.04 max	2.00 max	N08700
N08825	825 Alloy	38.0-46.0	—	19.5-23.5	1.5-3.0	bal	2.5-3.5	0.2 max	0.05 max	1.0 max	N08825
N08904	904L Alloy	23.0-28.0	—	19.0-23.0	1.0-2.0	bal	4.0-5.0	—	0.020 max	2.00 max	N08904
N08925	25-6Mo	24.0-26.0	—	19.0-21.0	0.05-1.50	bal	6.0-7.0	—	0.020 max	1.00 max	N08925
N09925	925 Alloy	38.0-46.0	—	19.5-23.5	1.50-3.00	22.0 min	2.50-3.50	0.10-0.50	0.03 max	1.00 max	N09925
N10002	C Alloy	bal	2.5 max	14.5-16.5	—	4.0-7.0	15.0-17.0	—	0.08 max	1.00 max	N10002
N10003	N Alloy	bal	0.20 max	6.0-8.0	0.35 max	5.00 max	15.0-18.0	0.50 max	0.04-0.08	1.00 max	N10003
N10004	W Alloy	bal	—	4.0-6.0	—	4.0-7.0	23.0-26.0	—	0.12 max	1.00 max	N10004
N10276	C-276 Alloy	bal	2.5 max	14.5-16.5	—	4.0-7.0	15.0-17.0	—	0.02 max	1.0 max	N10276

		Si	P	S	B	Ti	Cb	V	W	N	Zr	
N06002	X Alloy	1.00 max	0.040 max	0.030 max	—	—	—	—	0.20-1.0	—	—	N06002
N06007	G Alloy	1.0 max	0.04 max	0.03 max	—	—	—	—	1.0 max	—	—	N06007
N06022	C-22 Alloy	0.08 max	0.02 max	0.02 max	—	—	—	0.35 max	2.5-3.5	—	—	N06022
N06030	G-30 Alloy	0.8 max	0.04 max	0.02 max	—	—	—	—	1.5-4.0	—	—	N06030
N06110	Allcor	—	—	—	—	1.50 max	—	—	4.00 max	—	—	N06110
N06333	RA333 Alloy	0.75-1.50	0.030 max	0.030 max	—	—	—	—	2.50-4.00	—	—	N06333
N06455	C-4 Alloy	0.08 max	0.04 max	0.03 max	—	0.70 max	—	—	—	—	—	N06455
N06617	617 Alloy	1.00 max	—	0.015 max	0.006 max	0.60 max	—	—	—	—	—	N06617
N06625	625 Alloy	0.50 max	0.015 max	0.015 max	—	0.40 max	3.15-4.15	—	—	—	—	N06625
N06975	2550 Alloy	1.00 max	0.03 max	0.03 max	—	0.70-1.50	—	—	—	—	—	N06975
N06985	G-3 Alloy	1.00 max	0.04 max	0.03 max	—	—	0.50 max*	—	1.5 max	—	—	N06985
N07001	Waspaloy	0.75 max	0.030 max	0.030 max	0.003-0.01	2.75-3.25	—	—	—	—	0.02-0.12	N07001
N07031	Alloy 31	0.20 max	0.015 max	0.015 max	0.003-0.007	2.10-2.60	—	—	—	—	—	N07031
N07041	Rene 41	0.50 max	—	0.015 max	0.003-0.010	3.00-3.30	—	—	—	—	—	N07041
N07716	625 Plus	0.20 max	0.015 max	0.010 max	—	1.00-1.60	2.75-4.00	—	—	—	—	N07716
N07718	718 Alloy	0.35 max	0.015 max	0.015 max	0.006 max	0.65-1.15	4.75-5.50	—	—	—	—	N07718
N08020	20Cb-3	1.00 max	0.045 max	0.035 max	—	—	—	—	—	—	—	N08020
N08024	20Mo-4	0.50 max	0.035 max	0.035 max	—	—	—	—	—	—	—	N08024
N08026	20Mo-6	0.50 max	0.03 max	0.03 max	—	—	—	—	—	—	—	N08026
N08028	Sanicro 28	1.00 max	0.030 max	0.030 max	—	—	—	—	—	—	—	N08028
N08320	20 Mod	1.0 max	0.04 max	0.03 max	—	4xC min	—	—	—	—	—	N08320
N08366	AL-6X	1.00 max	0.030 max	0.030 max	—	—	—	—	—	—	—	N08366
N08367	AL-6XN	1.00 max	0.040 max	0.030 max	—	—	—	—	—	0.18-0.25	—	N08367
N08700	JS700	1.00 max	0.04 max	0.03 max	—	0.6-1.2	—	—	—	—	—	N08700
N08825	825 Alloy	0.5 max	—	0.03 max	—	—	—	—	—	—	—	N08825
N08904	904L Alloy	1.00 max	0.045 max	0.035 max	—	—	—	—	—	—	—	N08904
N08925	25-6Mo	0.50 max	0.045 max	0.030 max	—	—	—	—	—	0.10-0.20	—	N08925
N09925	925 Alloy	0.50 max	—	0.03 max	—	1.90-2.40	—	—	—	—	—	N09925
N10002	C Alloy	1.00 max	0.040 max	0.030 max	—	—	—	0.35 max	3.0-4.5	—	—	N10002
N10003	N Alloy	1.00 max	0.015 max	0.020 max	0.010 max	—	—	0.50 max	0.50 max	—	—	N10003
N10004	W Alloy	1.00 max	0.050 max	0.050 max	—	—	—	0.60 max	—	—	—	N10004
N10276	C-276 Alloy	0.05 max	0.030 max	0.030 max	—	—	—	0.35 max	3.0-4.5	—	—	N10276

*Cb + Ta
See also NICKEL ALLOYS, page 178

NICKEL ALLOYS - Mechanical Properties

UNS	Common Name	ASTM	Form	Heat Tr.	YS-ksi	TS-ksi	YS-MPa	TS-MPa	Hardness	UNS
N02200	Nickel 200	B161	Pipe,Tube	ANN	15 min	55 min	105 min	380 min	—	N02200
N02201	Nickel 201	B161	Pipe,Tube	ANN	12 min	50 min	80 min	345 min	—	N02201
N04400	400 Alloy	B165	Pipe,Tube	ANN	28 min	70 min	195 min	480 min	—	N04400
N04405	R-405 Alloy	B164	Bar	ANN	25 min	70 min	170 min	480 min	—	N04405
N06600	600 Alloy	B167	Pipe,Tube	HR	30 min	80 min	205 min	550 min	—	N06600
N06690	690 Alloy	B167	Pipe,Tube	HR	30 min	85 min	205 min	586 min	—	N06690
N07750	X-750 Alloy	B637	Forging	STA	130 min	185 min	896 min	1276 min	27-40 HRC	N07750
N08330	RA-330	B535	Pipe	ANN	30 min	70 min	207 min	483 min	—	N08330
N08800	800 Alloy	B407	Pipe,Tube	ANN	25 min	65 min	170 min	450 min	—	N08800
N08810	800H Alloy	B407	Pipe,Tube	ANN	25 min	65 min	170 min	450 min	—	N08810
N10001	B Alloy	B333	Plate	STQ	45 min	100 min	315 min	690 min	—	N10001
N10665	B-2 Alloy	B333	Plate	STQ	51 min	110 min	352 min	758 min	100 HRB max	N10665

CrMo NICKEL ALLOYS - Mechanical Properties

UNS	Common Name	ASTM	Form	Heat Tr.	YS-ksi	TS-ksi	YS-MPa	TS-MPa	Hardness	UNS
N06002	X Alloy	B435	Plate	STQ	35 min	95 min	240 min	660 min	—	N06002
N06007	G Alloy	B582	Plate	STQ	35 min	90 min	241 min	621 min	100 HRB max	N06007
N06022	C-22 Alloy	B575	Plate	STQ	45 min	100 min	310 min	690 min	100 HRB max	N06022
N06030	G-30 Alloy	B622	Pipe.Tube	STQ	35 min	85 min	241 min	586 min	—	N06030
N06333	RA333 Alloy	B722	Pipe.Tube	ANN	35 min	80 min	241 min	551 min	75-95 HRB	N06333
N06455	C-4 Alloy	B575	Plate	STQ	40 min	100 min	276 min	690 min	100 HRB max	N06455
N06625	625 Alloy	B444	Pipe.Tube	ANN	60 min	120 min	414 min	827 min	—	N06625
N06975	2550 Alloy	B582	Plate	STQ	32 min	85 min	221 min	586 min	100 HRB max	N06975
N06985	G-3 Alloy	B582	Plate	STQ	35 min	90 min	241 min	621 min	100 HRB max	N06985
N07718	718 Alloy	B637	Forging	STA	150 min	185 min	1034 min	1275 min	331 HB min	N07718
N08020	20Cb-3	B464	Pipe	ANN	35 min	80 min	241 min	551 min	—	N08020
N08024	20Mo-4	B464	Pipe	ANN	35 min	80 min	241 min	551 min	—	N08024
N08026	20Mo-6	B464	Pipe	STQ	35 min	80 min	241 min	551 min	—	N08026
N08028	Sanicro 28	B668	Tube	STQ	31 min	73 min	214 min	500 min	—	N08028
N08320	20 Mod	B622	Pipe.Tube	STQ	28 min	75 min	193 min	517 min	—	N08320
N08366	AL-6X	B690	Pipe	STQ	30 min	75 min	206 min	517 min	—	N08366
N08367	AL-6XN	B690	Pipe	STQ	46 min	104 min	317 min	717 min	—	N08367
N08700	JS700	B599	Plate	STQ	35 min	80 min	240 min	550 min	75-90 HRB	N08700
N08825	825 Alloy	B423	Pipe.Tube	ANN	25 min	75 min	172 min	517 min	—	N08825
N08904	904L Alloy	B677	Pipe.Tube	STQ	31 min	71 min	220 min	490 min	—	N08904
N08925	25-6Mo	B677	Pipe.Tube	STQ	43 min	87 min	300 min	600 min	—	N08925
N10003	N Alloy	B434	Plate	ANN	40 min	100 min	280 min	690 min	—	N10003
N10276	C-276 Alloy	B575	Plate	STQ	41 min	100 min	283 min	690 min	100 HRB max	N10276

COBALT ALLOYS - Composition, %

UNS	Common Name	Co	Ni	Cr	Fe	Mo	W	C	Mn	Si	UNS
R30003	Elgiloy	39.0-41.0	15.0-16.0	19.0-21.0	bal	6.0-8.0	—	0.15 max	1.5-2.5	—	R30003
R30004	Havar	41.0-44.0	12.0-14.0	19.0-21.0	bal	2.0-2.8	2.3-3.3	0.17-0.23	1.35-1.80	—	R30004
R30006	Stellite 6	bal	3.0 max	27.0-31.0	3.0 max	1.5 max	3.5-5.5	0.9-1.4	1.0 max	1.5 max	R30006
R30031	Stellite 31	bal	9.5-11.5	24.5-26.5	2.0 max	—	7.0-8.0	0.45-0.55	1.0 max	1.00 max	R30031
R30035	MP35N	bal	33.0-37.0	19.0-21.0	1.0 max	9.0-10.5	—	0.025 max	0.15 max	0.15 max	R30035
R30155	N-155	18.5-21.0	19.0-21.0	20.0-22.5	bal	2.5-3.5	2.0-3.0	0.08-0.16	1.0-2.0	1.00 max	R30155
R30188	HS-188	bal	20.0-24.0	20.0-24.0	3.0 max	—	13.0-16.0	0.05-0.15	1.25 max	1.00 max	R30188
R30260	Duratherm 2602	—	bal	11.7-12.3	9.8-10.4	3.7-4.3	3.6-4.2	0.05 max	0.4-1.1	0.20-0.50	R30260
R30556	HS-556	16.0-21.0	19.0-22.5	21.0-23.0	bal	2.0-3.5	—	0.05-0.15	0.5-2.0	0.20-0.80	R30556
R30605	L-605	bal	9.0-11.0	19.0-21.0	3.0 max	—	14.0-16.0	0.05-0.15	2.0 max	1.00 max	R30605

UNS	Common Name	Be	Cb	Cu	La	Sn	Ti	P	S	Other
R30003	Elgiloy	1.00 max	—	—	—	—	—	—	—	—
R30004	Havar	0.02-0.06	—	—	—	—	—	0.015 max	0.010 max	—
R30006	Stellite 6	—	—	—	—	—	—	—	—	—
R30031	Stellite 31	—	—	—	—	—	—	—	—	—
R30035	MP35N	—	—	—	—	—	—	—	—	—
R30155	N-155	—	0.75-1.25	—	—	—	1.00 max	0.040 max	0.030 max	—
R30188	HS-188	—	—	—	0.03-0.15	—	—	—	—	—
R30260	Duratherm 2602	0.20-0.30	0.10 max	0.30 max	—	—	—	—	—	—
R30556	HS-556	—	0.30 max	—	0.005-0.10	0.20-0.60	0.80-1.20	0.04 max	0.015 max	Al 0.10-0.50 B 0.02 max N 0.10-0.30 Ta 0.30-1.25 Zr 0.001-0.10
R30605	L-605	—	—	—	—	—	—	—	—	—

COBALT ALLOYS - Mechanical Properties

UNS	Common Name	ASTM	Form	Heat Tr.	YS-ksi	TS-ksi	YS-MPa	TS-MPa	Hardness	UNS
R30155	N-155	B639	Forging	STA	50 min	110 min	345 min	760 min	192 HB min	R30155
R30605	L-605	F90	Bar	ANN	45 min	125 min	310 min	860 min	—	R30605

REFRACTORY ALLOYS - Composition, %

UNS	Common Name	Cb	Mo	Ta	W	Zr	C	Fe	Ni	UNS
R03600	Molybdenum	—	bal	—	—	—	0.01-0.04	0.010 max	0.005 max	R03600
R03630	Molybdenum alloy	—	bal	—	—	—	0.01-0.4	0.010 max	0.005 max	R03630
R03650	Molybdenum, low C	—	bal	—	—	—	0.010 max	0.010 max	0.005 max	R03650
R04210	Columbium	bal	0.005 max	0.2 max	0.05 max	—	0.01 max	0.01 max	0.005 max	R04210
R05200	Tantalum	0.05 max	0.01 max	bal	0.03 max	0.01 max	0.01 max	0.01 max	0.01 max	R05200
R07005	Tungsten	—	—	—	99.95 min	—	.	—	—	R07005
R60702	Zr 702	—	—	—	—	+Hf99.2 min	0.05 max	+Cr0.2 max	—	R60702
R60704	Zr 704	2.0-3.0	—	—	—	+Hf97.5 min	0.05 max	+Cr0.20 0.40	—	R60704
R60705	Zr 705	—	—	—	—	+Hf95.5 min	0.05 max	—	—	R60705

UNS	Common Name	H	N	D	Si	Other	UNS
R03600	Molybdenum	—	0.0010 max	0.0030 max	0.010 max		R03600
R03630	Molybdenum alloy	—	0.001 max	0.003 max	0.010 max		R03630
R03650	Molybdenum, low C	—	0.001 max	0.003 max	0.010 max		R03650
R04210	Columbium	0.001 max	0.01 max	0.025 max	0.005 max	Hf 0.01 max	R04210
R05200	Tantalum	0.001 max	0.01 max	0.015 max	0.005 max		R05200
R07005	Tungsten	—	—	—	—	ea 0.01 max, lot 0.05 max	R07005
R60702	Zr 702	0.005 max	0.025 max	—	—	Hf 4.5 max	R60702
R60704	Zr 704	0.005 max	0.025 max	—	—	Hf 4.5 max; Sn 1.00-2.00	R60704
R60705	Zr 705	0.005 max	0.025 max	0.018 min	—	Hf 4.5 max	R60705

REFRACTORY ALLOYS - Mechanical Properties

UNS	Common Name	ASTM	Form	Heat Tr.	YS-ksi	TS-ksi	YS-MPa	TS-MPa	UNS
R03600	Molybdenum	B386	Plate	RX	25 min	55 min	170 min	380 min	R03600
R03630	Molybdenum alloy	B386	Plate	SRA	100 min	120 min	690 min	830 min	R03630
R03650	Molybdenum, low C	B386	Plate	SRA	80 min	100 min	550 min	690 min	R03650
R04210	Columbium	B394	Tube	ANN	12 min	18 min	85 min	125 min	R04210
R05200	Tantalum	B365	Rod	—	20 min	25 min	138 min	172 min	R05200
R60702	Zr 702	B523	Tube	ANN	30 min	55 min	207 min	380 min	R60702

TITANIUM ALLOYS - Composition, %

UNS	Common Name	Al	C	Fe	H	Mo	N	O	Other	UNS
R50250	Titanium, Gr 1	—	0.10 max	0.20 max	0.015 max	—	0.03 max	0.18 max	—	R50250
R50400	Titanium, Gr 2	—	0.10 max	0.30 max	0.015 max	—	0.03 max	0.25 max	—	R50400
R50550	Titanium, Gr 3	—	0.10 max	0.30 max	0.015 max	—	0.05 max	0.35 max	0.4 max	R50550
R50700	Titanium, Gr 4	—	0.10 max	0.50 max	0.015 max	—	0.05 max	0.40 max	0.4 max	R50700
R52250	Titanium, Gr 11	—	0.10 max	0.20 max	0.015 max	—	0.03 max	0.18 max	Pd 0.12-0.25	R52250
R52400	Titanium, Gr 7	—	0.10 max	0.30 max	0.015 max	—	0.03 max	0.25 max	Pd 0.12-0.25	R52400
R53400	Titanium, Gr 12	—	0.08 max	0.30 max	0.015 max	0.2-0.4	0.03 max	0.25 max	Ni 0.6-0.9	R53400
R54520	Titanium, Gr 6	4.0-6.0	0.10 max	0.50 max	0.020 max	—	0.05 max	0.20 max	Sn 2.0-3.0	R54520
R56260	Ti6Al6Mo2Sn4Zr	6.	—	—	—	6.	—	—	Sn 2.; Zr 4.	R56260
R56320	Titanium, Gr 9	2.5-3.5	0.05 max	0.25 max	0.013 max	—	0.02 max	0.12 max	V 2.0-3.0	R56320
R56400	Titanium, Gr 5	5.5-6.75	0.10 max	0.40 max	0.015 max	—	0.05 max	0.20 max	V 3.5-4.5	R56400
R58640	Beta-C	3.	—	—	—	4.	—	—	Cr 6; V 8.; Zr 4.	R58640

TITANIUM ALLOYS - Mechanical Properties

UNS	Common Name	ASTM	Form	Heat Tr.	YS-ksi	TS-ksi	YS-MPa	TS-MPa	UNS
R50250	Titanium, Gr 1	B338	Tube	ANN	25-45	35 min	170-310	240 min	R50250
R50400	Titanium, Gr 2	B338	Tube	ANN	40-65	50 min	275-450	345 min	R50400
R50550	Titanium, Gr 3	B338	Tube	ANN	55-80	65 min	380-550	450 min	R50550
R50700	Titanium, Gr 4	B265	Plate	ANN	70-95	80 min	485-655	550 min	R50700
R52250	Titanium, Gr 11	B338	Tube	ANN	25-45	35 min	170-240	240 min	R52250
R52400	Titanium, Gr 7	B338	Tube	ANN	40-65	50 min	275-450	345 min	R52400
R53400	Titanium, Gr 12	B338	Tube	ANN	50 min	70 min	345 min	483 min	R53400
R54520	Titanium, Gr 6	B265	Plate	ANN	115 min	120 min	795 min	830 min	R54520
R56320	Titanium, Gr 9	B338	Tube	CWK/SR	105 min	125 min	725 min	860 min	R56320
R56400	Titanium, Gr 5	B265	Plate	ANN	120 min	130 min	830 min	895 min	R56400

LEAD ALLOYS - Composition, %

UNS	Common Name	Pb	Ag	As	Bi	Cu	Fe	Sb	Sn	Zn	UNS
L50045	Common Lead	99.94 min	0.005 max	—	0.050 max	0.0015 max	0.002 max	—	—	0.001 max	L50045
L51120	Chemical Lead	99.90 min	0.002-0.02	—	0.005 max	0.04-0.08	0.002 max	—	—	0.001 max	L51120
L55030	50/50 Solder	50 nom	—	0.03 max	0.25 max	0.08 max	0.02 max	0.12 max	50 nom	0.005 max	L55030

MAGNESIUM ALLOYS - Composition, %

UNS	Common Name	Mg	Al	Cu	Fe	Mn	Ni	Zn	Other	UNS
M11311	Mg AZ31B	bal	2.5-3.5	0.05 max	0.005 max	0.20 min	0.005 max	0.6-1.4	Ca 0.04 max	M11311
M11914	Mg AZ91C	bal	8.1-9.3	0.10 max	—	0.13 min	0.01 max	0.40-1.0	tot 0.30 max	M11914
M12330	Mg EZ33A	bal	—	0.10 max	—	—	0.01 max	2.0-3.1	Rare Earths 2.5-4.0	M12330
M13310	Mg HK31A	bal	—	0.10 max	—	—	0.01 max	0.30 max	Th 2.5-4.0	M13310

MAGNESIUM ALLOYS - Mechanical Properties

UNS	Common Name	ASTM	Form	Temper	YS-ksi	TS-ksi	YS-MPa	TS-MPa	UNS
M11311	Mg AZ31B	B91	Forging	F	19 min	34 min	131 min	234 min	M11311
M11914	Mg AZ91C	B80	Casting	F	11 min	23 min	76 min	158 min	M11914
M12330	Mg EZ33A	B80	Casting	T5	14 min	20 min	96 min	138 min	M12330
M13310	Mg HK31A	B80	Casting	T6	13 min	27 min	89 min	186 min	M13310

PRECIOUS METALS AND ALLOYS - Composition, %

UNS	Common Name	Ag	Cu	Other	UNS
P00020	Refined gold	0.035 max	0.02 max	Au 99.95 min; Pd 0.02 max	P00020
P03980	Refined palladium	—	—	Ir 0.05 max; Pt 0.15 max; Rh 0.10 max; Ru 0.05 max	P03980
P04995	Refined platinum	0.005 max	0.01 max	0.01Au, Ru; 0.02Pd; 0.005Bi, Ca, Te; 0.015Ir; 0.03Rh max	P04995
P07015	Refined silver	99.95 min	0.04 max	Bi 0.001 max	P07015
P07931	Sterling silver	92.10-93.50	6.50-7.90	0.06 max	P07931

ZINC ALLOYS - Composition, %

UNS	Common Name	ASTM	MILSPEC	Al	Cd	Cu	Fe	Pb	Si	UNS
Z13000	Zinc Anode Type II	B418	—	0.005 max	0.003 max	—	0.0014 max	—	—	Z13000
Z32120	Zinc Anode Type I	B418	—	0.10-0.4	0.03-0.10	—	0.005 max	—	—	Z32120
Z32121	Zinc Anode Type III	B418	MIL-A-18001H	0.10-0.50	0.025-0.15	0.005 max	0.005 max	0.006 max	0.125 max	Z32121

API GRADES OF CASING AND TUBING - Composition, %

Common Name	C	Cr	Cu	Mn	Mo	Ni	P	S	Si	Common Name
H40	—	—	—	—	—	—	0.040 max	0.060 max	—	H40
J55	—	—	—	—	—	—	0.040 max	0.060 max	—	J55
K55	—	—	—	—	—	—	0.040 max	0.060 max	—	K55
N80	—	—	—	—	—	—	0.040 max	0.060 max	—	N80
C75 Type 1	0.50 max	(a)	(a)	1.90 max	0.15-0.40	(a)	0.040 max	0.060 max	0.45 max	C75 Type 1
C75 Type 2	0.43 max			1.50 max	—		0.040 max	0.060 max	0.45 max	C75 Type 2
C75 Type 3	0.38-0.48	0.80-1.10		0.75-1.00	0.15-0.25		0.040 max	0.040 max	—	C75 Type 3
C75 9Cr	0.15 max	8.0-10.0	0.25 max	0.30-0.60	0.90-1.10	0.5 max	0.020 max	0.010 max	1.0 max	C75 9Cr
C75 13Cr	0.15-0.22	12.0-14.0	0.25 max	0.25-1.00	—	0.5 max	0.020 max	0.010 max	1.0 max	C75 13Cr
L80 Type 1	0.43 max(d)	—	0.35 max	1.90 max	—	0.25 max	0.040 max	0.060 max	0.45 max	L80 Type 1
L80 9Cr	0.15 max	8.0-10.0	0.25 max	0.30-0.60	0.90-1.10	0.5 max	0.020 max	0.010 max	1.0 max	L80 9Cr
L80 13Cr	0.15-0.22	12.0-14.0	0.25 max	0.25-1.00	—	0.5 max	0.020 max	0.010 max	1.0 max	L80 13Cr
C90 Type 1	0.35 max	1.20 max	—	1.00 max	0.75 max	0.99 max	0.020 max	0.010 max	—	C90 Type 1
C90 Type 2	0.50 max	(b)	—	1.90 max	(b)	0.99 max	0.030 max	0.010 max	—	C90 Type 2
C95	0.45 max(c)	—	—	1.90 max	—	—	0.040 max	0.060 max	0.45 max	C95
P105	—	—	—	—	—	—	0.040 max	0.060 max	—	P105
P110	—	—	—	—	—	—	0.040 max	0.060 max	—	P110
Q125 Type 1	0.35 max	1.20 max	—	1.00 max	0.75 max	0.99 max	0.020 max	0.010 max	—	Q125 Type 1
Q125 Type 2	0.35 max	(b)	—	1.00 max	(b)	0.99 max	0.020 max	0.020 max	—	Q125 Type 2
Q125 Type 3	0.50 max	(b)	—	1.90 max	(b)	0.99 max	0.030 max	0.010 max	—	Q125 Type 3
Q125 Type 4	0.50 max	(b)	—	1.90 max	(b)	0.99 max	0.030 max	0.020 max	—	Q125 Type 4

(a)Cr + Ni + Cu shall not exceed 0.50%.
(b)No limit. Elements shown must be reported in product analysis.
(c)Carbon content may be increased to 0.50% max if product is oil quenched.
(d)Carbon content may be increased to 0.55% max if product is oil quenched.

Source: API Specification 5CT (March 15, 1988).

API GRADES OF CASING AND TUBING - Mechanical Properties

Common Name	Form	Mfr. Proc.	Heat Tr.	YS-ksi	TS-ksi	YS-MPa	TS-MPa	Hardness*	Common Name
H40	Casing, Tubing	SMLS, EW	None	40-80	60 min	276-552	414 min	—	H40
J55	Casing, Tubing	SMLS, EW	None, N, NT, QT	55-80	75 min	379-552	517 min	—	J55
K55	Casing	SMLS, EW	None, N, NT, QT	55-80	95 min	379-552	655 min	—	K55
N80	Casing	SMLS, EW	None, N, NT, QT	80-110	100 min	552-758	689 min	—	N80
N80	Tubing	SMLS, EW	N, NT, QT	80-110	100 min	552-758	689 min	—	N80
C75 Type 1	Casing, Tubing	SMLS, EW	NT (1150 F min)	75-90	95 min	517-620	655 min	—	C75 Type 1
C75 Type 2	Casing, Tubing	SMLS, EW	QT (1150 F min)	75-90	95 min	517-620	655 min	—	C75 Type 2
C75 Type 3	Casing, Tubing	SMLS, EW	NT (1150 F min)	75-90	95 min	517-620	655 min	—	C75 Type 3
C75 9Cr	Casing, Tubing	SMLS	QT (1100 F min)	75-90	95 min	517-620	655 min	22 HRC max	C75 9Cr
C75 13Cr	Casing, Tubing	SMLS	QT (1100 F min)	75-90	95 min	517-620	655 min	22 HRC max	C75 13Cr
L80 Type 1	Casing, Tubing	SMLS, EW	QT (1050 F min)	80-95	95 min	552-655	655 min	22 HRC max	L80 Type 1
L80 9Cr	Casing, Tubing	SMLS	QT (1100 F min)	80-95	95 min	522-655	655 min	23 HRC max	L80 9Cr
L80 13 Cr	Casing, Tubing	SMLS	QT (1100 F min)	80-95	95 min	522-655	655 min	23 HRC max	L80 13Cr
C90 Type 1	Casing, Tubing	SMLS	QT (1150 F min)	90-105	100 min	620-724	690 min	25.4 HRC max	C90 Type 1
C90 Type 2	Casing, Tubing	SMLS	QT (1150 F min)	90-105	100 min	620-724	690 min	25.4 HRC max	C90 Type 2
C95	Casing, Tubing	SMLS, EW	QT (1000 F min)	95-110	105 min	655-758	724 min	—	C95
P105	Tubing	SMLS	QT, NT	105-135	120 min	724-931	827 min	—	P105
P110	Casing	SMLS	QT, NT	110-140	125 min	758-965	862 min	—	P110
9125 Type 1	Casing	SMLS, EW	QT	125-150	135 min	860-1035	930 min	—	Q125 Type 1
Q125 Type 2	Casing	SMLS, EW	QT	125-150	135 min	860-1035	930 min	—	Q125 Type 2
Q125 Type 3	Casing	SMLS, EW	QT	125-150	135 min	860-1035	930 min	—	Q125 Type 3
Q125 Type 4	Casing	SMLS, EW	QT	125-150	135 min	860-1035	930 min	—	Q125 Type 4

*See Specification 5CT for allowable hardness variations.

Source: API Specification 5CT (March 15, 1988)

MAXIMUM ALLOWABLE STRESS IN TENSION FOR ALUMINUM ALLOY TUBES - ksi

Summarized from ASME Pressure Vessel Code, Section VIII, Table UNF-23.1 (1989)

(ksi x 6.895 = MPa)

UNS	Common Name	Temper	ASME	SMYS	100 F	150 F	200 F	250 F	300 F	350 F	400 F	UNS
A91060	Al 1060	O	SB210	2.5	1.7	1.7	1.6	1.5	1.3	1.1	0.8	A91060
A93003	Al 3003	O	SB210	5.	3.4	3.4	3.4	3.0	2.4	1.8	1.4	A93003
A95052	Al 5052	O	SB210	10.	6.2	6.2	6.2	6.2	5.6	4.1	2.3	A95052
A96061	Al 6061	T6	SB210	35.	10.5	10.5	10.5	9.9	8.4	6.3	4.5	A96061
A96063	Al 6063	T6	SB210	28.	8.3	8.3	7.9	7.4	5.5	3.4	2.0	A96063

MAXIMUM ALLOWABLE STRESS IN TENSION FOR COPPER ALLOY TUBES - ksi

Summarized from ASME Pressure Vessel Code, Section VIII, Table UNF-23.2 (1989)

(ksi x 6.895 = MPa)

UNS	Common Name	ASME	Temper	SMYS	100 F	150 F	200 F	250 F	300 F	350 F	400 F	450 F	500 F	550 F	600 F	650 F	700 F	UNS
C10200	OF Copper	SB111	H55	30.	9.0	9.0	9.0	9.0	8.7	8.5	8.2							C10200
C12200	DHP Copper	SB111	H04	40.	11.3	11.3	11.3	11.3	11.0	10.3	4.3							C12200
C23000	Red Brass	SB111	061	12.	8.0	8.0	8.0	8.0	8.0	7.0	5.0	2.0						C23000
C28000	Muntz Metal	SB111	061	20.	12.5	12.5	12.5	12.5	12.5	10.8	5.3							C28000
C44300	Adm. Brass, As	SB111	061	15.	10.0	10.0	10.0	10.0	10.0	9.8	3.5							C44300
C60800	Al Bronze, 6%	SB111	061	19.	12.5	12.4	12.2	11.9	11.6	10.0	6.0	4.0	2.0					C60800
C68700	Al Brass, As	SB111	061	18.	12.0	11.9	11.8	11.7	11.7	6.5	3.3	1.8						C68700
C70600	90-10 Cu-Ni	SB111	061	15.	10.0	9.7	9.5	9.3	9.0	8.7	8.5	8.2	8.0	7.0	6.0			C70600
C71500	70-30 Cu-Ni	SB111	061	18.	12.0	11.6	11.3	11.0	10.8	10.6	10.3	10.1	9.9	9.8	9.6	9.5	9.4	C71500

MAXIMUM ALLOWABLE STRESS IN TENSION FOR CARBON AND LOW ALLOY STEELS - ksi

Summarized from ASME Pressure Vessel Code, Section VIII, Table UCS-23 (1989)

(ksi x 6.895 = MPa)

UNS	Common Name	Form	ASME	SMYS	650 F	700 F	750 F	800 F	850 F	900 F	950 F	1000 F	1050 F	1100 F	1150 F	1200 F
K01800	A516-55	Plate	SA516-55	30.	13.8	13.3	12.1	10.2	8.4	6.5	4.5	2.5				
K02100	A516-60	Plate	SA516-60	32	15.0	14.1	13.0	10.8	8.7	6.5	4.5	2.5				
K02403	A516-65	Plate	SA516-65	35	16.3	15.5	13.9	11.4	9.0	6.5	4.5	2.5				
K02700	A516-70	Plate	SA516-70	38	17.5	16.6	14.8	12.0	9.3	6.5	4.5	2.5				
K02801	A285-C	Plate	SA285-C	30.	13.8	13.3	12.1	10.2	8.4	6.5						
K03005	A53-B	Pipe	SA53-B	35.	15.0	14.4	13.0	10.8	8.7	6.5						
K03006	A106-B	Pipe	SA106-B	35.	15.0	14.4	13.0	10.8	8.7	6.5	4.5	2.5				
K11522	C-0.5Mo	Pipe	SA335-P1	30.	13.8	13.8	13.8	13.5	13.2	12.7	8.2	4.8				
K11597	1.25Cr-0.5Mo	Pipe	SA335-P11	30.	15.0	15.0	14.8	14.4	14.0	12.1	9.3	6.3	4.2	2.8	1.9	1.2
K21590	2.25Cr-1Mo	Pipe	SA335-P22	30.		15.0	15.0	15.0	14.4	13.1	11.0	7.8	5.8	4.2	3.0	2.0
K41545	5Cr-0.5Mo	Pipe	SA335-P5	30.		13.7	13.2	12.8	12.1	10.9	8.0	5.8	4.2	2.9	1.8	1.0
S50400	9Cr-1Mo	Pipe	SA335-P9	30.		13.7	13.2	12.8	12.1	11.4	10.6	7.4	5.0	3.3	2.2	1.5

MAXIMUM ALLOWABLE STRESS IN TENSION FOR STAINLESS STEEL PIPE/TUBE - ksi

Summarized from ASME Pressure Vessel Code, Section VIII, Table UHA-23 (1989)

(ksi × 6.895 - MPa)

UNS	Common Name	ASME	SMYS	-20/100 F	200 F	300 F	400 F	500 F	600 F	650 F	700 F	750 F	800 F	850 F	900 F	Notes	UNS
S30400	304 SS	SA312	30.	18.8	17.8	16.6	16.2	15.9	15.9	15.9	15.9	15.6	15.2	14.9	14.7		S30400
S30403	304L SS	SA312	25.	16.7	16.5	15.3	14.7	14.4	14.0	13.7	13.5	13.3	13.0				S30403
S30451	304N SS	SA312	35.	20.0	20.0	19.0	18.3	17.8	17.4	17.3	17.1	16.9	16.6	16.3	15.9		S30451
S30900	309 SS	SA312	30.	18.8	17.2	16.4	15.9	15.5	15.3	15.2	15.1	15.0	14.9	14.6	13.9		S30900
S31000	310 SS	SA312	30.	18.8	17.2	16.4	15.9	15.5	15.3	15.2	15.1	15.0	14.9	14.6	13.9		S31000
S31254	254 SMO	SA312	44.	23.5	23.5	21.4	19.9	18.5	17.9	17.7	17.5	17.3					S31254
S31600	316 SS	SA312	30.	18.8	18.8	18.4	18.1	18.0	17.0	16.7	16.3	16.1	15.9	15.7	15.6		S31600
S31603	316L SS	SA312	25.	16.7	14.1	12.7	11.7	10.9	10.4	10.2	10.0	9.8	9.6	9.4			S31603
S31651	316N SS	SA312	35.	20.0	20.0	19.2	18.8	18.6	18.6	18.6	18.6	18.5	18.4	18.3	18.1		S31651
S32100	321 SS	SA312	30.	18.7	17.8	16.7	16.4	16.4	16.4	16.4	16.4	16.4	16.4	16.4	16.4		S32100
S34700	347 SS	SA312	30.	18.8	17.9	16.4	16.4	16.4	16.4	16.4	16.4	16.4	16.4	16.4	16.4		S34700
S40500	405 SS	SA268	30.	15.0	14.3	13.8	13.3	12.9	12.4	12.3	12.1	11.7	11.1	10.4	9.7	(a)	S40500
S41000	410 SS	SA268	30.	15.0	14.3	13.8	13.3	12.9	12.4	12.3	12.1	11.7	11.1	10.4	9.7		S41000
S43000	430 SS	SA268	35.	15.0	14.3	13.8	13.3	12.9	12.4	12.3	12.1	11.7	11.1	10.4	9.7	(a)	S43000
S44600	446 SS	SA268	40.	17.5	16.6	16.1	15.6	15.0	14.5	14.3							S44600
S31500	3RE60	SA790	64.	23.0	22.2	21.3	21.2	21.2	21.2	21.2	21.2	21.2				(a)	S31500
S31803	2205 alloy	SA790	65.	22.5	22.5	21.7	20.9	20.4	20.2							(b)	S31803
S32550	Ferralium 255	SA790	80.	27.5	27.4	25.7	24.7	24.7								(a)	S32550

Notes:
(a)This material may be expected to develop embrittlement after service at moderately elevated temperatures.
(b)This material may be expected to exhibit embrittlement at room temperature after service above 600°F.

Continued on Next Page

MAXIMUM ALLOWABLE STRESS IN TENSION FOR STAINLESS STEEL PIPE/TUBE - ksi (Cont)

UNS	Common Name	ASME	950 F	1000 F	1050 F	1100 F	1150 F	1200 F	1250 F	1300 F	1350 F	1400 F	1450 F	1500 F	UNS
S30400	304 SS	SA312	14.4	14.1	12.4	9.8	7.7	6.1	4.7	3.7	2.9	2.3	1.8	1.4	S30400
S30451	304N SS	SA312	15.6	15.0	12.4	9.8	7.7	6.1							S30451
S30900	309 SS	SA312	12.4	10.5	8.5	6.5	5.0	3.8	2.9	2.3	1.8	1.3	0.9	0.8	S30900
S31000	310 SS	SA312	12.5	11.0	7.1	5.0	3.6	2.5	1.5	0.8	0.5	0.4	0.3	0.2	S31000
S31600	316 SS	SA312	15.4	11.3	11.2	11.0	9.8	7.4	5.5	4.1	3.1	2.3	1.7	1.3	S31600
S31651	316N SS	SA312	17.8	13.2	12.7	12.2	9.8	7.4							S31651
S32100	321 SS	SA312	16.3	13.8	9.6	6.9	5.0	3.6	2.6	1.7	1.1	0.8	0.5	0.3	S32100
S34700	347 SS	SA312	16.4	14.4	12.1	9.1	6.1	4.4	3.3	2.2	1.5	1.2	0.9	0.8	S34700
S40500	405 SS	SA268	8.4	4.0	4.4	2.9	1.8	1.0							S40500
S41000	410 SS	SA268	8.4	6.4	4.5	3.2	2.4	1.8							S41000
S43000	430 SS	SA268	8.5	6.5											S43000

MAXIMUM ALLOWABLE STRESS IN TENSION FOR ANNEALED NICKEL ALLOY PIPE/TUBE - ksi

Summarized from ASME Pressure Vessel Code, Section VIII, Table UNF 23.3 (1989)

(ksi x 6.895 = MPa)

UNS	Common Name	ASME	SMYS	100 F	500 F	600 F	700 F	800 F	900 F	1000 F	1100 F	1200 F	1300 F	1400 F	1500 F	Notes	UNS
N02200	Nickel 200	SB163	15.	10.0	10.0	10.0											N02200
N04400	400 Alloy	SB163	28.	17.5	14.7	14.7	14.7	14.2	8.0								N04400
N06002	X Alloy	SB619	40.	19.8	14.0	13.3	12.7	12.5	12.3	12.1	12.1	9.6	6.5				N06002
N06022	C-22 Alloy	SB619	45.	21.2	18.0	17.1	16.3	15.8									N06022
N06030	G-30 Alloy	SB619	35.	18.1	13.9	13.4	12.9	12.4									N06030
N06455	C-4 Alloy	SB619	40.	21.2	17.8	17.1	16.7	16.2									N06455
N06600	600 Alloy	SB163	35.	20.0	20.0	20.0	19.6	19.1	16.0	7.0	3.0	2.0					N06600
N06625	625 Alloy	SB444	60.	30.0	27.0	26.4	26.0	26.0	26.0	26.0	26.0	13.2				(a)	N06625
N06985	G-3 Alloy	SB619	35.	19.1	13.8	13.1	12.4	11.9									N06985
N08020	20Cb-3	SB464	35.	17.0	15.5	15.1	14.7	14.3									N08020
N08028	Sanicro 28	SB668	31.	18.2	14.5	13.3											N08028
N08800	800 Alloy	SB163	30.	18.7	16.7	16.3	15.9	15.5	15.1	14.7	13.0	6.6	2.0	1.1			N08800
N08810	800H Alloy	SB163	25.	16.2	12.9	12.2	11.7	11.1	10.7	10.3	10.0	7.4	4.7	3.0			N08810
N08825	825 Alloy	SB163	35.	21.2	18.3	17.8	17.3	17.1	16.8	16.6							N08825
N10276	C-276 Alloy	SB619	41.	21.2	17.0	16.0	15.1	14.5	14.1	14.0	12.7	8.3					N10276
N10665	B-2 Alloy	SB619	51.	23.4	23.4	23.1	22.6	21.8							1.9		N10665

Notes:
(a)Alloy N06625 in the annealed condition is subject to severe loss of impact strength at room temperature after exposure in the range of 1000 F to 1400 F.

MAXIMUM ALLOWABLE STRESS IN TENSION FOR TITANIUM AND ZIRCONIUM ALLOY TUBES - ksi

Summarized from ASME Pressure Vessel Code, Section VIII, Tables UNF-23.4 and −23.5 (1989)

(ksi x 6.895 = MPa)

UNS	Common Name	ASME	SMYS	100 F	150 F	200 F	250 F	300 F	350 F	400 F	450 F	500 F	550 F	600 F	700 F	UNS
R50250	Titanium, Gr 1	SB338	25.	8.8	8.1	7.3	6.5	5.8	5.2	4.8	4.5	4.1	3.6	3.1		R50250
R50400	Titanium, Gr 2	SB338	40.	12.5	12.0	10.9	9.9	9.0	8.4	7.7	7.2	6.6	6.2	5.7		R50400
R50550	Titanium, Gr 3	SB338	55.	16.3	15.6	14.3	13.0	11.7	10.4	9.3	8.3	7.5	6.7	6.0		R50550
R52400	Titanium, Gr 7	SB338	40.	12.5	12.0	10.9	9.9	9.0	8.4	7.7	7.2	6.6	6.2	5.7		R52400
R53400	Titanium, Gr 12	SB338	50.	17.5	17.5	16.4	15.2	14.2	13.3	12.5	11.9	11.4	11.1	10.8		R53400
R60702	Zirconium 702	SB523	30.	13.0		11.0		9.3		7.0		6.1		6.0	4.8	R60702

CREEP STRENGTH OF METALS

Material	Form, Condition	Stress (ksi) for 0.01% Creep per 1000 h at Indicated Temp.					Stress (ksi) for 0.1% Creep per 1000 h at Indicated Temp.				
		300°F / 149°C	400 / 204	500 / 260	600 / 315	800 / 426	300 / 149	400 / 204	500 / 260	600 / 315	800 / 426
NONFERROUS METALS											
Coppers	Wrought (annealed)	3-8	1.5-5	0.4-2.6	—	—	—	—	—	—	—
Nonleaded Brasses	Wrought (annealed)	0.9-19	2-11	0.3-23	—	—	25	5-9	1-2	—	—
Bronzes	Wrought (annealed)	14-23	5-10	2-5	—	—	—	—	—	—	—
Cupro-Nickel	Wrought (water quenched, aged)	25-40	15-30	8-30	—	—	—	—	—	—	—
Aluminum 2024-T	Sheet	23	9.5	2.5	1.5	—	30	22	13	2	—
Aluminum 7075-T	Sheet	12	4	2.5	1.5	—	16	6	3	2	—
Titanium (commercial)	Sheet (annealed)	—	38	—	32	10	37	40	37	32	13
Ti-6Al-4V	Sheet (annealed)	—	—	—	—	—	—	—	—	80	—
Ti-7Al-4Mo	Bar or Forging (annealed)	—	—	—	—	—	—	—	—	85	18
		1000°F / 538°C	1100 / 593	12200 / 648	1500 / 816	1600 / 871	1000°F / 538°C	1100 / 593	1200 / 648	1500 / 816	1600 / 891
CARBON AND LOW ALLOY STEELS											
Low Carbon Steel	Wrought, Cast	1.8	—	0.1	—	—	3.3-5	—	0.5	—	—
Carbon-Molybdenum Steels	Wrought, Cast	5.7	3	1	—	—	10-12	4	2	—	—
Chromium-Molybdenum Steels (0.5-3%)	Wrought, Cast	6-12	2-4	1-2.5	—	—	10-20	3-8	2-4.5	—	—
Chromium Steels 4-6%	Wrought, Cast	6-7	2.5-3.5	1-2	—	—	8-11	5-6.5	2-3.5	—	—
Chromium Steels 6-10%	Wrought, Cast	5-9	2.5-4	1-2	—	—	8-12	4-6	2.5-3	—	—

CREEP STRENGTH OF METALS (Cont)

Material	Form, Condition	Stress (ksi) for 0.01% Creep per 1000 h at Indicated Temp.					Stress (ksi) for 0.1% Creep per 1000 h at Indicated Temp.				
		1000°F 538°C	1100 593	1200 648	1500 816	1600 871	1000 538	1100 593	1200 648	1500 816	1600°F 891°C
STAINLESS STEELS											
Martensitic Chromium Steels (403, 410, 416, 420, 440)	Wrought	8	3.5	1.3	—	—	9.2	4.2	2	—	—
Ferritic Chromium Steels (405, 430)	Wrought	4.2-7	2.3-4.5	1.0-1.6	—	—	6-8.5	3-5	1.5-2.2	—	—
Nickel-Chromium Steels											
304, 316, 321, 347	Wrought	12-17	7.5-11.5	4.5-7	1-2	—	17-25	12-18.2	7-12.7	1.2-2.8	—
309	Wrought	—	—	4	0.5	—	15.9	11.6	8	1.0	—
310, 314	Wrought	17	13	8	2	—	17	13-14	9	1-2.5	—
HEAT RESISTANT CAST HIGH ALLOYS											
Iron-Chromium Alloys (HA, HC, HD)	Cast	—	—	—	—	—	—	—	—	1.2-3.5[a]	0.7-1.9
Iron-Chromium-Nickel Alloys (HE, HF, HH, HI, HK, HL)	Cast	—	—	—	—	—	—	—	—	3.5-7[a]	2-4.3
Nickel-Chromium Alloys (HN, HT, HU, HW, HX)	Cast	—	—	—	—	—	—	—	—	6-8.5[a]	3-5

[a] At 1400°F (760°C).

Source: Materials in Design Engineering, p. 35, 1964.

TEMPER DESIGNATIONS FOR COPPERS AND COPPER ALLOYS

Cold Worked Tempers

H00	1/8 hard
H01	1/4 hard
H02	1/2 hard
H03	3/4 hard
H04	Hard
H06	Extra hard
H08	Spring
H10	Extra spring
H12	Special spring
H13	Ultra spring
H14	Super spring
H50	Extruded and drawn
H52	Pierced and drawn
H55	Light drawn; light cold rolled
H58	Drawn genereal purpose
H60	Cold heading; forming
H63	Rivet
H64	Screw
H66	Bolt
H70	Bending
H80	Hard drawn
H85	Medium-hard-drawn electrical wire
H86	Hard-drawn electrical wire

Cold Worked and Stress-relieved Tempers

HR01	H01 and stress relieved
HR02	H02 and stress relieved
HR04	H04 and stress relieved
HR06	H06 and stress relieved
HR08	H08 and stress relieved
HR10	H10 and stress relieved
HR50	Drawn and stress relieved

Cold Worked and Order-strengthened Tempers

HT04	H04 and order heat treated
HT06	H06 and order heat treated
HT08	H08 and order heat treated

As-manufactured Tempers

M01	As sand cast
M02	As centrifugal cast
M03	As plaster cast
M04	As pressure die cast
M05	As permanent mold cast
M06	As investment cast
M07	As continuous cast
M10	As hot forged and air cooled
M11	As hot forged and quenched
M20	As hot rolled
M30	As hot extruded
M40	As hot pierced
M45	As hot pierced and rerolled

Annealed Tempers(a)

O10	Cast and annealed(b)
O20	Hot forged and annealed
O25	Hot rolled and annealed
O30	Hot extruded and annealed
O40	Hot pierced and annealed
O50	Light annealed
O60	Soft annealed
O61	Annealed
O65	Drawing annealed
O68	Deep drawing annealed
O70	Dead soft annealed
O80	Annealed to temper—1/8 hard
O81	Annealed to temper—1/4 hard
O82	Annealed to temper—1/2 hard

Annealed Tempers(c)

OS005	Average grain size 0.005 mm
OS010	Average grain size 0.010 mm
OS015	Average grain size 0.015 mm
OS025	Average grain size 0.025 mm
OS035	Average grain size 0.035 mm
OS050	Average grain size 0.050 mm
OS070	Average grain size 0.070 mm
OS100	Average grain size 0.100 mm
OS120	Average grain size 0.120 mm
OS150	Average grain size 0.150 mm
OS200	Average grain size 0.200 mm

Continued on Next Page

TEMPER DESIGNATIONS FOR COPPER ALLOYS (Cont)

Solution-Treated Temper

TB00	Solution heat treated

Solution-Treated and Cold Worked Tempers

TD00	TB00 cold worked to 1/4 hard
TD01	TB00 cold worked to 1/4 hard
TD02	TB00 cold worked to 1/2 hard
TD03	Tb00 cold worked to 3/4 hard
TD04	TB00 cold worked to full hard

Precipitation-Hardedned Temper

TF00	TB00 and precipitation hardened

Cold Worked and Precipitation-Hardened Tempers

TH01	TD01 and precipitation hardened
TH02	TD02 and precipitation hardened
TH03	TD03 and precipitation hardened
TH04	TD04 and precipitation hardened

Precipitation-Hardened and Cold Worked Tempers

TL00	TF00 cold worked to 1/8 hard
TL01	TF00 cold worked to 1/4 hard
TL02	TF00 cold worked to 1/2 hard
TL04	TF00 cold worked to full hard
TL08	TF00 cold worked to spring
TL10	TF00 cold worked to extra spring
TR01	TL01 and stress relieved
TR02	TL02 and stress relieved
TR04	TL04 and stress relieved

Mill-Hardened Tempers

TM00	AM
TM01	1/4 HM
TM02	1/2 HM
TM04	HM
TM06	XHM
TM08	XHMS

Quench-Hardened Tempers

TQ00	Quench hardened
TQ50	Quench hardened and temper annealed
TQ75	Interrupted quench hardened

Tempers of Welded Tubing[d]

WH00	Welded and drawn to 1/8 hard
WH01	Welded and drawn to 1/4 hard
WM01	As welded from H01 strip
WM02	As welded from H02 strip
WM03	As welded from H03 strip
WM04	As welded from H04 strip
WM06	As welded from H06 strip
WM08	As welded from H08 strip
WM10	As welded from H10 strip
WM15	WM50 and stress relieved
WM20	WM00 and stress relieved
WM21	WM01 and stress relieved
WM22	WM02 and stress relieved
WM50	As welded from O60 strip
WO50	Welded and light annealed
WR00	WM00; drawn and stress relieved
WR01	WM01; drawn and stress relieved

[a]To produce specified mechanical properties. [b] Homogenization anneal. [c]To produce prescribed average grain size. [d]Tempers of fully finished tubing that has been drawn or annealed to produce specified mechanical properties or that has been annealed to produce a prescribed average grain size are commonly identified by the appropriate H, O or OS temper designation.

SOURCE: Metals Handbook, 9th ed., Volume 2, p. 249, ASM 1979.

TEMPER DESIGNATIONS FOR MAGNESIUM ALLOYS

F	As fabricated
O	Annealed
H10, H11	Slightly strain hardened
H23, H24, H26	Strain hardened and partially annealed
T4	Solution heat treated
T5	Artificially aged only
T6	Solution heat treated and artificially aged
T8	Solution heat treated, cold worked, and artificially aged

Source: Metals Handbook, 9th ed., Volume 2, p. 527, ASM 1979.

TEMPER DESIGNATIONS FOR ALUMINUM ALLOYS

Basic Temper Designations

F   **as fabricated.** Applies to the products of shaping processes in which no special control over thermal conditions or strain-hardening is employed. For wrought products, there are nomechanical property limits.

O   **annealed.** Applies to wrought products which are annealed to obtain the lowest strength temper, and to cast products which are annealed to improve ductility and dimensional stability. The O may be followed by a digit other than zero.

H   **strain-hardened (wrought products only).** Applies to products which have their strength increased by strain-hardening, with or without supplementary thermal treatments to produce some reduction in strength. The H is always followed by two or more digits.

W   **solution heat-treated.** An unstable temper applicable only to alloys which spontaneously age at room temperature after solution heat-treatment. This designation is specific only when the period of natural aging is indicated; for example: W 1/2 h.

T   **thermally treated to produce stable tempers other than F, O, or H.** Applies to products which are thermally treated, with or without supplementary strain-hardening, to produce stable tempers. The T is always followed by one or more digits.

Subdivision of H Temper: Strain-Hardened

The first digit following the H indicates the specific combination of basic operations, as follows:

H1   **strain-hardened only.** Applies to products which are strain-hardened to obtain the desired strength without supplementary thermal treatment. The number following this designation indicates the degree of strain-hardening.

Continued on Next Page

H2 **strain-hardened and partially annealed.** Applies to products which are strain-hardened more than the desired final amount and then reduced in strength to the desired level by partial annealing. For alloys that age-soften at room temperature, the H2 tempers have the same minimum ultimate tensile strength as the corresponding H3 tempers. For other alloys, the H2 tempers have the same minimum ultimate tensile strength as the corresponding H1 tempers and slightly higher elongation. The number following this designation indicates the degree of strain-hardening remaining after the product has been partially annealed.

H3 **strain-hardened and stabilized.** Applies to products which are strain-hardened and whose mechanical properties are stabilized either by a low temperature thermal treatment or as a result of heat introduced during fabrication. Stabilization usually improves ductility. This designation is applicable only to those alloys which, unless stabilized, gradually age-soften at room temperature. The number following this designation indicates the degree of strai-hardening remaining after the stabilization treatment.

The digit following the designations H1, H2, and H3 indicates the degree of strain-hardening.

Subdivision of T Temper: Thermally Treated

Numerals 1 through 10 following the T indicate specific sequences of basic treatments, as follows:

T1 **cooled from an elevated temperature shaping process and naturally aged to a substantially stable condition.** Applies to products which are not cold worked after cooling from an elevated temperature shaping process, or in which the effect of cold work in flattening or straightening may not be recognized in mechanical property limits.

T2 **cooled from an elevated temperature shaping process, cold worked, and naturally aged to a substantially stable condition.** Applies to products that are cold worked to improve strengfh after cooling from an elevated temperature shaping process, or in which the effect of cold work in flattening or straighting is recognized in mechanical property limits.

Continued on Next Page

T3   **solution heat-treated, cold worked, and naturally aged to a substantially stable condition.** Applies to products which are cold worked to improve strength after solution heat-treatment, or in which the effect of cold work in flattening or straightening is recognized in mechanical property limits.

T4   **solution heat-treated and naturally aged to a substantially stable condition.** Applies to products which are not cold worked after solution heat-treatment, or in which the effect of cold work in flattening or straightening may not be recognized in mechanical property limits.

T5   **cooled from an elevated temperature shaping process and then artifically aged.** Applies to products which are not cold worked after cooling from an elevated temperature shaping process, or in which the effect of cold work in flattening or straightening may not be recognized in mechanical property limits.

T6   **solution heat-treated and then artificially aged.** Applies to products which are not cold worked after solution heat-treatment, or in which the effect of cold work in flattening or straightening may not be recognized in mechanical property limits.

T7   **solution heat-treated and overaged/stabilized.** Applies to wrought products that are artificially aged after solution heat-treatment to carry them beyond a point of maximum strength to provide control of some significant characteristic. Applies to cast products that are artificially aged after solution heat-treatment to provide dimensional and strength stability.

T8   **solution heat-treated, cold worked, and then artificially aged.** Applies to products which are cold worked to improve strength, or in which the effect of cold work in flattening or straightening is recognized in mechanical property limits.

T9   **solution heat treated, artificially aged, and then cold worked.** Applies to products which are cold worked to improve strength.

T10   **cooled from an elevated temperature shaping process, cold worked, and then artificially aged.** Applies to products which are cold worked to improve strength, or in which the effect of cold work in flattening or straightening is recognized in mechanical property limits.

Source: Aluminum Association, Inc.

MELTING TEMPERATURES OF COMMON ALLOYS

UNS	Common Name	Melting Temperature	
		°C	°F
A24430	Cast Al B443.0	570-630	1065-1170
A91100	Al 1100	640-660	1190-1215
A95052	Al 5052	610-650	1125-1200
C11000	ETP Copper	1083	1980
C23000	Red Brass	990-1025	1810-1880
C28000	Muntz Metal	900-905	1650-1660
C44300	Admiralty Brass, As	900-935	1650-1720
C61400	Aluminum Bronze D	1045-1060	1910-1940
C70600	90-10 Copper-Nickel	1100-1150	2010-2100
C71500	70-30 Copper Nickel	1170-1240	2140-2260
C83600	Ounce Metal	854-1010	1510-1840
F10006	Gray Cast Iron	1150-1200	2100-2200
G10200	Carbon Steel	1520	2760
J94224	HK Cast SS	1400	2550
L13002	Tin	232	450
L51120	Chemical Lead	326	618
L55030	50/50 Solder	183-216	361-421
M11311	Mg AZ31B	605-632	1120-1170
M13310	Mg HK31A	589-651	1092-1204
N02200	Nickel 200	1435-1445	2615-2635
N04400	400 Alloy	1300-1350	2370-2460
N06600	600 Alloy	1350-1410	2470-2575
N10276	C-276 Alloy	1320-1370	2420-2500
N10665	B-2 Alloy	1300-1370	2375-2500
P00020	Gold	1063	1945
P03980	Palladium	1552	2826
P04995	Platinum	1769	3217
P07015	Silver	961	1761
R03600	Molybdenum	2610	4730
R04210	Niobium (Columbium)	2470	4470
R05200	Tantalum	2996	5425
R07005	Tungsten	3410	6170
R50250	Titanium, Gr 1	1705	3100
R56400	Titanium, Gr 5	1600-1660	2920-3020
R60702	Zr 702	1860	3380
S30400	304 SS	1400-1450	2550-2650
S31000	310 SS	1400-1450	2500-2650
S41000	410 SS	1480-1530	2700-2790
S44600	446 SS	1430-1510	2600-2750
S50200	5Cr-0.5Mo Steel	1480-1540	2700-2800
Z13001	Zinc	420	787

COEFFICIENTS OF THERMAL EXPANSION OF COMMON ALLOYS

UNS	Common Name	in/in/°F × 10^{-6}	mm/mm/°C × 10^{-6}	Range-°C
A24430	Cast Al B443.0	12.3	22.	20-100
A91100	Al 1100	13.1	24.	20-100
A95052	Al 5052	13.2	24.	20-100
C11000	ETP Copper	9.4	16.9	20-100
C23000	Red Brass	10.4	18.7	20-300
C28000	Muntz Metal	11.6	21.	20-300
C44300	Admiralty Brass, As	11.2	20.	20-300
C61400	Aluminum Bronze D	9.0	16.2	20-300
C70600	90-10 Copper-Nickel	9.5	17.1	20-300
C71500	70-30 Copper-Nickel	9.0	16.2	20-300
C83600	Ounce Metal	10.2	18.4	0-100
F10006	Gray Cast Iron	6.7	12.1	0-100
G10200	Carbon Steel	6.7	12.1	0-100
J94224	HK Cast SS	9.4	16.9	20-540
L13002	Tin	12.8	23.	0-100
L51120	Chemical Lead	16.4	30.	0-100
L55030	50/50 Solder	13.1	24.	0-100
M11311	Mg AZ31B	14.5	26.	20-100
M13310	Mg HK31A	14.5	26.	20-100
N02200	Nickel 200	7.4	13.3	20-90
N04400	400 Alloy	7.7	13.9	20-90
N06600	600 Alloy	7.4	13.3	20-90
N10276	C-276 Allooy	6.3	11.3	20-90
N10665	B-2 Alloy	5.6	10.1	20-90
R03600	Molybdenum	2.7	4.9	20-100
R05200	Tantalum	3.6	6.5	20-100
R50250	Titanium, Gr 1	4.8	8.6	0-100
R56400	Titanium, Gr 5	4.9	8.8	0-100
R60702	Zr 702	2.9	5.2	0-100
S30400	304 SS	9.6	17.3	0-100
S31000	310 SS	8.0	14.4	0-100
S41000	410 SS	6.1	11.0	0-100
S44600	446 SS	5.8	10.4	0-100
S50200	5Cr-0.5Mo Steel	7.3	13.1	20-540
Z13001	Zinc	18.	32.	0-100

IRON-CARBON EQUILIBRIUM DIAGRAM

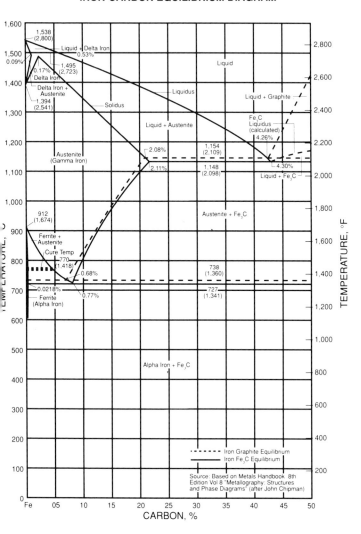

CRITICAL TRANSFORMATION TEMPERATURES FOR STEELS

Definitions

Transformation temperature range is that range of temperature within which austenite forms during heating and transforms during cooling. Transformation temperature is the temperature at which a change in phase occurs. The following symbols have been used:

Ac_1 The temperature at which austenite begins to form during heating.

Ac_3 The temperature at which the transformation of ferrite to austenite is completed during heating.

Ar_1 The temperature at which transformation of austenite to ferrite or to ferrite plus cementite is completed during cooling.

Ar_3 The temperature at which austenite begins to transform to ferrite during cooling.

Type of Steel	Thermal critical range							
	On heating				On cooling			
	Ac_1		Ac_3		Ar_3		Ar_1	
	°F	°C	°F	°C	°F	°C	°F	°C
Low Carbon	1330	723	1605	874	1540	838	1240	671
Medium Carbon	1350	732	1540	838	1470	799	1340	727
Carbon - 1/2 Mo	1340	727	1570	854	1480	804	1225	663
1 Cr 1/2 Mo	1420	771	1635	891	1550	843	1420	771
2 1/4 Cr 1 Mo	1480	804	1600	871	1510	821	1330	721
5 Cr 1/2 Mo	1505	819	1620	882	1445	785	1325	719
7 Cr 1/2 Mo	1520	827	1620	882	1450	788	1340	727
9 Cr 1 Mo	1490	810	1580	860	1420	771	1320	716
12 Cr - Type 410	1435	780	1545	841	1310	710	1130	609

Critical ranges were determined with a heating rate of 250°F per hour and a cooling rate of 50°F per hour.

TEMPER AND RADIATION COLOR OF CARBON STEEL

°F	Approx.°C	Temper Color
380-400	200	Pale yellow
420-440	220	Straw yellow
460-480	240	Yellowish brown
500-540	270	Bluish purple
540-560	285	Violet
560-580	300	Pale blue
600-640	325	Blue
		Radiation Color
1000	540	Black
1100	590	Faint dark red
1200	650	Cherry red (dark)
1300	700	Cherry red (med.)
1400	760	Red
1500	815	Light red
1600	870	Reddish orange
1700	930	Orange
1800	980	Changes
1900	1040	to
2000	1090	Pale orange lemon
2100	1150	Lemon
2200	1205	Light lemon
2300	1260	Yellow
2400	1315	Light yellow
2500	1370	Yellowish gray: "white"

ANNEALING TEMPERATURES FOR AUSTENITIC STAINLESS STEELS AND RELATED ALLOYS

Solution annealing consists of heating to temperature and cooling rapidly. Temperatures are from indicated ASTM standards. Consult alloy producers for details.

UNS	Name	ASTM	Form	°F	°C
J93370	CD-4MCu	A743	Casting	1900 min	1040 min
J95150	CN-7M	A743	Casting	2050 min	1120 min
N08020	20Cb-3	B464	Pipe	1800-1850	980-1010
N08024	20Mo-4	B464	Pipe	1925-1975	1050-1080
N08026	20Mo-6	B464	Pipe	2050-2200	1120-1205
N08028	Sanicro 28	B668	Tube	1975-2085	1080-1140
N08366	AL-6X	B675	Pipe	2200 min	1205 min
N08367	AL-6XN	B675	Pipe	2150 min	1175 min
N08700	JS700	B599	Plate	2000 min	1090 min
N08904	904L Alloy	B677	Pipe, Tube	1950-2100	1065-1150
N08925	25-6Mo	B677	Pipe, Tube	1950-2100	1065-1150
S20910	22-13-5	A312	Pipe	1900 min	1040 min
S24000	18-3 Mn	A312	Pipe	1900 min	1040 min
S30400	304 SS	A312	Pipe	1900 min	1040 min
S30403	304L SS	A312	Pipe	1900 min	1040 min
S30409	304H SS	A312	Pipe	1900 min	1040 min
S30451	304N SS	A312	Pipe	1900 min	1040 min
S30453	304LN SS	A312	Pipe	1900 min	1040 min
S30815	253MA	A312	Pipe	1900 min	1040 min
S30900	309 SS	A312	Pipe	1900 min	1040 min
S31000	310 SS	A312	Pipe	1900 min	1040 min
S31254	254 SMO	A312	Pipe	2100 min	1150 min
S31600	316 SS	A312	Pipe	1900 min	1040 min
S31603	316L SS	A312	Pipe	1900 min	1040 min
S31609	316H SS	A312	Pipe	1900 min	1040 min
S31651	316N SS	A312	Pipe	1900 min	1040 min
S31653	316LN SS	A312	Pipe	1900 min	1040 min
S31700	317 SS	A312	Pipe	1900 min	1040 min
S31703	317L SS	A312	Pipe	1900 min	1040 min
S31725	317LM SS	A312	Pipe	1900 min	1040 min
S31726	317L4 SS	A312	Pipe	1900 min	1040 min
S32100	321 SS*	A312	Pipe	1900 min	1040 min
S32109	321H SS*	A312	Pipe	CW 2000 min HR 1925 min	CW 1095 min HR 1050 min
S34700	347 SS*	A312	Pipe	1900 min	1040 min
S34709	347H SS*	A312	Pipe	CW 2000 min HR 1925 min	CW 1095 min HR 1050 min
S34800	348 SS*	A312	Pipe	1900 min	1040 min
S38100	18-18-2	A312	Pipe	1900 min	1040 min
S31200	44LN	A790	Pipe	1920-2010	1050-1100
S31260	DP-3	A790	Pipe	1870-2010	1020-1100
S31500	3RE60	A790	Pipe	1800-1900	980-1040
S31803	2205 Alloy	A790	Pipe	1870-2010	1020-1100
S32304	SAF 2304	A790	Pipe	1800-1900	930-1040
S32550	Ferralium 255	A790	Pipe	1900 min	1040 min
S32950	7 Mo Plus	A790	Pipe	1820-1880	990-1025

*A stabilization heat treatment after solution anneal improves resistance to intergranular corrosion.

ANNEALING TEMPERATURES AND PROCEDURES FOR MARTENSITIC STAINLESS STEELS

UNS	Process (subcritical) annealing(a)		Full annealing(b) (c)		Isothermal annealing(c)	
	°C	Hardness	°C	Hardness	Procedure(d)	Hardness
S40300, S41000	650-760	82-92 HRB	830-885	75-85 HRB	Heat to 830 to 885°C; hold to 6 h at 705°C	85 HRB
S41400	650-730	99 HRB-24 HRC	Not recommended		Not recommended	
S41600, S41623	650-760	86-92 HRB	830-885	75-85 HRB	Heat to 830 to 885°C; hold 2 h at 720°C	85 HRB
S42000	675-760	94-97 HRB	830-885	86-95 HRB	Heat to 830 to 885°C; hold 2 h at 705°C	95 HRB
S43100	620-705	99 HRB-30 HRC	Not recommended		Not recommended	
S44002	675-760	90 HRB-22 HRC	845-900	94-98 HRB	Heat to 845 to 900°C; hold 4 h at 690°C	98 HRB
S44003	675-760	98 HRB-23 HRC	845-900	95 HRB-20 HRC	Same as S44002	20 HRC
S44004	675-760	98 HRB-23 HRC	845-900	98 HRB-25 HRC	Same as S44002	25 HRC

(a) Air cool from temperature; maximum softness is obtained by heating to temperature at high end of range. (b) Soak thoroughly at temperature within range indicated; furnace cool at 790 °C; continue cooling at 15 to 25 °C/h to 595°C; air cool to room temperature; (c) Recommended for applications in which full advantage may be taken of the rapid cooling to the transformation temperature and from it to room temperature; (d) Preheating to a temperature within the process annealing range is recommended for thin-gage parts, heavy sections, previously hardened parts, parts with extreme variatsions in section or with sharp re-entrant angles, and parts that have been straightened or heavily ground or machined to avoid cracking and minimize distortion, particularly for S42000, S43100, S44002, S44003, and S44004.

ANNEALING TREATMENTS FOR FERRITIC STAINLESS STEELS

UNS	Common Name	Treatment temperature	
		°C	°F
Conventional ferritic grades			
S40500	405	650-815	1200-1500
S40900	409	870-900	1600-1650
S43000	430	705-790	1300-1450
S43020	430F	705-790	1300-1450
S43400	434	705-790	1300-1450
S44600	446	760-830	1400-1525
Low-interstitial ferritic grades			
S43035	439	870-925	1600-1700
S44400	444	955-1010	1750-1850
S44626	26-1Ti	760-955	1400-1750
S44660	SC-1	1010-1065	1850-1950
S44735	29-4C	1010-1065	1850-1950
S44800	29-4-2	1010-1065	1850-1950
S44635	26-4-4	1010-1065	1850-1950

Note: Postweld heat treating of low-interstitial ferritic stainless steels is generally unnecessary and frequently undesirable. Any annealing of these grades should be followed by water quenching or very rapid cooling.

Source: Metals Handbook, Desk Edition, p. 28-61, ASM 1985.

SCHOEFER DIAGRAM FOR ESTIMATING
THE AVERAGE FERRITE CONTENT IN AUSTENITIC
IRON-CHROMIUM-NICKEL ALLOY CASTINGS

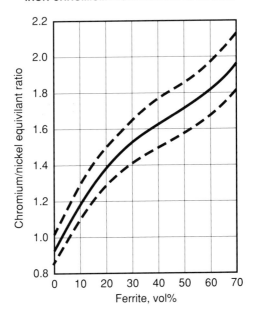

*Source: Metals Handbook, 9th ed.,
Volume 13*, p. 577, ASM 1987.

DELTA FERRITE CONTENT OF STAINLESS STEEL
WELD METALS

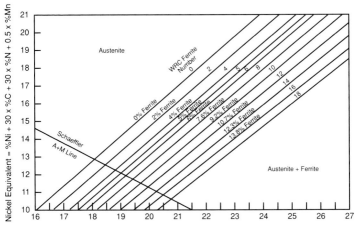

Note: The actual nitrogen content is preferred. If this is not available, the following applicable nitrogen value shall be used: GMAW welds–0.08% (except self-shielding flux cored electrode GMAW welds–0.12%); welds of other process–0.06%.

Source: ASME Pressure Vessel Code, Section III, Figure NB-2433.1-1 (1977)

OVERVIEW OF JOINING PROCESSES

Material	Thickness	SMAW	SAW	GMAW	FCAW	GTAW	PAW	ESW	EGW	RW	FW	OFW	DFW	FRW	EBW	LBW	TB	IFB	RB	DB	IRB	DFB	S
																	Brazing						
Carbon Steel	S	X	X	X		X	X		X	X	X	X			X	X	X	X	X	X	X	X	X
	I	X	X	X	X	X	X		X	X	X	X			X	X	X	X	X	X		X	X
	M	X	X	X	X				X	X	X	X			X	X	X	X				X	
	T	X	X	X	X			X	X		X	X			X							X	
Low Alloy Steel	S	X	X	X		X	X		X	X	X	X	X	X	X	X	X	X	X	X	X	X	X
	I	X	X	X	X	X	X		X	X	X		X	X	X	X	X	X	X			X	X
	M	X	X	X	X		X		X		X		X	X	X	X	X	X				X	
	T	X	X	X	X		X		X		X		X	X	X	X	X				X		
Stainless Steel	S	X	X	X		X	X		X	X	X	X	X	X	X	X	X	X	X	X	X	X	X
	I	X	X	X	X	X	X		X	X	X		X	X	X	X	X	X				X	X
	M	X	X	X	X		X		X		X		X	X	X	X	X	X				X	
	T	X	X	X	X		X	X	X		X		X	X	X	X						X	
Cast Iron	I	X	X	X	X							X					X	X	X	X	X	X	X
	M	X	X	X	X				X			X					X	X				X	X
	T	X							X			X						X				X	
Nickel and Alloys	S	X	X	X		X	X		X	X	X	X	X	X	X	X	X	X	X	X	X	X	X
	I	X	X	X	X	X	X		X	X	X		X	X	X	X	X	X		X		X	X
	M	X	X	X	X		X		X		X		X	X	X	X	X	X				X	
	T	X			X		X		X		X			X	X	X	X					X	

Continued on Next Page

OVERVIEW OF JOINING PROCESSES (Cont)

Material	Thickness	SMAW	SAW	GMAW	FCAW	GTAW	PAW	ESW	EGW	RW	OFW	DFW	FRW	EBW	LBW	TB	FB	IB	RB	DB	IRB	DFB	S
Aluminum and Alloys	S	×		×	×	×	×			×	×			×	×	×	×	×	×	×	×		×
	I	×		×	×	×	×			×		×		×	×	×	×	×	×		×	×	×
	M	×		×	×	×						×		×	×	×	×			×		×	
	T	×		×		×		×	×			×		×			×					×	
Titanium and Alloys	S	×		×		×	×			×		×		×	×	×	×	×				×	
	I	×		×		×						×		×	×	×	×	×			×		
	M	×		×		×						×		×	×	×	×	×					
	T	×		×								×		×			×						
Copper and Alloys	S	×		×		×	×			×			×	×	×	×	×	×	×			×	×
	I	×		×		×	×			×			×	×	×	×	×	×	×		×	×	×
	M	×		×		×							×	×	×	×	×	×				×	
	T	×		×										×		×	×	×				×	
Magnesium and Alloys	S	×		×		×				×			×	×	×	×	×	×	×			×	
	I	×		×		×				×			×	×	×	×	×	×	×		×	×	
	M	×		×		×								×	×	×	×	×				×	
	T	×		×										×			×					×	
Refractory Alloys	S	×		×		×	×			×			×	×	×	×	×	×		×		×	×
	I	×		×		×	×			×			×	×	×	×	×	×		×		×	×
	M					×				×			×				×						
	T																						

This table presented as a general survey only. In selecting processes to be used with specific alloys, the reader should refer to other appropriate sources of information.

Continued on Next Page

OVERVIEW OF JOINING PROCESSES (Cont)

LEGEND

Process Code

SMAW - Shielded Metal Arc Welding-
SAW-Submerged Arc Welding
GMAW-Gas Metal Arc Welding
FCAW-Flux Cored Arc Welding
GTAW-Gas Tungsten Arc Welding
PAW-Plasma Arc Welding
ESW-Electroslag Welding
EGW-Electrogas Welding
RW-Resistance Welding
FW-Flash Welding
OFW-Oxyfuel Gas Welding
DFW-Diffusion Welding

FRW-Friction Welding
EBW-Electron Beam Welding
LBW-Laser Beam Welding
Brazing
TB-Torch Brazing
FB-Furnace Brazing
IB-Induction Brazing
RB-Resistance Brazing
DB-Dip Brazing
IRB-Infrared Brazing
DFB-Diffusion Brazing
S-Soldering

Thickness

S-Sheet up to 3 mm (1/8 in.)
I-Intermediate 3 to 6 mm (1/8 to 3/4 in.)
M-Medium 6 to 19 mm (1/4 to 3/4 in.)
T-Thick 19 mm (3/4 in.) and up

X-Commercial Process

Source: Welding Handbook, 8th ed., volume 1, p. 3, AWS 1987.

PREHEAT TEMPERATURES FOR WELDING CARBON AND ALLOY STEELS

Base Metal P-No.[1]	Weld Metal Analysis A-No.[2]	Base Metal Group	Nominal Wall Thickness in.	Nominal Wall Thickness mm	Min. Specified Tensile Strength, Base Metal ksi	Min. Specified Tensile Strength, Base Metal MPa	Min. Temperature Required °F	Min. Temperature Required °C	Min. Temperature Recommended °F	Min. Temperature Recommended °C
1	1	Carbon steel	<1	<25.4	≤71	≤490	…	…	50	10
			≥1	≥25.4	All	All	…	…	175	79
			All	All	>71	>490	…	…	175	79
3	2, 11	Alloy steels, Cr ≤ 1/2%	<1/2	<12.7	≤71	≤490	…	…	50	10
			≥1/2	≥12.7	All	All	…	…	175	79
			All	All	>71	>490	…	…	175	79
4	3	Alloy steels 1/2% < Cr ≤ 2%	All	All	All	All	300	149	…	…
5	4, 5	Alloy steels, 2 1/4% ≤ Cr ≤ 10%	All	All	All	All	350	177	…	…
6	6	High alloy steels martensitic	All	All	All	All	…	…	300[3]	149[3]
7	7	High alloy steels ferritic	All	All	All	All	…	…	50	10
8	8, 9	High alloy steels austenitic	All	All	All	All	…	…	50	10
9A, 9B	10	Nickel Alloy steels	All	All	All	All	…	…	200	93

Continued on Next Page

PREHEAT TEMPERATURES FOR WELDING CARBON AND ALLOY STEELS (Cont)

| Base Metal P-No.[1] | Weld Metal Analysis A-No.[2] | Base Metal Group | Nominal Wall Thickness | | Min. Specified Tensile Strength, Base Metal | | Min. Temperature | | | |
| | | | | | | | Required | | Recommended | |
			in.	mm	ksi	MPa	°F	°C	°F	°C
10	...	Cr-Cu steel	All	All	All	All	300-400	149-204	...	...
10A	...	Mn-V steel	All	All	All	All	...	...	175	79
10E	...	27Cr steel	All	All	All	All	300[4]	149[4]	...	...
11A SG 1	...	8Ni, 9Ni steel	All	All	All	All	...	...	50	10
11A SG 2	...	5Ni steel	All	All	All	All	50	10	...	...
21-52	...	...	All	All	All	All	...	...	50	10

Notes:
[1] P-Number from BPV Code, Section IX, Table QW-422. Special P-Numbers (SP-1, SP-2, SP-3, SP-4, and SP-5) require special consideration. The required thermal treatment for Special P-Numbers shall be established by the engineering design and demonstrated by the welding procedure qualification.
[2] A-Number from BPV Code, Section IX, Table QW-442.
[3] Maximum interpass temperature 600°F (315°C).
[4] Maintain interpass temperature between 350°F-450°F (177°C-232°C).

Source: ASME B31.3-1990 EDITION, TABLE 330.1.

POSTWELD HEAT TREATMENT REQUIREMENTS FOR CARBON AND ALLOY STEELS

Base Metal P-Number[1]	Weld Metal Analysis A-Number[2]	Base Metal Group	Nominal Wall Thickness in.	Nominal Wall Thickness mm	Min. Specified Tensile Strength, Base Metal ksi	Min. Specified Tensile Strength, Base Metal MPa	Metal Temperature Range °F	Metal Temperature Range °C	Holding Time hr/in. Nominal Wall[3]	Holding Time Min. Time, hr	Brinell Hardness,[4] Max.
1	1	Carbon steel	≤ 3/4	≤19	All	All	None	None	...	...	...
			>3.4	>19	All	All	1100-1200	593-649	1	1	...
3	2, 11	Alloy steels, Cr ≤ 1/2%	≤3/4	≤19	≤71	≤490	None	None	...	...	...
			>3/4	>19	All	All	1100-1325	593-718	1	1	225
			All	All	>71	>490	1100-1325	593-718	1	1	225
4	3	Alloy steels, 1/2% < Cr ≤ 2%	≤1/2	≤12.7	≤71	≤490	None	None	...	...	...
			>1/2	>12.7	All	All	1300-1375	704-746	1	2	225
			All	All	>71	>490	1300-1375	704-746	1	2	225
5	4,5	Alloy steels, 2 1/4% ≤ Cr ≤ 10% ≤3% Cr & ≤0.15% C & >3% Cr or >0.15% C or	≤1/2	≤12.7	All	All	None	None	...	...	...
			>1/2	>12.7	All	All	1300-1400	704-760	1	2	241
6	6	High alloy steels martensitic A 240 Gr. 429	All	All	All	All	1350-1450	732-788	1	2	241
			All	All	All	All	1150-1225	621-663	1	2	241
7	7	High alloy steels ferritic	All	All	All	All	None	None	...	...	...

Base Metal P-Number[1]	Weld Metal Analysis A-Number[2]	Base Metal Group	Nominal Wall Thickness		Min. Specified Tensile Strength, Base Metal		Metal Temperature Range		Holding Time		Brinell Hardness,[4] Max.
			in.	mm	ksi	MPa	°F	°C	hr/in. Nominal Wall[3]	Min. Time, hr	
8	8, 9	High alloy steels austenitic	All	All	All	All	None	None			
9A, 9B	10	Nickel alloy steels	≤3/4	≤19	All	All	None	None			
			>3/4	>19	All	All	1100–1175	593–635	1/2	1	
10		Cr-Cu steel	All	All	All	All	1400–1500[5]	760–816[5]	1/2	1/2	
10A		Mn-V steel	≤3/4	≤19	≤71	≤490	None	None			
			>3/4	>19	All	All	1100–1300	593–704	1	1	225
			All	All	>71	>490	1100–1300	593–704	1	1	225
10E		27Cr steel	All	All	All	All	1225–1300[6]	663–704[6]	1	1	
10H		Cr-Ni-Mo steel	All	All	All	All	See Note[7]		1/2	1/2	
11A SG 1		8Ni, 9Ni steel	≤2	≤51	All	All	None	None			
			>2	>51	All	All	1025–1085[8]	552–585[8]	1	1	
11A SG 2		5Ni steel	>2	>51	All	All	1025–1085[8]	552–585[8]	1	1	

Notes:
(1) P-Number from BPV Code, Section IX, Table QW-422. Special P-Numbers (SP-1, SP-2, SP-3, SP-4, and SP-5) require special consideration. The required thermal treatment for Special P-Numbers shall be established by the engineering design and demonstrated by the welding procedure qualification.
(2) A-Number from BPV Code, Section IX, Table QW-442.
(3) For SI equivalent, h/mm, divide hr/in. by 25.   (4) See 331.1.7.   (5) Cool as rapidly as possible after the hold period.
(6) Cooling rate to 1200°F (650°C) shall be less than 100°F (55°C)/hr; thereafter, the cooling rate shall be fast enough to prevent embrittlement.
(7) Postweld heat treatment is neither required nor prohibited, but any heat treatment applied shall be performed at 1800°F–1900°F, (982°C–1038°C) followed by rapid cooling.
(8) Cooling rate shall be > 300°F (167°C)/hr to 600°F (316°C).

Source: ASME B31.3-1990 EDITION, TABLE 331.1.1

FILLER METALS SUITABLE FOR WELDING JOINTS BETWEEN DISSIMILAR AUSTENITIC STAINLESS STEELS

Suitable filler metals (listed in no preferred order, prefix ER omitted)
Base metal B (type of steel being welded to base metal listed in the first column)

Base metal A	304L	308	309	309S	310	310S	316, 316H	316	317	321, 321 H	347, 347H 348, 348H
304, 304H, 305	308L	308L	308, 309	308, 309	308, 309, 310	308, 309, 310	308, 316	308, 316	308, 316, 317	308	308
304L		308L	308, 309	308, 309	308, 309, 310	308, 309, 310	308, 316	308L, 316L	308, 316, 317	308L, 347	308L, 347
308			308, 309	308, 309	308, 309, 310	308, 309, 310	308, 316	308, 316	308, 316, 317	308	308, 347
309				309	309, 310	309, 310	309, 316	309, 316	309, 316	309, 347	309, 347
309S					309, 310	309S, 310S	309, 316	309S, 316L	309, 316	309, 347	309, 347
310							310, 316	310, 316	310, 317	308, 310	308, 310
310S							316	316	316, 317	308, 310	308, 310
316, 316H								316	317	308, 316	308, 316, 347
316L									316, 317	316L	316L, 347
317										308, 317	308, 317, 347
321, 321H											308L, 347

Source: Metals Handbook, 9th ed., Volume 6, p.335, ASM 1983

ELECTRODES AND FILLER METALS FOR DISSIMILAR JOINTS BETWEEN NICKEL ALLOYS AND OTHER METALS

UNS	Common Name	Carbon and Low Alloy Steels		300 Series Stainless Steels		Copper	
		Electrode	Filler Metal	Electrode	Filler Metal	Electrode	Filler Metal
N02200	Nickel 200	Weld A	82	Weld A	82	190	60
N04400	400 Alloy	190	61	Weld A	82	190	60
N05500	K-500 Alloy	190	61	Weld A	82	190	60
N06002	X Alloy	X	X	X	X		
N06022	C-22 Alloy	C-22	C-22	C-22	C-22		
N06030	G-30 Alloy	C-22	C-22	C-22	C-22		
N06455	C-4 Alloy	C-22	C-22	C-22	C-22		
N06600	600 Alloy	Weld A	82	Weld A	82	141	61
N06625	625 Alloy	Weld A	82	Weld A	82	141	61
N06985	G-3 Alloy	C-22	C-22	C-22	C-22		
N07718	718 Alloy	Weld A	82	Weld A	82	141	61
N07750	X-750 Alloy	Weld A	82	Weld A	82	141	61
N08020	20Cb-3	C-22	C-22	C-22	C-22		
N08366	AL-6X	C-22	C-22	C-22	C-22		
N08800	800 Alloy	Weld A	82	Weld A	82	141	61
N08825	825 Alloy	112	625	112	625	141	61
N08904	904L Alloy	C-22	C-22	C-22	C-22		
N10276	C-276 Alloy	C-276	C-276	C-276	C-276		
N10665	B-2 Alloy	B-2	B-2	B-2	B-2		

Designations for Nickel-Base Electrodes and Filler Metals:

Common	UNS	AWS
60	N04060	A5.14 (ERNiCu-7)
61	N02061	A5.14 (ERNi-1)
82	N06082	A5.14 (ERNiCr-3)
141	W82141	A5.11 (ENi-1)
190	W84190	A5.11 (ENiCu-7)
Weld A	W86133	A5.11 (ENiCrFe-2)

Source: This partial listing is adapted from Inco Alloys International ''Joining'' and Haynes International ''Welding Filler Material Information.''

TYPICAL PROPERTY RANGES FOR PLASTICS

Thermosets[a]	Specific gravity	Tensile Strength ksi	(MPa)	Modulus of Elast., Tension 10² ksi	(10² MPa)	Impact Strength, Izod[c] ft-lb	(J)	Max use temp. (no load) °F	(°C)	HDT at 254 psi[d] °F	(°C)	Weather res	Chem. res.[e] Weak acid	Strong acid	Weak alkali	Strong alkali	Solvents
Alkyds																	
Glass filled	2.12-2.15	4-9.5	(28-66)	20-28	(138-193)	0.6-10	(0.8-14)	450	(230)	400-500	(200-260)	R	A	A	A	A	A
Mineral filled	1.60-2.30	3-9	(21-62)	5-30	(34-207)	0.3-0.5	(0.4-0.7)	300-450	(150-230)	350-500	(180-260)	R	R	A	A	D	A
Asbestos filled	1.65	4.5-7	(31-48)	—		0.4-0.5	(0.6-0.7)	450	(230)	315	(160)	R	R	S	R	S	R
Syn. fiber filled	1.24-2.10	4.5-7	(31-48)	20	(138)	0.5-4.5	(0.7-6.1)	300-430	(150-220)	245-430	(120-220)	R	R	S	R	S	A
Alkyl diglycol carbonate	1.30-1.40	5-6	(34-41)	3.0	(21)	0.2-0.4	(0.3-0.5)	212	(100)	140-190	(60-90)	R	R	A[*]	R	R-S	R
Diallyl phthalates																	
Glass filled	1.61-1.78	6-11	(41-76)	14-22	(97-152)	0.4-15	(0.5-20)	300-400	(150-200)	330-540	(165-280)	R	R	S	R	S	R
Mineral filled	1.65-1.68	5-9	(34-62)	12-22	(83-152)	0.3-0.5	(0.4-1)	300-400	(150-200)	320-540	(160-280)	R	R	S	R-S	S	R
Asbestos filled	1.55-1.65	7-8	(48-55)	12-22	(83-152)	0.4-0.5	(0.5-0.7)	300-400	(150-200)	320-540	(160-280)	R	R	S	R-S	S	R
Epoxies (bis A)																	
No filler	1.06-1.40	4-13	(28-90)	2.15-5.2	(15-36)	0.2-1.0	(0.3-1.4)	250-500	(120-260)	115-500	(45-260)	R	R	A	R	S	R-S
Graphite fiber reinf	1.37-1.38	185-200	(1280-1380)	118-120	(814-827)	—	—	—		—		S	R	R	R	R	R-S
Mineral filled	1.6-2.0	5-15	(34-103)	—		0.3-0.4	(0.4-0.5)	300-500	(150-260)	250-500	(120-260)	S	R	R	R	R	R-S
Glass filled	1.7-2.0	10-30	(69-207)	30	(207)	10-30	(14-41)	300-500	(150-260)	250-500	(120-260)	S	R	R-S	R	R	R-S
Epoxies (novolac)																	
No filler	1.12-1.24	5-11	(34-76)	2.15-5.2	(15-36)	0.3-0.7	(0.4-0.9)	400-500	(200-260)	450-500	(230-260)	R	R	A	R	R	R
Epoxies (cycloaliphatic)																	
No filler	1.12-1.18	10-17.5	(69-121)	5-7	(34-48)	—	—	480-550	(250-290)	500-550	(260-290)	R	R	R-A	R	R-A	R
Melamines																	
Cellulose filled	1.45-1.52	5.9	(34-62)	11	(76)	0.2-0.4	(0.3-0.5)	250	(120)	270	(130)	S	R-S	D	R	D	R
Flock filled	1.50-1.55	7-9	(48-62)	—		0.4-0.5	(0.5-0.7)	250	(120)	270	(130)	S	R-S	D	R	D	R-S
Asbestos filled	1.70-2.0	5-7	(34-48)	20	(138)	0.3-0.4	(0.4-0.5)	250-400	(120-200)	265	(130)	S	R-S	D	S	S	R
Fabric filled	1.5	8-11	(55-76)	14-16	(97-110)	0.6-1.0	(0.8-1.4)	250	(120)	310	(150)	S	R	D	R	A	R-S
Glass filled	1.8-2.0	5-10	(34-69)	24	(165)	0.6-18	(0.8-24)	300-400	(150-200)	400	(200)	S	R	D	R	R-S	R
Phenolics																	
Woodflour filled	1.34-1.45	5-9	(34-62)	8-17	(55-117)	0.2-0.6	(0.3-0.8)	300-350	(150-180)	300-370	(150-190)	S	R-S	S-D	S-D	A	R-S
Asbestos filled	1.45-2.00	4.5-7.5	(31-52)	10-30	(69-207)	0.2-0.4	(0.3-0.5)	350-500	(180-260)	300-500	(150-260)	S	R-S	S-D	S-D	A	R-S
Mica filled	1.65-1.92	5.5-7	(38-48)	25-50	(172-345)	0.3-0.4	(0.4-0.5)	250-300	(120-150)	300-350	(150-180)	S	R-S	S-D	S	A	R-S
Glass filled	1.69-1.95	5-18	(34-124)	19-33	(131-228)	0.3-18	(0.4-24)	350-550	(180-290)	300-600	(150-320)	S	R-S	S-D	S-D	A	R-S
Fabric filled	1.36-1.43	3-9	(21-62)	9-14	(62-97)	0.8-8	(1.1-11)	220-250	(100-120)	250-330	(120-170)	S	R-S	S-D	S-D	A	R-S

TYPICAL PROPERTY RANGES FOR PLASTICS (Cont)

Thermosets[a]	Specific gravity	Tensile Strength, ksi	(MPa)	Modulus of Elast., Tension 10^2 ksi	(10^2 MPa)	Impact Strength, Izod[c] ft-lb	(J)	Max use temp. (no load) °F	(°C)	HDT at 254 psi[d] °F	(°C)	Weather res	Weak acid	Strong acid	Weak alkali	Strong alkali	Solvents
Polybutadienes																	
Very high vinyl (no filler)	1.00	8	(55)	2	(14)	1.1	(1.5)	500	(260)	—	—	S	R	R	R	R	R
Polyesters																	
Glass filled BMC	1.7-2.3	4-10	(28-69)	16-25	(110-172)	1.5-16	(2.0-22)	300-350	(150-180)	400-450	(200-230)	R-E	R-A	S-A	S-A	S-D	A-D
Glass filled SMC	1.7-2.1	8-20	(55-138)	16-25	(110-172)	8-22	(11-30)	300-350	(150-180)	400-450	(200-230)	R-E	R-A	S-A	S-A	S-D	A-D
Glass cloth reinf.	1.3-2.1	25-50	(172-345)	19-45	(131-310)	5-30	(7-41)	300-350	(150-180)	400-450	(200-230)	R-E	R-A	S-A	S-A	S-D	A-D
Silicones																	
Glass filled	1.7-2.0	4-6.5	(28-45)	10-15	(69-103)	3-15	(4-20)	600	(320)	600	(320)	R-S	R-S	R-S	S	S-A	R-A
Mineral filled	1.8-2.8	4-6	(28-41)	13-18	(90-124)	0.3-0.4	(0.4-0.5)	600	(320)	600	(320)	R-S	R-S	R-S	S	S-A	R-A
Ureas																	
Cellulose filled	1.47-1.52	5.5-13	(38-90)	10-15	(69-103)	0.2-0.4	(0.3-0.5)	170	(80)	260-290	(130-140)	S	R-S	A-D	S-A	D	R-S
Urethanes																	
No filler	1.1-1.5	0.2-10	(1-69)	1-10	(7-69)	5-NB	(7)	190-250	(90-120)	—	—	R-S	S	A	S	S-A	R-S

Continued on Next Page

TYPICAL PROPERTY RANGES FOR PLASTICS (Cont)

Thermoplastics		Specific gravity	Tensile Strength ksi	(MPa)	Modulus of Elast., Tension 10² ksi	(10² MPa)	Impact Strength, Izod[f] ft-lb	(J)	Max use temp. (no load) °F	(°C)	HDT at 66 psi °F	(°C)	HDT at 264 psi °F	(°C)	Weather res	Weak acid	Strong acid	Weak alkali	Strong alkali	Solvents
ABS	GP	1.05-1.07	5.9	(41)	3.1	(21)	6	(8)	160-200	(70-90)	210-225	(100-110)	190-206	(90-95)	R:E	R	A[x]	R	R	A[m]R
	Hi. imp.	1.01-1.06	4.8	(33)	2.4	(17)	7.5	(10)	140-210	(60-100)	210-225	(100-110)	188-211	(85-100)	R:E	R	A[x]	R	R	A[m]R
	Ht. res.	1.06-1.08	7.4	(51)	3.9	(27)	2.2	(3.0)	190-230	(90-110)	225-252	(110-120)	226-240	(110-115)	R:E	R	A[x]	R	R	A[m]R
	Trans.	1.07-1.20	5.6-6.0	(39-41)	2.9-3.2	(20-22)	5.3-2.5	(7.1-3.4)	130-130	(55-55)	180-210	(80-100)	165-195	(75-90)	R:E R:E	R R	A[x] A[x]	R R	R R	A[m]R A[m]R
Acetals	Homo	1.42	10	(69)	5.2	(36)	1.4-1.6	(1.9-2.2)	195-180	(90-80)	338	(170)	255	(125)	R[j]	R	A	R	A-D	R
	Copol	1.41	8.8	(61)	4.1	(28)	1.2	(1.6)	212	(100)	316	(160)	230	(110)	R[j]	R	A	R	R	R
Acrylics	GP	1.11-1.19	5.6-11.0	(39-76)	2.25-4.65	(16-32)	0.3-0.4	(0.4-3.1)	130-230	(55-110)	175-225	(80-110)	165-210	(75-100)	R	R	A[x]	R	A	A[m]R
	Hi. imp.	1.12-1.16	5.8-8.0	(40-55)	2.3-3.3	(16-23)	0.8-2.3	(1.1-3.1)	140-195	(60-90)	180-205	(80-95)	165-190	(75-90)	R	R	A[x]	R	R	A[m]R
		1.21-1.28	8.0-12.5	(55-86)	3.5-4.8	(24-33)	0.3-0.4	(0.4-0.5)	125-200	(50-90)	170-200	(75-95)	155-205	(70-95)	R	R	A[x]	R	A	A[m]R
	Cast	1.18-1.28	9.0-12.5	(62-86)	3.7-5.0	(26-34)	0.4-1.5	(0.5-2.0)	140-200	(60-90)	165-235	(75-115)	160-215	(70-100)	R	R	A[x]	R	A	A[m]R
	Multi poly-mer	1.09-1.14	6-8	(41-55)	3.1-4.3	(21-30)	1-3	(1-4)	165-175	(75-80)	—	—	185-195	(85-90)	E	R	A[x]	R	S	A[m]
Cellulosics	Acetate	1.23-1.34	3.0-8.0	(21-55)	1.05-2.55	(7-18)	1.1-6.8	(1.5-9)	140-220	(60-105)	120-209	(50-100)	111-195	(45-90)	S	S	D	S	D	D-S
	Butyrate	1.15-1.22	3.0-6.9	(21-48)	0.7-1.8	(5-12)	3.0-10.0	(4-14)	140-220	(60-105)	130-227	(55-110)	113-202	(45-95)	S	S	D	S	D	D-S

Continued on Next Page

TYPICAL PROPERTY RANGES FOR PLASTICS (Cont)

Thermoplastics		Specific gravity	Tensile Strength ksi	(MPa)	Modulus of Elast., Tension 10² ksi	(10³ MPa)	Impact Strength, Izod[c] ft-lb	(J)	Max use temp. (no load) °F	(°C)	HDT at 66 psi °F	(°C)	HDT at 264 psi °F	(°C)	Weather res	Weak acid	Strong acid	Weak alkali	Strong alkali	Solvents
Cellulosics	E. cellulose	1.10-1.17	3-8	(21-55)	0.5-3.5	(3-24)	1.7-7.0	(2.3-9.5)	115-185	(45-85)	—	—	115-190	(45-90)	S	S	D	S	S	D
	Nitrate	1.35-1.40	7-8	(48-55)	1.9-2.2	(13-15)	5-7	(7-9)	140	(60)	—	—	140-160	(60-70)	E	S	D	S	D	D
	Propionate	1.19-1.22	4.0-6.5	(28-45)	1.1-1.8	(8-12)	1.7-9.4	(2.3-13)	155-220	(70-105)	147-250	(65-120)	111-228	(45-110)	S	S	D	S	D	D-S
Ch. polyether		1.4	5.4	(37)	1.5	(10)	0.4	(0.5)	290	(140)	285	(140)	—	—	R-S	R	A[k]	S	D	R
Eth. copolymers	EEA	0.93	2.0	(14)	0.05	(0.3)	NB	—	190	(90)	—	—	—	—	S	R	A[k]	R	R	A-D
	EVA	0.94	3.6	(25)	0.02-0.12	(0.14-0.8)	NB	—	—	—	140-147	(60-65)	93	(35)	S	R	A	R	R	A-D
Fluoropolymers	FEP	2.14-2.17	2.5-3.9	(17-27)	0.5-0.7	(3-5)	NB	—	400	(208)	158	(70)	—	—	R	R	R	R	R	R
	PTFE	2.1-2.3	1.4	(7-28)	0.38-0.65	(2.6-4.5)	2.5-4.0	(3.4-5.4)	550	(290)	250	(120)	—	—	R	R	R	R	R	R
	CTFE	2.10-2.15	4.6-5.7	(32-39)	1.8-2.0	(12-14)	3.5-3.6	(4.7-4.9)	350-390	(180-200)	258	(125)	—	—	R	R	R	R	R	S[m]
	PVF$_2$	1.77	7.2	(50)	1.7	(12)	3.8	(5.2)	300	(150)	300	(150)	195	(90)	S	R	A[l]	R	R	R
	ETFE & ECTFE	1.68-1.70	6.5-7.0	(45-48)	2-2.5	(14-17)	NB	—	300	(150)	220	(105)	160	(70)	R	R	R	R	R	R
Methylpentene		0.83	3.3-3.6	(23-25)	1.3-1.9	(10-13)	0.95-3.8	(1.3-5.2)	275	(135)	—	—	—	—	E	R	A[l]	R	R	A
Nylons	6/6	1.13-1.15	9-12	(62-83)	3.85	(27)	2.0	(2.7)	180-300	(80-150)	360-470	(180-240)	150-220	(65-105)	R	R	A	R	R	R-D[o]
	6	1.14	12.5	(86)	—	—	1.2	(1.6)	180-250	(80-120)	300-365	(150-185)	140-155	(60-70)	R	R	A	R	R	R-A[o]
	6/10	1.07	7.1	(49)	2.8	(19)	1.6	(2.2)	180	(80)	300	(150)	—	—	R	R	A	R	R	R-A[o]
	8	1.09	3.9	(27)	—	—	>16	(>22)	175-260	(80-125)	—	—	120-130	(50-55)	R	R	A	R	R	R-A[o]
	12	1.01	6.5-8.5	(45-59)	1.7-2.1	(12-14)	1.2-4.2	(1.6-5.7)	180-250	(80-120)	—	—	130-350	(55-180)	R	R	A	R	R	R-A[o]
	Copolymers	1.08-1.14	7.5-11.0	(52-76)	—	—	1.5-19	(2.26)	180-250	(80-120)	—	—	—	—	R	R	A	R	R	R-A[o]

Continued on Next Page

TYPICAL PROPERTY RANGES FOR PLASTICS (Cont)

Thermoplastics		Specific gravity	Tensile Strength ksi (MPa)	Modulus of Elast., Tension 10²ksi (10²MPa)[c]	Impact Strength, Izod[c] ft-lb (J)	Max use temp. (no load) °F (°C)	HDT at 66 psi °F (°C)	HDT at 264 psi °F (°C)	Weather res	Weak acid	Strong acid	Weak alkali	Strong alkali	Solvents
Polyesters	PET	1.37	10.4 (72)	—	0.8 (1.1)	175 (80)	240 (115)	185 (85)	R	R	A[a]	R	A	R-A[a]
	PBT	1.31	8.0-8.2 (55-57)	3.6 (25)	1.2-1.3 (1.6-1.8)	280 (140)	310 (155)	130 (55)	R	R	R	R	A	R
	PTMT Copol.	1.31	7.3 (50)	—	1.0 (1.4)	270 (130)	302 (150)	122 (50)		R	R	R	A	R
Polyaryl ether		1.2	8.2 (57)	3.2 (22)	1.0 (1.4)	250 (120)	320 (160)	154 (70)	R	R	R	R	A	A
Polyaryl sulfone		1.14	7.5 (52)	3.7 (26)	10 (14)	500 (260)	—	300 (150)	E	R	R	R	R	R
		1.36	13 (90)	—	2 (2.7)	225 (105)	—	525 (275)	Darkens	R	R	R	R	R
Polybutylene		0.910	3.8 (26)	0.26 (1.8)	NB	250 (120)	215 (100)	130 (55)	E	R	A[a]	R	R	A
Polycarbonate		1.2	9 (62)	3.45 (24)	12-16 (16-22)	220 (105)	290 (130)	265-285 (130-140)	R	R	A[a]	A	A	A
PC/ABS		1.14	8.2 (57)	3.7 (26)	10 (14)	180 (80)	235 (115)	220 (105)	R-E	R	A[a]	R	S	A
Polyethylenes*	LD	0.91-0.93	0.9-2.5 (6-17)	0.20-0.27 (1.4-1.9)	NB	212 (100)	100-120 (40-50)	90-105 (30-40)	E	R	A[a]	R	R	R
	HD	0.95-0.96	2.9-5.4 (20-37)	—	0.4-14 (0.5-19)	175-250 (80-120)	140-190 (60-90)	110-130 (45-55)	E	R	A[a]	R	R	R
	HMW	0.945	2.5 (17)	1 (7)	NB	160-180 (70-80)	155 (70)	105-180 (40-80)	E	R	R	R	R	R
Ionomer		0.94-0.95	3.4-4.5 (23-31)	0.3-0.7 (2-5)	6-NB (8-)	175 (80)	110 (45)	100-120 (40-50)	E	A	A[a]	R	R	R
Phenylene oxide based mtls		1.06-1.10	7.8-9.6 (54-66)	3.5-3.8 (24-26)	5.0 (6.8)	220 (105)	230-280 (110-140)	212-265 (100-130)	R	R	R	R	R	R-A
Polyphenylene sulfide		1.34	10 (69)	4.8 (33)	0.3 (0.4)	500 (260)	200-230 (95-110)	278 (135)	R	R	R	R	A	R
Polyimide		1.43	5.7-7.5 (34-52)	5.4 (37)	5.7 (7.9)	225-300 (105-150)	—	680 (360)	—	R	A[a]	A	A	A
Polypropylenes	GP	0.90-0.91	4.8-5.5 (33-38)	1.6-2.2 (11-15)	0.4-2.2 (0.5-3.0)	200-250 (95-120)	160-200 (70-95)	125-140 (50-60)	E	R	A[a]	A	R	R
	Hi. imp.	0.90-0.91	3.5-5 (21-34)	1.3 (9)	1.5-12 (2-16)	190-240 (90-115)	185-230 (85-110)	120-135 (50-60)	E	R	A[a]	R	R	R
Propylene copolymer		0.91	4 (28)	1.0-1.7 (7-12)	1.1 (1.5)	—	—	115-140 (45-60)	E	R	A[a]	R	R	R

*Polyethylene may stress crack in HF service.

Continued on Next Page

TYPICAL PROPERTY RANGES FOR PLASTICS (Cont)

Thermoplastics		Specific gravity	Tensile Strength, ksi	(MPa)	Modulus of Elast., Tension 10³ ksi	(10² MPa)	Impact Strength., Izod[c] ft-lb	(J)	Max use temp. (no load) °F	(°C)	HDT at 66 psi °F	(°C)	HDT at 264 psi °F	(°C)	Weather res	Weak acid	Strong acid	Weak alkali	Strong alkali	Solvents
Polystyrenes	GP	1.04-1.07	6.0-7.3	(41-50)	4.5	(31)	0.3	(0.4)	150-170	(65-80)	—	—	180-220	(80-105)	S	R	A[k]	R	R	D
	Hi. imp.	1.04-1.07	2.8-4.6	(20-32)	2.9-4.0	(20-28)	0.7-1.0	(0.9-1.4)	140-175	(60-80)	—	—	175-210	(80-100)	S	R	A[k]	R	R	D
Polysulfone		1.24	10.2	(70)	3.6	(25)	1.2	(1.6)	300	(150)	360	(180)	345	(175)	S	R	R	R	R	R-A
Polyurethanes		1.11-1.25	4.5-8.4	(31-58)	0.1-3.5	(0.7-24)	NB	(—)	190	(90)	—	—	—	—	R-S	S-D	S-D	S-D	S-D	R
Vinyl Rigid		1.3-1.5	5-8	(34-55)	3.5	(21-34)	0.5-20	(0.7-27)	150-175	(65-80)	135-180	(60-80)	130-175	(55-80)	R	R	R-S	R	R	R-A
Vinyl Flexible		1.2-1.7	1-4	(7-28)	—	—	0.5-20	(0.7-27)	140-175	(60-80)	—	—	—	—	S	R	R-S	R	R	R-A
Rigid CPVC		1.49-1.58	7.5-9.0	(52-62)	3.6-4.7	(25-32)	1.0-5.6	(1.4-7.6)	230	(110)	215-245	(100-120)	200-235	(95-115)	R	R	R	R	R	R
PVC/acrylic		1.30-1.35	5.5-6.5	(38-45)	2.75-3.35	(19-23)	15	(20)	—	—	180	(80)	170	(80)	R	R	S	R	R	A
PVC/ABS		1.10-1.21	2.6-6.0	(18-41)	0.8-3.4	(6-23)	10-15	(14-20)	—	—	—	—	—	—	S	R	R-S	R	R	R-D
SAN		1.08	10-12	(69-83)	5.0-5.6	(34-39)	0.4-0.5	(0.5-0.7)	140-200	(60-95)	—	—	190-220	(90-105)	S-E	R	A	R	R	A

[a]All values at room temperature unless otherwise listed. [b]Heat deflection temperature. [c]Per ASTM. [d]Notched samples. [e]Ac is acid and Al is alkali; R is resistant; A is attacked; S is slight effects; E is embrittles and D is decomposes. [f]Chalks slightly. [g]By oxidizing acids. [h]By turning sulfuric. [i]By ketones, esters, and chlorinated and aromatic hydrocarbons. [j]Halogenated solvents cause swelling. [k]Dissolved by phenols and formic acid.

Source: *Plastics Engineering Handbook*, 4th Edition.

PROPERTIES OF ELASTOMERS

Property	NR Natural rubber (cis-polyisoprene)	SBR Butadiene-styrene (GR-S)	IR Synthetic (polyisoprene)	COX Butadiene-acrylonitrile (nitrile)	CR Chloroprene (neoprene)	ITR Butyl (isobutylene-isoprene)	BR Polybutadiene	T Poly-sulfide	Silicone (poly-siloxane)
Physical Properties:									
Specific gravity (ASTM D792)	0.93	0.91	0.93	0.98	1.25	0.90	0.91	1.35	1.1-1.6
Thermal conductivity Btu/(h)(ft²)(°F/ft) (ASTM C 177)	0.082	0.143	0.082	0.143	0.112	0.053	—	—	0.13
Coefficient of thermal expansion (cubical), 10^{-5} per °F (ASTM D 696)	37	37	—	39	34	32	37.5	—	45
Electrical insulation	Good	Good	Good	Fair	Fair	Good	Good	Fair	Excellent
Flame resistance	Poor	Poor	Poor	Poor	Good	Poor	Poor	Poor	Good
Min. recommended service temp. °F	−60	−60	−60	−60	−40	−50	−150	−60	−178
Max. recommended service temp. °F	180	180	180	300	240	300	200	250	600
Mechanical properties									
Tensile strength, lb/in.²									
Pure gum (ASTM D 412)	2,500-3,500	200-300	2,500-3,500	500-900	3,000-4,000	2,500-3,000	200-1,000	250-400	600-1,300
Black (ASTM D 412)	3,500-4,500	2,500-3,500	3,500-4,500	3,000-4,500	3,000-4000	2,500-3,000	2,000-3,000	1,000	—
Elongation, %									
Pure gum (ASTM D 412)	750-850	400-600	—	300-700	800-900	750-950	400-1,000	450-650	100-500
Black (ASTM D 412)	550-650	500-600	300-700	300-650	500-600	650-850	450-600	150-450	—
Hardness (durometer)	A30-90	A40-90	A40-80	A40-95	A20-95	A40-90	A40-90	A40-85	A30-90

Continued on Next Page

PROPERTIES OF ELASTOMERS (Cont)

Property	NR Natural rubber (cis-polyisoprene)	SBR Butadiene-styrene (GR-S)	IR Synthetic (polyisoprene)	COX Butadiene-acrylonitrile (nitrile)	CR Chloroprene (neoprene)	ITR Butyl (isobutylene-isoprene)	BR Polybutadiene	T Polysulfide	Silicone (polysiloxane)
Mechanical properties									
Rebound:									
Cold	Excellent	Good	Excellent	Good	Very good	Bad	Excellent	Good	Very good
Hot	Excellent	Good	Excellent	Good	Very good	Very good	Excellent	Good	Very good
Tear resistance	Excellent	Fair	Excellent	Good	Fair to good	Good	Fair	Poor	Fair
Abrasion resistance	Excellent	Good to excellent	Excellent	Good to excellent	Good	Good to excellent	Excellent	Poor	Poor
Chemical resistance:									
Sunlight aging	Poor	Poor	Fair	Poor	Very good	Very good	Poor	Very good	Excellent
Oxidation	Good	Good	Excellent	Good	Excellent	Excellent	Good	Very good	Excellent
Heat aging	Good	Very good	Good	Excellent	Excellent	Excellent	Good	Fair	Excellent
Solvents:									
Aliphatic hydrocarbons	Poor	Poor	Poor	Excellent	Good	Poor	Poor	Excellent	Fair
Aromatic hydrocarbons	Poor	Poor	Poor	Good	Fair	Poor	Poor	Excellent	Poor
Oxygenated, alcohols	Good	Good	Good	Good	Very good	Very good	—	Very good	Excellent
Oil, Gasoline	Poor	Poor	Poor	Excellent	Good	Poor	Poor	Excellent	Poor
Animal, vegetable oils	Poor to good	Poor to good	—	Excellent	Excellent	Excellent	Poor to good	Excellent	Excellent
Acids:									
Dilute	Fair to good	Fair to good	Fair to good	Good	Excellent	Excellent	—	Good	Very good
Concentrated	Fair to good	Fair to good	Fair to good	Good	Good	Excellent	—	Good	Good
Permeability to gases	Low	Low	Low	Very low	Low	Very low	Low	Very low	High
Water-swell resistance	Fair	Excellent	Excellent	Excellent	Fair to Excellent	Excellent	Excellent	Excellent	Excellent

Continued on Next Page

PROPERTIES OF ELASTOMERS (Cont)

Property	ECO, CO Epichlorohydrin homopolymer and copolymer	Fluorosilicone	EPDM Ethylene propylene	CSM Chloro-sulfonated polyethylene	FPM Fluorocarbon elastomers
Physical properties:					
Specific gravity	1.32-1.49	1.4	0.86	1.11-1.26	1.4-1.95
Thermal conductivity, Btu/(h)(ft^2)(°F/ft)	—	0.13	—	0.065	0.13
Coefficient of thermal expansion, 10^{-5}/°F	—	45	—	27	8.8
Flame resistance	Fair	Poor	Poor	Good	Excellent
Colorability	Good	Good	Excellent	Excellent	Good
Mechanical properties:					
Hardness (Shore A)	30-95	40-70	30-90	45-95	65-90
Tensile strength, 1,000 lb./in.^{2}					
Pure gum	—	1	<1	4	<2
Reinforced	2-3	<2	0.8-3.2	1.5-2.5	1.5-3
Elongation, %					
Reinforced	320-350	200-400	200-600	250-500	100-450
Resilience	Poor to excellent	Good to fair	Good	Good	Fair
Compression-set resistance	Very good	—	Good	Fair to good	Good to excellent
Hysteresis resistance	Good	Good	Good	Good	Good
Flex-cracking resistance	Very good	Good	Good	Good	Good
Slow rate	Very good	Good	Good	Good	Good
Fast rate	Good	Good	Good	Good	Good
Tear strength	Good	Fair	Poor to fair	Fair to good	Poor to fair
Abrasion resistance	Fair to good	Poor	Good	Excellent	Good
Electrical properties					
Dielectric strength	Fair	Good	Excellent	Excellent	Good
Electrical insulation:	Fair	Good	Very good	Good	Fair to good
Thermal properties:					
Service temp. °F:					
Min for continuous use	−15 to −80	−90	−60	−40	−10
Max for continuous use	300	400	<350	<325	<500

Continued on Next Page

PROPERTIES OF ELASTOMERS (Cont)

Property	ECO, CO Epichlorohydrin homopolymer and copolymer	Fluorosilicone	EPDM Ethylene propylene	CSM Chloro-sulfonated polyethylene	FPM Fluorocarbon elastomers
Corrosion resistance:					
Weather	Excellent	Excellent	Excellent	Excellent	Excellent
Oxidation	Very good	Excellent	Excellent	Excellent	Outstanding
Ozone	Good to excellent	Excellent	Excellent	Excellent	Excellent
Radiation	—	Good	Excellent	Fair to Good	Fair to Good
Water	Good	Excellent	Good to excellent	Good	Good
Acids	Good	Very good to excellent	Good to excellent	Excellent	Good to excellent
Alkalies	Good	Very good	Good to excellent	Excellent	Poor to good
Aliphatic hydrocarbons	Excellent	Excellent	Poor	Fair	Excellent
Aromatic hydrocarbons	Very good	Excellent	Fair	Poor to fair	Excellent
Halogenated hydrocarbons	Good	—	Poor	Poor to fair	Good
Alcohol	Good	—	Good	Very good	Excellent
Synthetic lubricants (diester)	Fair to good	Excellent	Poor to fair	Poor	Fair to good
Hydraulic fluids:					
Silicates	Very good	Excellent	Fair to good	Good	Good
Phosphates	Poor to fair	Excellent	Good to excellent	Poor to fair	Poor

Source: C.H. Harper, Handbook of Plastics & Elastomers, Table 35, McGraw Hill

OXYGEN AND WATER PERMEABILITY IN PLASTIC FILMS

Plastic	Permeability to O_2 om^3-mil/100 in^2/ 24 hr. 25°C ASTM D1434	Rate of H_2O vapor trans. g-mil/100 in^2/ 24 hr. 38°C ASTM E96
Acrylonitrite-Butadiene-Styrene (ABS)	50-70	—
Ethylene-Chlorotrifluoroethylene Copolymer	25	0.6
Ethylene-Tetrofluoroethylene Copolymer	100	1.7
Fluorinated Ethylene-Propylene Copolymer	750	0.4
Polyvinyl Fluoride	3	3.2
Polycarbonate	300	11
Polyester (PE Terephthalate)	3-6	1.0-1.3
Nylon 6/6	5	3-8
Nylon 11	34	0.32-0.85
Nylon 12	52-92	0.07
Polyethylene, low dens.	500	1.0-1.5
Polyethylene, med. dens.	250-535	0.7
Polyethylene, high dens.	185	0.3
Polyimide	25	5.4
Polypropylene	150-240	0.7
Polystyrene	350	7-10
Polysulfone	230	18
Vinyl Chloride-Acetate Copolymer (Non-plasticized)	15-20	4
Vinyl Chloride-Acetate Copolymer (Plasticized)	20-150	5-8
Vinylidene Chloride-Vinyl Chloride Copolymer	0.8-6.9	0.2-0.6
Polyvinyl Chloride-(Non-plasticized)	5-20[a]	0.9-5.1
Polyvinyl Chloride-(Plasticized)	2-4000	5-30
Polyurethane Elastomer	75-327	40-75

[a]50% R.H.

Adapted from Modern Plastics Encyclopedia, 1985-1986, pp.481-485, McGraw Hill.

DIMENSIONS OF POLYETHYLENE LINE PIPE (API)

(All Dimensions are in Inches)
(inches × 25.40 = mm)

Nominal pipe size	Outside diameter	Minimum Wall Thickness								
		SDR 32.5	SDR 26	SDR 21	SDR 17	SDR 13.5	SDR 11	SDR 9	SDR 7.3	SDR 7
1/2	0.840	—	—	0.062	0.062	0.062	0.076	0.083	0.115	—
3/4	1.050	—	—	0.062	0.062	0.078	0.095	0.117	0.144	—
1	1.315	—	—	0.062	0.077	0.097	0.119	0.146	0.180	—
1 1/4	1.660	—	—	0.079	0.098	0.123	0.151	0.184	0.227	—
1 1/2	1.900	—	—	0.090	0.112	0.141	0.173	0.211	0.260	—
2	2.375	—	—	0.113	0.140	0.176	0.216	0.264	0.325	—
3	3.500	—	—	0.167	0.206	0.259	0.318	0.389	0.479	0.500
4	4.500	—	—	0.214	0.264	0.333	0.409	0.500	0.616	0.643
5	5.563	—	—	0.265	0.328	0.413	0.506	0.618	0.762	0.795
6	6.625	0.204	0.255	0.316	0.390	0.491	0.603	0.736	0.908	0.946
8	8.625	0.265	0.332	0.410	0.508	0.639	0.785	0.985	1.182	1.232
10	10.750	0.331	0.413	0.511	0.633	0.797	0.978	1.194	1.473	1.536
12	12.750	0.392	0.490	0.608	0.745	0.945	1.160	1.417	1.747	1.821

See API Spec 15LE for dimensions of pipe sizes 14-36 in.

Source: API Spec 15LE, Second Edition (1987).

POLYETHYLENE LINE PIPE (API)

MECHANICAL PROPERTIES

Property	Method of Test ASTM	Minimum Strength	
		psi	MPa
Short-Term Hydrostatic Hoop Strength	D1599*	2520	17.4
Long-Term Hydrostatic Hoop Strength (PE2406, 3406)	D1598	1320	9.1
	(PE3408)	1600	11.0
Elevated Temperature, Hydrostatic Hoop Strength	D1598	(see table below)	

*Ring Tensile ASTM D2290 may be used as alternate on sizes above 4″ OD.

176°F (80°C) SUSTAINED PRESSURE

REQUIREMENTS FOR WATER PIPE

Pipe Test Category	Base Resin Melt Index, D1238 (g/10 min)	Base Resin Density, D1505 (g/cc)	Minimum Average Hours to Failure		
			S = 725 psi (5 MPa)	S = 580 psi (4 MPa)	S = 435 psi (3 MPa)
C1	<0.05	0.941-0.948	100	200	—
C2	<0.05	0.935-0.940	100	200	—
C3	0.05-0.25	0.941-0.948	60	150	—
C4	0.05-0.25	0.935-0.940	60	150	—
C5	>0.25	0.941-0.948	45	100	—
C6	>0.25	0.935-0.940	45	100	—
C7	>0.50	0.926-0.940	—	80	150

Source: API Spec 15LE, Second Edition (1987).

DIMENSIONS OF POLY (VINYL CHLORIDE)
AND CHLORINATED POLY (VINYL CHLORIDE) LINE PIPE (API)

(All Dimensions are in Inches)
(inches × 25.40 = mm)

Nominal pipe size	Outside diameter	Minimum Wall Thickness							
		Schedule 40	Schedule 80	SDR 32.5	SDR 26	SDR 21	SDR 17	SDR 13.5	SDR 11
1/2	0.840	0.109	0.147	—	—	0.062	0.062	0.062	0.076
3/4	1.050	0.113	0.154	—	—	0.090	0.090	0.090	0.095
1	1.315	0.133	0.179	—	—	0.090	0.090	0.097	0.119
1 1/4	1.660	0.140	0.191	—	—	0.090	0.098	0.123	0.151
1 1/2	1.900	0.145	0.200	—	—	0.090	0.112	0.141	0.173
2	2.375	0.154	0.218	—	—	0.113	0.140	0.176	0.216
2 1/2	2.875	0.203	0.276	—	—	0.137	0.169	0.213	0.261
3	3.500	0.216	0.300	—	—	0.167	0.206	0.259	0.318
3 1/2	4.000	0.226	0.318	—	—	0.190	0.236	0.296	0.363
4	4.500	0.237	0.337	—	—	0.214	0.264	0.333	0.409
5	5.563	0.258	0.375	—	—	0.265	0.328	0.413	0.506
6	6.625	0.280	0.432	0.204	0.255	0.316	0.390	0.491	0.603
8	8.625	0.322	0.500	0.265	0.332	0.410	0.508	0.639	0.785
10	10.750	0.365	0.593	0.331	0.413	0.511	0.633	0.797	0.978
12	12.750	0.406	0.687	0.392	0.490	0.608	0.745	0.945	1.160

Source: API Specification 15LP, Sixth Edition (1987)

PVC AND CPVC LINE PIPE (API)

MECHANICAL PROPERTIES

Property	Method of Test	PVC and CPVC psi	MPa
Short-Term Hydrostatic Hoop Strength, min.	ASTM D 1599	6,400	44
*Long-Term Hydrostatic Hoop Strength, min.	ASTM D 1598	4,200	29
Ring Tensile Strength, min.	ASTM D 2513	6,400	44

*Test specimens for long-term hydrostatic hoop strength shall include representative fittings in the center of each specimen. Minimum strength listed is for the 1,000-hour test.

MINIMUM IMPACT STRENGTH,
ft-lbs (Joules) at 32-35°F (0-2°C)*

Nominal Pipe Size	Schedules 40 and 80	SDR 17	SDR 21	SDR 26
1/2	16 (21.7)			
3/4	20 (27.1)			
1	20 (27.1)			
1 1/4	20 (27.1)	20 (27.1)		
1 1/2	30 (40.7)	30 (40.7)	30 (40.7)	
2	40 (54.2)	40 (54.2)	40 (54.7)	40 (54.7)
2 1/2	40 (54.2)	40 (54.2)	40 (54.7)	40 (54.7)
3	40 (54.2)	40 (54.2)	40 (54.7)	40 (54.7)
4	40 (54.2)	40 (54.2)	40 (54.7)	40 (54.7)
6	55 (74.5)	55 (74.5)	55 (74.5)	55 (74.5)
8	60 (81.3)	60 (81.3)	60 (81.3)	60 (81.3)
Fittings all Sizes and Types	5 (6.8)			

*Test Method ASTM D 2444.

Source: API Spec 15LP, Sixth Edition (1987).

GLASS FIBER REINFORCED THERMOSETTING RESIN LINE PIPE - DIMENSIONS

API Specification

Nominal Size (inches)	Outside[1] Dia. (inches)	Inside[2] Dia. (inches) Minimum
2	2.375	2.00
2 1/2	2.875	2.40
3	3.500	3.00
4	4.500	4.00
6	6.625	5.80
8	8.625	8.20
10	10.750	10.30
12	12.750	11.90
14	[3]	13.50
16	[3]	15.40

Notes:

[1] The outer diameters are applicable to

A. Sizes 2″ through 5″ with 300 psi cyclic pressure ratings.

B. Sizes 8″ through 12″ with 150 psi cyclic pressure ratings.

C. All centrifugal cast pipe.

D. Other outside diameters shall be permitted by agreement between purchaser and manufacturer.

[2] The minimum inside diameters are applicable to all filament wound pipe with cyclic pressure ratings greater than pipe covered by Note (1).

[3] The minimum inside diameters are applicable to all pressure ratings of 14″ and 16″ diameter pipe.

Source: API Specification 15LR, 5th ed. (1986).

Typical Values

Nominal Pipe Size		Pipe OD	Pipe ID	Nominal Wall Thickness*	Pipe Weight
(in)	(mm)	(in)	(in)	(in)	(lb/ft)
2	50	2.37	2.09	0.157	0.8
3	80	3.50	3.22	0.157	1.2
4	100	4.50	4.14	0.203	2.0
6	150	6.62	6.26	0.203	3.0
8	200	8.62	8.22	0.226	4.3
10	250	10.75	10.35	0.226	5.4
12	300	12.75	12.35	0.226	6.4

*Determined in accordance with ASTM D2996.

Source: Ameron (1985).

GLASS FIBER REINFORCED THERMOSETTING RESIN LINE PIPE

Typical Physical Properties

Pipe Property	Units	Value	Method ASTM
Thermal conductivity	Btu·in/(h·ft²·°F)	1.7	C177
Thermal expansion (linear)	10^{-6} in/in/°F	8.5	D696
Flow coefficient	Hazen-Williams	150	—
Absolute roughness	10^{-6} ft	50	—
Specific gravity	—	1.81	D792
Bore of hardness	Impressor 934-1	65	D2583

Typical Mechanical Properties

Property	Units	Value*	Method ASTM
Tensile strength			
Longitudinal	10^3 psi	25.0	D2105
Circumferential	10^3 psi	50.0	D1599
Tensile modulus			
Longitudinal	10^6 psi	2.0	D2105
Circumferential	10^6 psi	3.0	—
Compressive strength			
Longitudinal	10^3 psi	25.0	—
Compressive modulus			
Longitudinal	10^6 psi	2.0	—
Long-term hydrostatic design basis			
Static	10^3 psi	31.5	D2992(B)
Poisson's ratio**			
ν_{yx}	—	0.11	—
ν_{xy}	—	0.19	—

*Based on structural wall thickness.

**The first subscript denotes the direction of contraction and the second that of the applied stress.
  x denotes longitudinal direction.
  y denotes circumferential direction.

Ameron (1985)

TYPES OF PORTLAND CEMENT

Type	Non-air-entraining
I	For general concrete construction when special properties specified for the other types are not required. When no type is indicated, it is assumed to be Type I.
II	For general concrete construction exposed to moderate sulfate action or where moderate heat of hydration is required.
III	For construction when high early strength is required.
IV	For construction when a low heat of hydration is required. This type is not generally carried in stock.
V	For construction when high sulfate resistance is required. This type is not generally carried in stock.

	Air-entraining
IA IIA IIIA	Same as corresponding Types I, II, and III of non-air-entraining cement, but imparting to the concrete properties of greatly improved resistance to (1) severe weathering, and (2) the deleterious effect of applications of sodium or calcium salts to pavement surfaces for snow and ice removal.

CHEMICAL REQUIREMENTS FOR PORTLAND CEMENTS*

Cement Type	I and IA	II and IIA	III and IIIA	IV	V
Silicon dioxide (SiO_2), min, %	...	20.0	...	...	...
Aluminum oxide (Al_2O_3), max, %	...	6.0	...	...	...
Ferric oxide (Fe_2O_3), max, %	...	6.0	...	6.5	...
Magnesium oxide (MgO), max, %	6.0	6.0	6.0	6.0	6.0
Sulfur trioxide (SO_3), max, %					
When (C_3A) is 8% or less	3.0	3.0	3.5	2.3	2.3
When (C_3A) is more than 8%	3.5		4.5		
Loss on ignition, max, %	3.0	3.0	3.0	2.5	3.0
Insoluble residue, max, %	0.75	0.75	0.75	0.75	0.75
Tricalcium silicate (C_3S) max, %	...	...	...	35	...
Dicalcium silicate (C_2S) min, %	...	...	...	40	...
Tricalcium aluminate (C_3A) max, %	...	8	15	7	5
Tetracalcium aluminoferrite plus twice the tricalcium aluminate ($C_4AF + 2(C_3A)$), or solid solution ($C_4AF + C_2F$), as applicable, max, %	...	...	...	...	25

OPTIONAL CHEMICAL REQUIREMENTS

Cement Type	I and IA	II and IIA	III and IIIA	IV	V	Remarks
Tricalcium aluminate (C_3A), max, %	...	...	8	...	...	for moderate sulfate resistance
Tricalcium aluminate (C_3A), max, %	...	...	5	...	...	for high sulfate resistance
Sum of tricalcium silicate and tricalcium aluminate, max, %	...	58	...	...	...	for moderate heat of hydration
Alkalies ($Na_2O + 0.658K_2O$), max, %	0.60	0.60	0.60	0.60	0.60	low-alkali cement

The expressing of chemical limitations by means of calculated assumed compounds does not necessarily mean that the oxides are actually or entirely present as such compounds. When expressing compounds, C = CaO, S = SiO_2, A = Al_2O_3, F = Fe_2O_3. For example, C_3A = $3CaO \cdot Al_2O_3$.

*Summarized from ASTM C150. For additional details see the current edition of that standard.

HYDRAULIC CEMENTS

Some typical properties of mortars (1 cement 3 sand - water/cement: 0.32)	Portland cement	High-alumina cement
Specific gravity	2.2	2.2
Tensile strength, MPa*	3.5-4.1	4.0-4.4
Compressive strength, MPa*	38-50	68-70
Modulus of elasticity, MPa	14,100-15,200	
Linear coeff. of thermal expansion/°C	about 11×10^{-6}	
Maximum working temperature, °C	Above about 100°C, decrease in tensile and compressive strength. However, useful in concrete up to 300-400°C depending on composition and preparation of the concrete.	used in refractory concrete

*After hardening for about one month in water.

Chemical resistance	Portland cement	High-alumina cement
Water	+	+
Acids	minimum pH 6.5	minimum pH 5.5
Phenol	−	+
Alkalis	+	5% NaOH: −
Salt solutions:		
sodium chloride	+	+
sodium sulfate 0.07%	+	
sodium sulfate 0.7%	+	
sodium sulfate 2% or more	−	
calcium chloride 5%	−	−
magnesium chloride 1%	+	5-15%: +
Hydrocarbons		
aliphatic	+	+
aromatic	+	+
chlorinated	+	+
Alcohols	−	+
Esters	+	+
Ketones	+	+

High-alumina cement has a normal setting time, but shows extremely rapid hardening and development of high strength (24 hrs) as compared with Portland cement. Concrete to be used in seawater is made with blast furnace cement.

+ Resistant
− Not resistant

CHEMICAL RESISTANT MORTARS AND GROUTS

	Furan	Epoxy	Polyester	Vinyl Ester	Phenolic	Silicate	Sulfur
Density, lb./ft^3	95-120	110-125	100-125	110-125	95-120	130	135
Tensile strength, psi	1,000	1,800	2,300	2,300	1,200	500	600
Flexural strength, psi	2,800	3,800	4,800	4,200	2,800	900	1,300
Compressive strength, psi	8,800	12,000	10,000	10,000	9,000	5,800	8,000
Bond strength, psi	280	*		280	175	280	175	120
Thermal resistance, max F	380	200	225	180	380	2,000	200
Chemical resistance (ASTM C267)							
Acetic Acid, Glacial	R	NR	NR	NR	—	—	NR
Acetone	R	NR	NR	NR	—	—	NR
Chlorine Dioxide Solution	NR	NR	R	R	—	—	R
Chromic Acid, 10%	NR	NR	R	R	—	—	NR
Dichloracetic Acid, 10%	R	NR	R	R	—	—	NR
Ethyl Alcohol	R	R	R	R	—	—	R
Formic Acid, 20%	R	C	C	C	—	—	NR
Gasoline	R	R	R	R	—	—	NR
Hydrochloric Acid, 20%	R	R	R	R	—	—	R
Lactic Acid, 15%	R	C	R	R	—	—	R
Nitric Acid, 20%	NR	NR	R	R	—	—	R
Phosphoric Acid, 30%	R	R	R	R	—	—	R
Sodium Choride	R	R	R	R	—	—	R
Sodium Hydroxide, 25%	R	R	R	R	—	—	NR
Sodium Hypochlorite, 10%	NR	NR	R	R	—	—	NR
Sulfuric Acid, 20%	R	R	R	R	—	—	R
Trichloroethylene	R	NR	C	C	—	—	NR
Xylene	R	NR	C	C	—	—	NR

R Recommended   NR Not Recommended   C Conditional

	Furan	Epoxy	Polyester	Vinyl Ester	Phenolic	Silicate	Sulfur
Oxidizing Acids	P	P	E	E	P	E	E
Nonoxidizing Acids	E	E	E	E	E	E	E
Alkali	E	E	G	G	P	P	P
Alkaline Hypochlorite	P	P	E	E	P	P	P
Solvents	E	G	G	G	E	E	P

E Excellent   G Good   P Poor

*Brick failed

Source: Adapted from A.A. Boova, Process Industries Corrosion, The Theory and the Practice, p. 657, NACE 1986.

PROPERTIES OF GRAPHITE AND SILICON CARBIDE

	Graphite	Impervious Graphite	Impervious Silicon Carbide
Specific gravity	1.4-1.8	1.75	3.10
Tensile strength, psi (MPa)	400-1400 (3-10)	2600 (18)	20,650 (143)
Compressive strength psi (MPa	2000-6000 (14-42)	10,500 (72)	150,000 (1000)
Flexural strength, psi (MPa)	750-3000 (5-21)	4700 (32)	—
Modulus of elasticity ($\times 10^6$), psi (MPa)	0.5-1.8 (0.3-12$\times 10^4$)	2.3 (1.6$\times 10^4$)	56 (39$\times 10^4$)
Thermal expansion, in/in/°F$\times 10^{-6}$ (mm/mm/°C)	0.7-2.1 (1.3-3.8)	2.5 (4.5)	1.80 (3.4)
Thermal conductivity, Btu/hr/ft^2/°F/ft (Watts/m, K)	15-97 (85-350)	85 (480)	60 (340)
Max. working temp (inert atm) °F (°C)	5000 (2800)	350 (180)	4200 (2300)
Max. working temp (oxidizing atm) °F (°C)	660 (350)	350 (180)	3000 (1650)

Source: Carborundum Co.

PROPERTIES OF GLASS AND SILICA

	Pyroceram	96% silica	Borosilicate	Glass lining
Specific gravity, 77°F	2.60	2.18	2.23	2.56
Water absorption, %	0.00	0.00	0.00	
Gas permeability	Gastight	Gastight	Gastight	
Softening temp., °F (°C)	2282 (1250)	2732 (1500)	1508 (820)	
Specific heat, 77°F, B.t.u./lb.) (°F) (Joules/kg/K)	0.185 (775)	0.178 (746)	0.186 (779)	
Mean specific heat (77°-752°F)	0.230	0.224	0.233	
Thermal conductivity, mean temp. 77°F, B.t.u./(sq. ft.)(hr.)(°F)/(in.) (Watts/m, K)	25.2 (3.6)	—	7.5 (1.1)	
Linear thermal expansion, per °F, (77°-572°F)(per °C) $\times 10^{-6}$	3.2 (5.8)	0.44 (0.79)	1.8 (3.2)	
Modulus of elasticity, ksi (MPa) $\times 10^3$	17.3 (119)	9.6 (66)	9.5 (66)	6-9 (40-60)
Poisson's ratio	0.245	0.17	0.20	
Modulus of rupture, ksi (MPa)	20 (140)	5-9 (35-63)	6-10 (42-70)	
Knoop hardness, 100 g	698	532	481	480
Knoop hardness, 500 g	619	477	442	
Adhesion strength ksi (MPa)	—	—	—	5-10 (35-70)
Max. operating temp., °F (°C)	—	—	—	500 (260)
Thermal shock resistance, temp. diff., °F (°C)	—	—	—	305 (152)

PROPERTIES OF HIGH TEMPERATURE REFRACTORIES

Item	Magnesia[1] F	Magnesia[1] C	Mullite F	Mullite C	Silicon Stabilized Carbide F	Silicon Stabilized Carbide C	Bonded Zirconia F	Bonded Zirconia C	99%Al$_2$O$_3$ F	99%Al$_2$O$_3$ C
Fusion Point	4800	2650	3300	1815	—	—	4700	2600	3650	2010
Use Limit, oxid.	4170	2300	3000	1650	3000	1650	4400	2430	3300	1815
Modulus of rupture, psi MPa	2500		1500 10		2000[3] 14		1900 13		2000 14	
Moh's hardness[2]	6		6.5		9.6		7		9	
Thermal shock resist.	Poor		Good		Good		Fair		Fair	
Relative Cost	2.8		1		2.1		10		3.1	

[1]Basic refractories have poor resistance to hot acids.
[2]Scale 1 to 10. Talc = 1, low carbon steel = 4, diamond = 10.
[3]At 2500°F (1371°C).

Source: NACE. *Basic Corrosion Course.*

TYPICAL PROPERTIES OF CERAMIC BRICKS AND CHEMICAL STONEWARE

Composition	Regular Acid Brick	High Temp. Acid Brick	Chemical Stoneware	Silica Brick
SiO_2	68	66	71	98.-99.6
Al_2O_3	26	28	23	0.2-0.5
Fe_2O_3	1.3	1.3	0.6	0.02-0.3
TiO_2	7.5	1.5	0.9	—
CaO	0.2	0.2	0.4	0.02-0.03
MgO	0.5	0.5	1.1	0.02-0.1
$Na_2O + K_2O + Li_2O$	2.8	2.5	2.3	0.01-0.2
Density, g/cm^3	2.2	2.2	2.3	1.8-2.0
Apparent porosity, %	7-10	6-9	—	7-16
Water absorption, %	4.5-5.0	2.5-3.5	0.5-2.5	3-14
Acid solubility, %w	8.5-10.5	8.-10.5		1-4
Mod. of rupture, ksi	2.5-2.8	3.0-3.2	6-12	0.5-2
Compressive strength, ksi	6-10	7-12	70-80	2-12
MPa	40-70	50-80	500-550	14-80
Mod. of elasticity, ksi ($\times 10^3$)	—	—	4-10	1-5
MPa ($\times 10^3$)	—	—	30-70	7-35
Coefficient of thermal expansion				
in/in/°F ($\times 10^{-6}$)	—	—	2	0.2-3
mm/mm/°C ($\times 10^{-6}$)	—	—	4	0.4-5

Source: Adapted from: "Corrosion and Chemical Resistant Masonry Materials Handbook" (1985, Noyes Publication) and "Process Industries Corrosion", p. 644 (1986, NACE).

SURFACE PREPARATION STANDARDS

Three standards used in industry to describe surface preparation are the National Association of Corrosion Engineers Standards (NACE), Steel Structures Painting Council (SSPC) "Surface Preparation Specifications" and Swedish Pictorial Standards (SA)

		Comparable Standards
NACE No. 1 "For Tank Linings"	White Metal Blast—This is defined as removing all rust, scale, paint, etc. to clean white metal which has a uniform gray-white appearance. Streaks and stains of rust or other contaminants are not allowed.	SSPC 5 SA-3*
NACE No. 2 "For Some Tank Linings and Heavy Maintenance"	Near-White Metal Blast—This provides a surface about 95% as clean as white metal. Light shadows and streaks are allowed.	SSPC 10 SA-2 1/2*
NACE No. 3 "For Maintenance"	Commercial Blast—This type of blast is more difficult to describe. It essentially amounts to about 2/3 of a SA-2* white metal blast which allows for very slight residues of rust and paint in the form of staining.	SSPC 6
NACE No. 4 "For Very Light Maintenance"	Brush Off Blast—This preparation calls for removal of loose paint, scale, rust, etc. Tightly adherent paint rust, scale is permitted to remain.	SSPC 7 SA-1*

*Swedish standards are not exactly the same as the NACE and SSPC standards. It is advisable to check the wording of the Swedish standards prior to use.

ABRASIVE/PROFILE COMPARATIVE CHART

The following chart should be used only for approximating abrasive size required to obtain a specified anchor pattern. This information can be used for centrifugal wheel as well as pressure blasting. Pressure blasting should be done using 90-100 psi nozzle pressure. The depth of anchor pattern used in this chart is an average and not a minimum or maximum depth obtainable.

1 Mil Profile
30/60 Mesh Silica Sand
G-80 Steel Grit
S-110 Steel Shot*
80 Mesh Gamet
100 Aluminum Oxide
Clemtex #4

2 Mil Profile
16/35 Mesh Silica Sand
G-40 Steel Grt
S-230 Steel Shot*
36 Mesh Gamet
36 Grit Aluminum Oxide
Clemtex #3
Black Beauty BB-50 or BB-2040

3-4 Mils Profile
8/20 Mesh Silica Sand
G-25 Steel Grit
S-330 or 390 Steel Shot*
16 Mesh Gamet
16 Grit Aluminum Oxide
Clemtex #2
Black Beauty BB-40 or BB-25

1.5 Mil Profile
16/35 Mesh Silica Sand
G-50 Steel Grit
S-170 SteelShot*
36 Mesh Gamet
50 Grit Aluminum Oxide
Clemtex#3

2.5 Mils Profile
8/35 Mesh Silica Sand
G-40 Steel Grit
S-280 Steel Shot*
16 Mesh Gamet
24 Grit Aluminum Oxide
Clemtex #2
Black Beauty BB-400

*The steel shot alone will not give a good angular anchor pattern and should be used in combination with steel grit for best results.

SUMMARY OF SSPC SURFACE PREPARATION SPECIFICIATIONS

SSPC Specification	Description
SP 1, Solvent Cleaning	Removal of oil, grease, dirt, soil, salts, and contaminants by cleaning with solvent, vapor, alkali, emulsion, or steam.
SP 2, Hand Tool Cleaning	Removal of loose rust, loose mill scale, and loose paint to degree specified, by hand chipping, scraping, sanding, and wire brushing.
SP 3, Power Tool Cleaning	Removal of loose rust, loose mill scale, and loose paint to degree specified, by power tool chipping, descaling, sanding, wire brushing, and grinding.
SP 5, White Metal Blast Cleaning	Removal of all visible rust, mill scale, paint, and foreign matter by blast cleaning by wheel or nozzle (dry or wet) using sand, grit, or shot. (For very corrosive atmospheres where high cost of cleaning is warranted.)
SP 6, Commercial Blast Cleaning	Blast cleaning until at least two-thirds of the surface area is free of all visible residues. (For rather severe conditions of exposure.)
SP 7, Brush-Off Blast Cleaning	Blast cleaning of all except tightly adhering residues of mill scale, rust, and coatings, exposing numerous evenly distributed flecks of underlying metal.
SP 8, Pickling	Complete removal or rust and mill scale by acid pickling, duplex pickling, or electrolytic pickling.
SP 10, Near-White Blast Cleaning	Blast cleaning nearly to White Metal Cleanliness, until at least 95% of the surface area is free of all visible residues. (For high humidity, chemical atmosphere, marine, or other corrosive environments.)

Source: Steel Structures Painting Manual, Vol. 2, p. 12, SSPC 1982.

COMPARATIVE MAXIMUM HEIGHTS OF PROFILE OBTAINED WITH VARIOUS ABRASIVES

Blast Cleaning of Mild Steel Plates Using 80 psig Air and 5/16 in. Diameter Nozzle

Abrasive	Size[1]	Max. Height of Profile in Mils
Large River Sand	Through U.S. 12, on U.S. 50	2.8
Medium Ottawa Silica Sand	Through U.S. 18, on U.S. 40	2.5
Fine Ottawa Silica Sand	Through U.S. 30, on U.S. 80	2.0
Very Fine Ottawa Silica Sand	Through U.S. 50.80% through U.S. 100	1.5
Black Beauty (crushed slag)	Estimated at minus 80 mesh	1.3
Crushed Iron Grit	G-50	3.3
Crushed Iron Grit	G-40	3.6
Crushed Iron Grit	G-25	4.0
Crushed Iron Grit	G-16	8.0
Crushed Iron Grit	G-16	8.0
Chilled Iron Shot	S-230	3.0
Chilled Iron Shot	S-330	3.3
Chilled Iron Shot	S-390	3.6

[1]Sizes listed are U.S. Seive Series Screen sizes or SAE grit or shot sizes.

Source: Steel Structures Painting Manual, Vol. 1, p. 20, SSPC 1966.

PROPERTIES OF ABRASIVES

Abrasive	Type	Major Component	Shape	Hardness	Specific Gravity	Reusable	Recommended Use
Steel Shot	Metallic	Iron	Round	Hard*	7.2	Yes	Peening
Steel Grit	Metallic	Iron	Angular	Hard*	7.6	Yes	Metal Etching
Iron Grit	Metallic	Iron	Angular	Hard	7.4	Yes	Metal Etching
Alum. Oxide	Oxide	Alumina	Angular	Hard	3.8	Yes	Metal Etching
Silicon Carbide	Oxide	Silicon Carbide	Angular	Hard	3.8	Yes	Metal Etching
Garnet	Oxide	Iron Silica	Irregular	Hard	4.0	Yes	Metal Etching
Mineral Slag	Conglomerate	Iron Alum-Silica	Irregular	Hard	2.8	Yes	Metal Etching
Flint	Silica	Silica	Sharp	Medium	2.7	Yes	Metal Etching
Sand	Silica	Silica	Irregular	Medium	2.7	No	Metal Etching
Limestone	Oxide	$CaCO_3$	Irregular	Soft	2.4	No	Light Cleaning No Etch
Walnut Shell	Vegetable	Cellulosic	Irregular	Soft	1.3	Yes	Light Cleaning No Etch
Corn Cob Grit	Vegetable	Cellulosic	Irregular	Soft	1.2	Yes	Light Cleaning No Etch
Glass Beads	Oxide	Silica	Round	Medium	2.7	Yes	Light Cleaning No Etch

*Various hardnesses available

Source: Steel Structures Painting Manual, Vol. 1, p.58, SSPC 1982

PROTECTIVE COATING CLASSIFICATIONS

Basic Coating Formation	Generic Coating Material
Natural Air-Oxidizing Coatings	Drying Oils
	Tung Oil Phenolic Varnish
Synthetic Air-Oxidizing Coatings	Alkyds
	Vinyl Alkyds
	Epoxy Esters
	Silicone Alkyds
	Uralkyds
Solvent Dry Lacquers	Nitrocellulose
	Polyvinylchloride-acetate
	Copolymers
	Acrylic Polymers
	Chlorinated Rubber
	Coal Tar Curback
	Asphalt Cutback
Coreactive Coatings	Epoxy
	Coal Tar Epoxy
	Polyurethane
	Polyesters
	Silicone
Emulsion-Type (Coalescent) Coatings	Vinyl Acetate
	Vinyl Acrylic
	Acrylic
	Epoxy
Heat-Condensing Coatings	Pure Phenolic
	Epoxy Phenolic
100% Solid Coatings	Coal Tar Enamel
	Asphalt
	Polyesters
	Epoxy Powder Coatings
	Vinyl Powder Coatings
	Plastisols
	Furan Materials
	Polyurethane

Source: Charles G. Munger, Corrosion Prevention by Protective Coatings, p.90, NACE 1984

ALKYD COATINGS - PROPERTIES

Property	Medium Oil Alkyd	Vinyl Alkyd	Silicone Alkyd	Uralkyd	Epoxy Ester
Physical Properties	Flexible	Tough	Tough	Hard Abrasion Resistant	Hard
Water Resistance	Fair	Good	Good	Fair	Good
Acid Resistance	Fair	Best of group	Fair	Fair	Fair-Good
Alkali Resistance	Poor	Poor	Poor	Poor	Fair
Salt Resistance	Fair	Good	Good	Fair	Good
Solvent Resistance	Poor-Fair	Fair	Fair	Fair-Good	Fair-Good
Weather Resistance	Good	Very Good	Very Good Excellent gloss retention	Fair	Poor
Temperature Resistance	Good	Fair-Good	Excellent	Fair-Good	Good
Age Resistance	Good	Very Good	Very Good	Good	Good
Best Characteristic	Application	Weather Resistance	Weather & Heat Resistance	Abrasion Resistance	Alkali Resistance
Poorest Characteristic	Chemical Resistance	Alkali Resistance	Alkali Resistance	Chemical Resistance	Weathering
Recoatability	Excellent	Difficult	Fair	Difficult	Fair
Primary Coating Use	Weather-Resistant Coating	Corrosion-Resistant Coating	Corrosion-Resistant Coating	Abrasion-Resistant Coating	Machinery Enamel

Source: Charles G. Munger, Corrosion Prevention by Protective Coatings, p.96, NACE 1984

SOLVENT DRY LACQUERS - PROPERTIES

Properties	Vinyl Chloride Acetate Copolymer	Vinyl Acrylic Copolymer	Chlorinated Rubber Resin Modified	Alkyd Modified	Acrylic Lacquers	Coal Tar	Asphalt
Physical Property	Tough Strong	Tough	Hard	Tough	Hard-Flexible	Soft Adherent	Soft Adherent
Water Resistance	Excellent	Good	Very Good	Good	Good	Very Good	Good
Acid Resistance	Excellent	Very Good	Very Good	Fair	Good	Very Good	Very Good
Alkali Resistance	Excellent	Fair-Good	Very Good	Poor-Fair	Fair	Good	Good
Salt Resistance	Excellent	Very good	Very Good	Good	Good	Very Good	Very Good
Solvent (Hydrocarbon) Aromatic Aliphatic Oxygenated	Poor Good Poor	Poor Good Poor	Poor Okay Poor	Poor Okay Poor	Poor Okay Poor	Poor Fair Poor	Poor Poor Poor
Temperature Resistance	Fair 65°C (150°F)	Fair 65°C (150°F)	Fair	Fair 60°C (140°F)	Fair	Depends on softening point	Depends on softening point
Weather Resistance	Very Good	Excellent	Good	Very Good	Excellent	Poor	Good
Age Resistance	Excellent	Excellent	Very Good	Good	Very Good	Good	Good
Best Characteristic	Broad Chemical Resistance	Weather Resistance	Water Resistance	Drying Speed	Clear Color Retention Gloss Retention	Easy Application	Easy Application
Poorest Characteristic	Critical Application	Critical Application	Spray Application	Chemical Resistance	Solvent Resistance	Black Color	Black Color
Recoatability	Easy	Easy	Easy	Easy	Easy	Easy	Easy
Primary Coating Use	Chemical-Resistant Coatings	Exterior Chemical-Resistant	Maintenance Coatings	Weather-Resistant Coatings	Weather-Resistant Coatings	Water-Resistant Coatings	Chemical-Resistant Coatings

Source: Charles G. Munger, Corrosion Prevention by Protective Coatings, p. 104, NACE 1984

EPOXY COATINGS - PROPERTIES

Properties	Aliphatic Amine Cure	Polyamide Cure	Aromatic Amine Cure	Phenolic Epoxy	Silicone Epoxy	Coal Tar Epoxy Amine Cure	Coal Tar Epoxy Polyamide Cure	Water Based Epoxy
Physical Property	Hard	Tough	Hard	Hard	Medium-Hard	Hard (brittle)	Tough	Tough
Water Resistance	Good	Very Good	Very Good	Excellent	Good-Excellent	Excellent	Excellent	Fair-Good
Acid Resistance	Good	Fair	Very Good	Excellent	Good	Good	Good	Fair
Alkali Resistance	Good	Very Good	Very Good	Excellent	Good	Good	Very Good	Fair
Salt Resistance	Very Good	Very Good	Very Good	Excellent	Very Good	Very Good	Very Good	Fair-Good
Solvent Resistance (Hydrocarbons)								
Aromatic	Very Good	Fair	Very Good	Very Good	Good	Poor	Poor	Poor-Fair
Aliphatic	Very Good	Good	Very Good	Very Good	Very Good	Good	Good	Good
Oxygenated	Fair	Poor	Good	Very Good	Fair	Poor	Poor	Poor
Temperature Resistance	95°C	95°C	120°C	120°C	120°C	95°C	95°C	95°C
Weather Resistance	Fair, Chalks	Good, Chalks	Good	Fair	Very Good, Chalk Resistant	Fair	Fair	Good
Age Resistance	Very Good	Very Good	Very Good	Very Good	Very Good	Very Good	Very Good	Good
Best Characteristics	Strong Corrosion Resistance	Water and Alkali Resistance	Chemical Resistance	Chemical Resistance	Water and Weather Resistance	Water Resistance	Water Resistance	Ease of Application
Poorest Characteristics	Recoatability	Recoatability	Slow Cure	Very Slow Air Cure	Recoatability	Black Color Recoatability	Poor Recoatability Black Color	Proper Coalescence
Recoatability	Difficult	Difficult	Difficult	Difficult	Difficult	Difficult	Difficult	Difficult
Primary Coating Use	Chemical Resistance	Water Immersion	Chemical Coating	Chemical Lining	Weather Resistance	Water Immersion	Water Immersion	Atmospheric Corrosion

Source: Charles G. Munger, Corrosion Prevention by Protective Coatings, p. 110, NACE 1984

100% SOLIDS COATINGS - PROPERTIES

Properties	Coal Tar Enamel	Asphalt Enamel	Polyester	Epoxy Powder	100% Solids Liquid Epoxy	Vinyl Powder	Vinyl Plastisol	Furfural Alcohol Resins
Physical Properties	Hard	Somewhat Resistant	Hard	Hard	Hard-Tough	Hard-Tough	Soft-Rubbery	Hard-Brittle
Water Resistance	Excellent	Very Good	Good	Good	Good	Good	Good	Excellent
Acid Resistance	Very Good	Very Good	Excellent	Good	Good	Good	Very Good	Excellent
Alkali Resistance	Good	Fair-Good	Poor	Very Good	Very Good	Good	Good	Excellent
Salt Resistance	Very Good	Very Good	Very Good	Very Good	Very Good	Very Good	Very Good	Excellent
Solvent Resistance (Hydrocarbons)								
Aliphatic	Good	Poor	Very Good	Very Good	Very Good	Very Good	Good	Excellent
Aromatic	Poor	Poor	Fair	Very Good	Very Good	Fair	Poor	Excellent
Oxygenated	Poor	Poor	Poor	Good	Good	Poor	Poor	Good
Temperature Resistance	60 C (140 F)	50 C (122 F)	65 C (150 F)	93 C (200 F)	93 C (200 F)	60 C (140 F)	60 C (140 F)	120 C (210 F)
Weather Resistance	Poor	Good	Good	Good (Chalks)	Good (Chalks)	Very Good	Very Good	Good
Age Resistance	Excellent	Very Good	Good	Good	Good	Very Good	Very Good	Good
Best Characteristic	Water Resistance	Water and Weather Resistance	Acid and Oxidizing Chemical Resistance	General Water and Alkali Resistance	General Chemical and Alkali Resistance	General Chemical Resistance	Water and Chemical Resistance	Temperature and Acid Resistance
Poorest Characteristic	Weather Resistance Black	Solvent Resistance. Black	Alkali Resistance	Critical Application	Critical Application	Critical Application	Adhesion	Adhesion Brittleness
Recoatability	Poor	Good	Fair	Poor	Difficult	Good	Good	Good
Principal Use	Lining and Coating Pipe	Waterproofing Structures	Tank Lining	Exterior Pipe Coating	Chemical Lining	Chemical Resistant Product Finish	Pipe Lining	Cement for Acid-Proof Brick

Source: Charles G. Munger. Corrosion Prevention by Protective Coatings. p. 127. NACE 1984.

URETHANE COATINGS - PROPERTIES[1]

Properties	Type 1 Oil Modified	Type 2 Moisture Cure	Type 3 Blocked	Type 4 Prepolymer Catalyst	Type 5 Two Component	Aliphatic Isocyanate Cure (Non-Yellowing)
Physical Property	Very Tough	Very Tough Abrasion Resistant	Tough Abrasion Resistant	Tough Abrasion Resistant	Tough-Hard Rubbery	Tough-Rubbers
Water Resistance[2]	Fair	Good	Good	Fair	Good	Good
Acid Resistance[2]	Poor	Fair	Fair	Poor-Fair	Fair	Fair
Alkali Resistance[2]	Poor	Fair	Fair	Poor	Fair	Fair
Salt Resistance[2]	Fair	Fair	Fair	Fair	Fair	Fair
Solvent Resistance (Hydrocarbon)						
Aromatic	Fair	Good	Good	Poor	Good	Good
Aliphatic	Fair	Good	Good	Fair	Good	Good
Oxygenated	Poor	Fair	Fair	Fair	Good	Fair
Temperature Resistance	Good 100°C	Good 120°C	Good 120°C	Good 100°C	Good 120°C	Good 120°C
Weather Resistance	Good, Yellows	Good, Yellows	Good, Yellows	Good, Yellows	Good, some yellowing, chalk	Excellent, good color and gloss retention
Age Resistance	Good	Good	Good	Good	Good	Good
Best Characteristic	Exterior, Wood Coating	Abrasion, Impact	Abrasion, Impact	Speed of cure	Abrasion, Impact	Weather resistance, color and gloss retention
Poorest Characteristic	Oil Base Chemical Resistance	Dependent on humidity for cure	Heat required for cure	Chemical Resistance	Two package	—
Recoatability	Fair	Difficult	Difficult	Difficult	Difficult	Difficult
Primary Coating Use	Clear Wood Coating	Abrasion Resistance, Floors	Product Finish	Abrasion Resistance	Abrasion Resistance, Impact	Exterior Coatings

[1]The properties of urethanes vary over a wide range due to the many and varied basic polyols and isocyanates. The above listings are only indicative. Manaufacturers must be contacted for specific properties of specific materials. Harder coatings are more resistant than softer, more rubbery types.
[2]Resistances are for nonimmersion conditions.

Source: Charles G. Munger, Corrosion Prevention by Protective Coatings, p. 113, NACE 1984.

HEAT-CONDENSING COATINGS - PROPERTIES

Properties	Phenolic	Epoxy Phenolics
Physical Property	Very Hard	Hard-Tough
Water Resistance	Excellent 100 C	Excellent 100 C
Acid Resistance	Excellent	Good
Alkali Resistance	Poor	Excellent
Salt Resistance	Excellent	Excellent
Solvent Resistance Hydrocarbon		
Aliphatic	Excellent	Excellent
Aromatic	Excellent	Excellent
Oxygenated	Very Good	Good
Temperature Resistance	120 C (250 F)	120 C (250 F)
Weather Resistance	Good (darkens)	Good
Age Resistance	Excellent	Excellent
Best Characteristics	Acid and Temperature Resistance	Alkali and Temperature Resistance
Worst Characteristics	Brittle, poor recoatability	Poor Recoatability
Recoatability	Poor	Poor
Principal Use	Chemical and Food Lining	Chemical Lining

Source: Charles G. Munger, Corrosion Prevention by Protective Coatings, p. 122, NACE 1984.

COALESCENT-EMULSION COATINGS - PROPERTIES

Properties	Vinyl Acetate	Vinyl Acrylic	Acrylic	Epoxy	Alkyd
Physical Property	Scour Resistant	Scour Resistant	Scour Resistant	Tough	Pliable
Water Resistance	Fair	Good	Good	Good	Good
Acid Resistance	NR[1]	NR	NR	Fair	NR
Alkali Resistance	NR	NR	NR	Good	NR
Salt Resistance	Fair	Fair	Fair	Good	Fair
Solvent Resistance Aliphatic hydrocarbon Aromatic hydrocarbon Oxygenated hydrocarbon	Fair NR NR	Good NR NR	Fair NR NR	Good Good NR	Good NR NR
Temperature Resistance	60 C (140 F)	60 C (140 F)	60 C (140 F)	70 C (158 F)	60 C (140 F)
Weather Resistance	Good	Very Good	Very Good	Fair	Good
Age Resistance	Good	Good	Good	Good	Good
Best Characteristic	Weather Resistant	Weather Resistant	Weather Resistant	Reasonable Corrosion Resistance	Weather Resistant, Easy Application
Poorest Characteristic	Porosity	Porosity	Porosity	More porous than solvent base	Poor Alkali Resistance
Recoatability	Good	Good	Good	Fair-Good	Good
Principal Use	Decorative Topcoat	Decorative Topcoat	Decorative Topcoat	Topcoat	Exterior Wood

[1]NR = Not Recommended.

These coatings are primarily for decorative purposes over primed steel or stucco. Alkyd formulations good for exterior wood.

Source: Charles G. Munger, Corrosion Prevention by Protective Coatings, p. 118, NACE 1984.

ZINC COATINGS - SUMMARY OF PROPERTIES

| | Organic Zinc Rich | | Inorganic Zinc | |
	Chlorinated Rubber Base	Epoxy Base	Water Base	Solvent Base
Physical Property	Tough	Medium hard	Hard-metallic	Medium hard
Water Resistance	Very good	Very good	Excellent	Very good
Acid Resistance	Poor	Poor	Poor	Poor
Alkali Resistance	Fair	Fair	Fair	Fair
Salt Resistance	Good	Good	Very good	Very good
Solvent Resistance				
Aliphatic Hydrocarbon	Very good	Very good	Excellent	Very good
Aromatic Hydrocarbon	Poor	Good	Excellent	Good
Oxygenated Hydrocarbon	Poor	Fair	Excellent	Good
Temperature Resistance	80 C (180°F) Dry	90 C (200°F) Dry	370 C (700°F) Dry	370 C (700°F) Dry
Weather Resistance	Very Good	Good	Excellent	Excellent
Age Resistance	Good	Good	Excellent	Excellent
Best Characteristic	Fast dry	Good adhesion	Hard abrasion resistant	Easier Application
Poorest Characteristic	Solvent resistance	Application Weather resistance	More difficult application	Poor cure, dry condition
Reactability	Good	Fair	Good	Good
Principal Use	Touch up-galvanize inorganic zinc	Touch up topcoated zinc systems. Corrosion resistant base coat	Corrosion resistance-abrasion resistance	Corrosion resistant base coat

Source: C. G. Munger.

ZINC COATINGS - PROPERTIES

	Inorganic zinc, postcured	Inorganic zinc, self-cured, water-based	Inorganic zinc, self-cured, water-based, ammonium	Inorganic zinc, self-cured, solvent-based	Organic zinc, one-package	Organic zinc, two-package	Modified inorganic zinc primer
Cure type	Inorganic	Inorganic	Inorganic	Hydrolyzable organic silicate	Lacquer	Coreacting	Coreacting
Effect of sunlight	Unaffected	Unaffected	Unaffected	Unaffected	Surface chalking	Surface chalking	Very slow; chalk
Wet or humid environments	Outstanding	Outstanding	Outstanding	Outstanding	Very good	Very good	Very good
Industrial-atmosphere contaminants:							
Acids	Requires topcoat	Requires topcoat	Requires topcoat	Requires topcoat	Requires topcoat	Requires topcoat	Requires topcoat
Alkalies	Requires topcoat	Requires topcoat	Requires topcoat	Requires topcoat	Requires topcoat	Requires topcoat	Requires topcoat
Oxidizing	Requires topcoat	Requires topcoat	Requires topcoat	Requires topcoat	Requires topcoat	Requires topcoat	Requires topcoat
Solvents	Outstanding	Outstanding	Outstanding	Outstanding	Limited	Very good	Very good
Spillage and splash of industrial compounds:							
Acids	Not recommended	Not recommended	Not recommended	Not recommended	Not recommended	Not recommended	Not recommended
Alkalies	Not recommended	Not recommended	Not recommended	Not recommended	Not recommended	Not recommended	Not recommended
Oxidizing	Not recommended	Not recommended	Not recommended	Not recommended	Not recommended	Not recommended	Not recommended
Solvents	Outstanding	Outstanding	Outstanding	Outstanding	Limited	Very good	Good

ZINC COATINGS - PROPERTIES (Cont)

	Inorganic zinc, postcured	Inorganic zinc, self-cured, water-based	Inorganic zinc, self-cured, water-based, ammonium	Inorganic zinc, self-cured, solvent-based	Organic zinc, one-package	Organic zinc, two-package	Modified inorganic zinc primer
Water immersion	With suitable topcoat system on ship hulls and other marine structures				Not recommended	With epoxy top-coat on marine structures	See manual-user's literature
Tank linings	Marine cargo and ballast tanks, fuel storage, including floating root tanks				Not used	With suitable top-coat in marine cargo and ballast	New manual-user's literature
Physical properties:							
Abrasion resistance	Outstanding	Outstanding	Outstanding	Outstanding	Good	Good	Excellent
Heat stability	Outstanding	Outstanding	Outstanding	Outstanding	Good	Good	Good
Hardness	Outstanding	Outstanding	Outstanding	Outstanding	Good	Good	Excellent
Gloss	None	None	None	None	Flat	Flat	None
Colors	Gray or tints of gray	Gray or tints of gray	Gray or tints of gray	Gray or tints of gray	Gray or tints of gray	Gray or tints of gray	Gray or tints of gray
Additional notes	As primers for organic systems, provide greatly extended service life. Special application technique or use of tie coat may be required to avoid solvent bubbling in organic topcoat. Alkyds always require tie coat				Can be used to touch up inorganic primers compatible with topcoats	Choice of top-coat is critical	In some applications, satisfactory over mechanically cleaned surfaces. Excellent for touch-up of in-organic zincs

Source: G. E. Weismantel, Paint Handbook, p. 7-30, McGraw-Hill 1981.

COMPATABILITY OF COATING MATERIALS WITH VARIOUS PRIMERS

Primers	Alkyd	Alkyd-phe-nolic	Vinyl-alkyd	Vinyl	Vinyl-acrylic	Epoxy cata-lyzed	Epoxy ester	Coal-tar epoxy	Chlo-rinated-rubber	Phe-nolic oleo-resinous	Poly-ure-thane	Poly-ester flake or glass
Alkyd	R	R	NR	NR	NR	NR	NR	NR	NR	R	NR	NR
Bituminous (aluminum)[a]	NR	NR	NR	NR	NR	NR	NR	NR	NR	NR	NR	NR
Vinyl-alkyd[b]	R	R	R	R*	R*	NR	R	NR	NR	NR	NR	NR
Vinyl[b]	R	R	R	R	R	NR	NR	NR	R	NR	NR	NR
Epoxy ester	R	R	R	NR	NR	NR	R	NR	R	R	NR	NR
Epoxy catalyzed[c]	NR	NR	R*	R*[d]	R*	R*	R*	R*	X	NR	R*	R*
Epoxy noncatalyzed[c]	R	R	R	X	X	R	R	X	X	R	NR	R*
Epoxy-organic zinc[c]	NR	NR	NR	R*	NR	R	R	R	R	NR	NR	NR
Phenolic oleoresinous	R	R	NR	NR	NR	NR	R	NR	NR	R	NR	NR
Vinyl-phenolic	R	R	R	R	R	NR	R	NR	R	R	X	NR
Coal-tar epoxy	NR	NR	NR	NR[d]	NR	R*	R*	R*	NR	NR	NR	NR
Inorganic zinc, postcure[e]	NR	NR	NR	R*	NR	R*	R*	R*	R*	NR	NR	NR
Inorganic zinc, self-cure[c]; water-base[e]	NR	NR	NR	R*	NR	R*	R*	R*	R*	NR	NR	NR
Inorganic zinc, self-cure[c]; solvent-base[e]	NR	NR	NR	R*	NR	R*	R*	R*	R*	NR	NR	NR
Chlorinated-rubber	R	R	R	NR	R	NR	R	NR	R	R	NR	NR

Note: R = known compatibility; normal practice. R* = known compatibility with special surface preparation and/or application. NR = not recommended. This is defined as meaning that it is not common practice to apply this topcoat over the specified primer, although certain products, if properly formulated, may be compatible. Attention is called also to the fact that certain combinations marked NR may be used, provided a suitable tie coat is applied between the two. X = not recommended because of insufficient data.

[a] Topcoated with itself or with an antifouling coating.　[b] Vinyl wash primer required.　[c] May be used as an after-blast primer.

[c] Vinyl antifouling coating such as MIL-P-15931 may be applied.　[d] May be used without topcoat.

Source: G. E. Weismantel, Paint Handbook, p. 14-29, McGraw-Hill 1981.

RESISTANT PROPERTIES OF BINDERS FOR COATINGS
Rating: E = Excellent, G = Good, F = Fair, P = Poor

Binder Type	Generic Type	Resistant Properties				Temperature	Primary Use
		Alkali	Acid	Water	Weather		
Lacquer	Copolymer-Vinyl Chloride-Vinyl Acetate	E	E	E	E	to 65 C (150 F)	Resistant intermediate and topcoats
	Polyacrylates	F	F	F	E	to 65 C (150 F)	Resistant topcoats
	Chlorinated Rubber	E	E	E	G	to 60 C (140 F)	Resistant intermediate
Co-reacting	Epoxy-Amine Cure	E	G	G	F	to 93 C (200 F)	Resistant coatings and linings
	Epoxy-Polyamide	E	F	G	G	to 93 C (200 F)	Resistant coatings and linings
	Urethane (2 package)	G	F	G	F	to 120 C (250 F)	Abrasion-resistant coatings
	Urethane (moisture cure)	G	F	G	G	to 120 C (250 F)	Abrasion-resistant coatings
	Urethane-Aliphatic Isocyanate	G	F	G	E	to 120 C (250 F)	Weather- and abrasion-resistant topcoats
Condensation (requires added heat to cure)	Phenolic	P	E	E	F	up to C (250 F)	Chemical- and food-resistant lining
	Epoxy Phenolic	F	E	E	F	to 120 C (250 F)	Chemical- and food-resistant lining
	Epoxy-Powder Coating (requires high heat to fuse and cure)	G	G	G	F	to 93 C (200 F)	Pipe coating and lining
Inorganic	Zinc Silicate	P	P	G	E	to 315 C (600 F)	Permanent primer or single coat weather-resistant coating
	Glass (fused to metallic substrate)	F	E	E	E	to 260 C (500 F)	Chemical- and Food-resistant lining

Source: Kirk-Othmer Encyclopedia of Chemical Technology, Munger, C. G., Coatings Resistant, Vol. 6, 3rd Ed., John Wiley & Sons, pp. 456–578, 1979.

PROPERTIES OF GENERIC COATINGS FOR ATMOSPHERIC SERVICE

APPLICATION PROPERTIES

	Alkyd	2-Can Epoxy	Acrylic Latex	Linseed Oil	Phenolic	Chlorinated Rubber	Aliphatic Urethane	Vinyl
Solvents	Aliphatic or Aromatic	Lacquer	Water	Aliphatic	Aromatic	Aromatic	Lacquer	Lacquer
Min. Surface Preparation*	SP 3	SP 6	SP 6	SP 2	SP 6	SP 6	***	SP 6
Stability During Use	EX	F	EX	EX	EX	EX	F	EX
Brushability	G	F	EX	VG	G	F	G	P
Method of Cure	Oxid.	Chem.	Coal.	Oxid.	Oxid.	Evap.	Chem.	Evap.
Speed of Cure								
50°F-90°F**	G	G	EX	F	G	EX	EX	EX
35°F-50°F**	F	NR	NR	P	F	G	G	G
Film Build per Coat	G	VG	F	G	G	G	VG	G
Use in Primers	G	EX	F	EX	G	G	G	G
Use on Damp Surfaces	P	G	VG	P	P	P	G	G

APPEARANCE PROPERTIES

	Alkyd	2-Can Epoxy	Acrylic Latex	Linseed Oil	Phenolic	Chlorinated Rubber	Aliphatic Urethane	Vinyl
Use as Clear Finish (Varnish)	VG	F	P	NR	VG	NR	EX	NR
Use in Ready Mixed Aluminum Paint	G	F	NR	F	EX	F	F	G
Pipe Color	VG	G	EX	G	P	VG	EX	EX
Ability to Produce High Gloss	EX	EX	F	G	EX	VG	EX	F

PERFORMANCE PROPERTIES

	Alkyd	2-Can Epoxy	Acrylic Latex	Linseed Oil	Phenolic	Chlorinated Rubber	Aliphatic Urethane	Vinyl
Hardness	G	VG	F	P	VG	VG	EX	G
Adhesion	G	EX	F	VG	G	VG	VG	F
Flexibility	G	G	EX	VG	F	VG	VG	EX
Resistance To—								
Abrasion	F	VG	F	P	G	VG	EX	VG
Water	G	EX	F	P	EX	EX	VG	EX
Strong Solvents	P	EX	F	P	G	P	EX	P
Acid	F	VG	F	P	EX	EX	EX	EX
Alkali	P	EX	G	P	G	EX	VG	EX
Heat—200°F	G	G	F	F	G	NR	G	NR

Continued on Next Page

PROPERTIES OF GENERIC COATINGS FOR
ATMOSPHERIC SERVICE (Cont)

DURABILITY

	Alkyd	2-Can Epoxy	Acrylic Latex	Linseed Oil	Phenolic	Chlorinated Rubber	Aliphatic Urethane	Vinyl
Moisture Permeability	Mod	Low	High	Mod	Low	Low	Low	Low
Normal Exposure	VG	VG	VG	G	VG	EX	EX	EX
Marine Exposure	F	EX	F	F	G	EX	EX	EX
Corrosive Exposure	F	EX	F	NR	G	VG	EX	EX
Color Retention	G	P	VG	F	P	G	EX	VG
Gloss Retention	G	P	EX	P	G	G	EX	VG
Chalk Resistance	G	P	VG	P	G	G	EX	VG

CODES
EX—Excellent
VG—Very Good
G—Good
F—Fair
P—Poor
NR—Not Recommended

SOLVENTS
Aliphatic—Mineral spirits
Aromatic—Xylene, toluene, etc.
Lacquer—Aromatic plus ketone,
  ester, or ether solvents
  (See Solvents)

ABBREVIATIONS
Oxid.—Oxidative polymerization or
  oxidation
Chem.—Chemical reaction (two
  component)
Coal.—Coalescence (latex)
Evap.—Solvent evaporation
  (lacquer)
Min.—Minimum

*SSPC Surface Preparation Specifications
**Painting should not be done above 90°F or below 34°F
***Usually used in topcoats

Source: Steel Structures Painting Manual, Vol. 1, pp. 121-122, SSPC 1982.

TEMPERATURE LIMITS OF COATINGS

Coating	Immersion	Nonimmersion
Vinyl Copolymer	38 C (100 F)	65 C (150 F)
Chlorinated Rubber	38 C (100 F)	60 C (140 F)
Coal Tar	50 C (120 F)	65 C (150 F)
Coal Tar Epoxy	50 C (120 F)	95 C (200 F)
Epoxy	50 C (120 F)	95 C (200 F)
Urethane	38 C (100 F)	120 C (250 F)
Epoxy Phenolic	82 C (180 F)	120 C (250 F)
Baked Phenolic	82 C (180 F)	120 C (250 F)
Inorganic Zinc	—	370 C (700 F)
Silicone	—	370 C (700 F)

Source: Kirk-Othmer Encyclopedia of Chemical Technology, Munger, C.G., Coatings Resistant, Vol. 6, 3rd Ed., John Wiley & Sons, New York, NY 1979

RADIATION TOLERANCE OF COATINGS

Severe Exposure = Greater than 4.5×10^9 Rads
Moderate Exposure = 5×10^8 to 4.5×10^9 Rads
Light Exposure = Less than 5×10^8 Rads

Coating	Maximum Allowable Radiation Dose in Air	
	On Steel	On Concrete
Chlorinated Rubber	1×10^8 Rads[1]	1×10^8 Rads
Epoxy-Amine	1×10^9	1×10^9
Epoxy Coal Tar	5×10^8	5×10^8
Epoxy-Polyamide	1×10^{10}	NA
Inorganic Silicate Finish	1×10^{10}	1×10^{10}
Inorganic Zinc	2.2×10^{10}	NA
Epoxy Phenolic	1×10^{10}	1×10^{10}
Silicone (Baked)	1×10^{10}	NA
Urethane	5×10^8	6×10^9
Vinyl	1×10^8	—

[1]Rad: The unit of absorbed radiation. For most organic material, one rentgen = 1 Rad (ANSI N 5.12, 1973).

Source: Kirk-Othmer Encyclopedia of Chemical Technology, Munger, C. G., Coatings Resistant, Vol. 6, 3rd Ed., John Wiley & Sons, New York, NY 1979.

COEFFICIENT OF FRICTION—SLIP FACTOR FOR VARIOUS SURFACE FINISHES AND COATINGS

Surface Treatment	No. of Tests	Slip Factor		
		Mean	Max.	Min.
Plain Steel				
Mill Scale	352	0.32	0.60	0.17
Rusted	15	0.43	0.55	0.41
Flame cleaned	88	0.48	0.75	0.31
Blast cleaned	183	0.57	0.81	0.32
Coated Steel				
Red lead paint	6	0.07	—	0.05
Rust preventive paint	3	0.11	—	0.07
Hot-dip galvanized	95	0.19	0.36	0.08
Lacquer-varnish	17	0.24	0.30	0.10
Blast cleaned vinyl wash primer	24	0.28	0.34	0.22
Galvanized and grit blasted	12	0.49	0.55	0.42
Grit blasted and inorganic zinc rich paint	48	0.51	0.65	0.38
Grit blasted and zinc sprayed	42	0.65	0.99	0.42

Source: Transportation Research Board #112, National Research Council, Dec. 1984.

CHEMICAL RESISTANCE OF COATINGS FOR IMMERSION SERVICE
(Room Temperature)

Chemical resistance data are for coatings only. Thin coatings generally are not suitable for substrates such as carbon steel which are corroded significantly (e.g., > 20 mpy) in the test environment.

	Asphalt, Unmodified	Coal Tar Hot Applied	Coal Tar Cold Applied	Coal Tar -Epoxy	Coal Tar -Urethanes	Epoxy-Phenolic Baked	Epoxy-Amine Cured	Epoxy Ester	Furfuryl Alcohol	Phenolics, Baked	Polyesters (Unsaturated)
Acids											
Sulfuric, 10%	R	LR	NR	R	R	R	R	LR	R	R	R
Sulfuric, 80%	–	NR	NR	–	NR	NR	NR	NR	LR	NR	R
Hydrochloric, 10%	R	LR	NR	LR	LR	R	R	LR	R	R	R
Hydrochloric, 35%	R	NR	NR	–	–	LR	NR	NR	R	R	R
Nitric, 10%	NR	LR	NR	–	–	R	NR	NR	NR	NR	R
Nitric, 50%	NR	NR	NR	–	–	NR	NR	NR	NR	NR	NR
Acetic, 100%	–	NR	NR	NR	NR	NR	NR	NR	LR	LR	NR
Water											
Distilled	R	R	R	R	LR	R	R	R	R	R	R
Salt Water	R	R	R	R	LR	R	R	R	R	R	R
Alkalies											
Sodium Hydroxide, 10%	R	R	LR	R	R	R	R	NR	R	NR	R
Sodium Hydroxide, 70%	–	NR	NR	–	LR	R	R	NR	LR	NR	NR
Ammonium Hydroxide, 10%	R	R	LR	R	R	R	LR	LR	R	NR	R
Sodium Carbonate, 5%	R	R	R	–	–	R	–	–	–	–	–
Gases											
Chlorine	R	NR	NR	LR	NR	LR	LR	LR	NR	NR	R
Ammonia	–	LR	LR	NR	NR	LR	LR	R	R	NR	NR
Hydrogen Sulfide	–	R	R	R	–	R	R	–	R	R	–

Continued on Next Page

CHEMICAL RESISTANCE OF COATINGS FOR IMMERSION SERVICE (Cont)
(Room Temperature)

Chemical resistance data are for coatings only. Thin coatings generally are not suitable for substrates such as carbon steel which are corroded significantly (e.g., > 20 mpy) in the test environment.

	Asphalt, Unmodified	Coal Tar Hot Applied	Coal Tar Cold Applied	Coal Tar -Epoxy	Coal Tar -Urethanes	Epoxy-Phenolic Baked	Epoxy-Amine Cured	Epoxy Ester	Furfuryl Alcohol	Phenolics, Baked	Polyesters (Unsaturated)
Organics											
Alcohols	R	LR	LR	NR	NR	R	R	LR	R	R	R
Aliphatic Hydrocarbons	NR	LR	LR	LR	LR	R	R	R	R	R	R
Aromatic Hydrocarbons	NR	NR	NR	NR	LR	R	R	R	R	R	R
Ketones	NR	NR	NR	NR	NR	LR	LR	NR	LR	R	NR
Ethers	NR	NR	NR	—	—	LR	LR	NR	LR	R	—
Esters	NR	NR	NR	NR	NR	LR	LR	NR	R	R	LR
Chlorinated Hydrocarbons	NR	NR	NR	NR	LR	LR	LR	NR	LR	R	NR
Max. Temp. (dry conditions) °F	150	—	—	200	200	250	250	250	300	250-300	—
Max. Temp. (wet conditions) °F	—	120	120	150	150	150	150	150	190	160-250	250

R = Recommended
LR = Limited recommendation
NR = No recommendation

Continued on Next Page

CHEMICAL RESISTANCE OF COATINGS FOR IMMERSION SERVICE (Cont)
(Room Temperature)

Chemical resistance data are for coatings only. Thin coatings generally are not suitable for substrates such as carbon steel which are corroded significantly (e.g., > 20 mpy) in the test environment.

	Polyvinyl Chlor-acetates	Vinyl Ester	Urethanes Air-Dry	Urethanes Bake	Vinylidene Chloride	Chlorinated Rubber
Acids						
Sulfuric, 10%	R	R	LR	LR	R	R
Sulfuric, 80%	NR	R	NR	LR	LR	R
Hydrochloric, 10%	R	R	LR	LR	R	R
Hydrochloric, 35%	LR	R	LR	LR	R	R
Nitric, 10%	R	R	LR	LR	R	R
Nitric, 50%	NR	—	NR	NR	LR	NR
Acetic, 100%	NR	LR	NR	NR	NR	NR
Water						
Distilled	R	R	LR	LR	—	R
Salt Water	R	R	LR	LR	R	R
Alkalies						
Sodium Hyroxide, 10%	R	R	LR	LR	LR	R
Sodium Hydroxide, 70%	LR	R	LR	LR	NR	R
Ammonium Hydroxide, 10%	R	R	LR	R	NR	R
Sodium Carbonate, 5%	R	R	R	R	R	R
Gases						
Chlorine	LR	R	LR	R	LR	R
Ammonia	LR	R	LR	R	NR	NR
Hydrogen Sulfide	LR	R	R	R	R	R

Continued on Next Page

CHEMICAL RESISTANCE OF COATINGS FOR IMMERSION SERVICE (Cont)
(Room Temperature)

Chemical resistance data are for coatings only. Thin coatings generally are not suitable for substrates such as carbon steel which are corroded significantly (e.g., > 20 mpy) in the test environment.

	Polyvinyl Chlor-acetates	Vinyl Ester	Urethanes		Vinylidene Chloride	Chlorinated Rubber
			Air-Dry	Bake		
Organics						
Alcohols	R	R	NR	R	R	LR
Aliphatic Hydrocarbons	R	R	R	R	R	LR
Aromatic Hydrocarbons	NR	LR	R	R	LR	NR
Ketones	NR	NR	NR	R	NR	NR
Ethers	NR	LR	R	R	NR	NR
Esters	NR	LR	NR	R	NR	NR
Chlorinated Hydrocarbons	LR	LR	LR	R	LR	NR
Max. Temp. (dry conditions) °F	160	350	—	—	—	160
Max. Temp. (wet conditions) °F	150	210	—	—	150	140

R = Recommended
LR = Limited recommendation
NR = No recommendation

Source: NACE, TPC 2 *Coatings & Linings for Immersion Service*

TYPICAL PHYSICAL PROPERTIES OF SURFACE COATINGS FOR CONCRETE[1]

| | Concrete | Polyester | | Epoxy | | Urethane[4] |
		Isophthalic	Bisphenol	Polyamide	Amine	
Tensile Strength						
(ASTM C307) psi	200-400	1200-2500	1200-2500	600-4000	1200-2500	200-1200
MPa	1.4-2.8	8.3-17	8.3-17	4.0-28	8.3-17	1.4-8.3
Thermal coefficient						
of expansion (ASTM C531)						
max. in/in/F	6.5×10^6	20×10^6	20×10^6	40×10^6	40×10^6	[2]
mm/mm/C	11.7×10^6	36×10^6	36×10^6	72×10^6	72×10^6	
Compressive strength,						
(ASTM C579) psi	3,500	10,000	10,000	4,000	6,000	[2]
MPa	24	70	70	28	42	
Abrasion resistance						
Taber Abraser—Wt loss						
in milligrams 1000 gram						
load/1000 cycles		15 to 27	15 to 27	15 to 27	15 to 27	[2]
Shrinkage, ASTM C531 %		2 to 4	2 to 4	0.25 to 0.75	0.25 to 0.75	0 to 2
Work life, minutes		15 to 45	15 to 45	30 to 90	30 to 90	15 to 60
Traffic limitations,	Light	16	16	24	24	24
hours after application	Heavy	36	36	48	48	48
	Ready for service	48	48	72	72	72

Continued on Next Page

TYPICAL PHYSICAL PROPERTIES OF SURFACE COATINGS FOR CONCRETE[1] (Cont)

	Concrete	Polyester		Polyamide	Epoxy Amine	Urethane[4]
		Isophthalic	Bisphenol			
Adhesion characteristics[3]		Poor	Fair	Excellent	Good	Fair
Flexural Strength (ASTM C580) psi		1500	1500	1000	1500	(2)
MPa		10	10	7	10	

[1] All physical values depend greatly on reinforcing. Values are for ambient temperatures.

[2] Urethanes not shown because of great differences in physical properties, depending on formulations. Adhesion characteristics should be related by actual test data. Any system which shows concrete failure when tested for surfacing adhesion should be rated excellent with decreasing rating for systems showing failure in cohesion or adhesion below concrete failure.

[3] Adhesion to concrete: primers usually are used under polyesters and urethanes to improve adhesion.

[4] Type of urethane used is one of three: (1) Type II Moisture Cured. (2) Type IV Two Package Catalyst or (3) Type V Two Package Polyol. Ref. ASTM C16.

Source: RP 03-76. "Monolithic Organic Corrosion Resistant Floor Surfacings." NACE. 1976.

DRY FILM THICKNESS (IN MILS) OF COATINGS AS A FUNCTION OF SOLIDS CONTENT AND COVERAGE RATE

Film Forming Solids by Volume, Percent	Gallons to Coat 1000 Sq. Ft.																				
	10.	8.0	6.7	5.7	5.0	4.4	4.0	3.6	3.3	3.1	2.9	2.7	2.5	2.4	2.2	2.1	2.0	1.9	1.8	1.7	1.7
	Coverage Rates, Sq. Ft. Per Gallon																				
	100	125	150	175	200	225	250	275	300	325	350	375	400	425	450	475	500	525	550	575	600
10	1.6	1.3	1.1	.9	.8	.7	.6	.6	.4												
15	2.4	1.9	1.6	1.4	1.2	1.1	1.	.9	.8	.7	.7	.6	.6	.6	.5						
20	3.2	2.6	2.1	1.8	1.6	1.4	1.3	1.2	1.1	1.	.9	.8	.8	.8	.7	.7	.6	.6	.6	.6	.5
25	4.	3.2	2.7	2.3	2.	1.8	1.6	1.5	1.4	1.2	1.1	1.1	1.	1.	.9	.8	.8	.8	.7	.7	.7
30	4.8	3.8	3.2	2.7	2.4	2.1	1.9	1.8	1.6	1.5	1.4	1.3	1.2	1.1	1.1	1.	1.	.9	.8	.8	.8
35	5.6	4.5	3.7	3.2	2.8	2.5	2.2	2.1	1.9	1.7	1.6	1.5	1.4	1.3	1.2	1.1	1.1	1.1	1.	1.	.9
40	6.2	5.1	4.3	3.6	3.2	2.8	2.6	2.4	2.2	2.	1.8	1.7	1.6	1.5	1.4	1.3	1.3	1.2	1.2	1.1	1.1
45	7.2	5.8	4.8	4.1	3.6	3.2	2.9	2.7	2.4	2.2	2.	2.	1.8	1.7	1.6	1.5	1.4	1.4	1.3	1.3	1.2
50	8.	6.4	5.4	4.6	4.	3.6	3.2	2.9	2.7	2.4	2.3	2.1	2.	1.9	1.8	1.7	1.6	1.5	1.5	1.4	1.4
55	8.8	7.	5.9	5.	4.4	3.9	3.5	3.2	3.	2.7	2.5	2.3	2.2	2.1	1.9	1.8	1.8	1.7	1.6	1.5	1.5
60	9.6	7.7	6.4	5.5	4.8	4.3	3.8	3.5	3.2	2.9	2.7	2.5	2.4	2.3	2.1	2.	1.9	1.8	1.7	1.7	1.6
65	10.4	8.3	7.	5.9	5.2	4.6	4.2	3.8	3.5	3.2	2.9	2.7	2.6	2.5	2.3	2.1	2.1	2.	1.9	1.8	1.8
70	11.2	9.	7.5	6.4	5.6	5.	4.5	4.1	3.8	3.4	3.2	2.9	2.8	2.7	2.5	2.3	2.2	2.1	2.	2.	1.8
75	12.	9.5	8.	6.8	6.	5.1	4.8	4.4	4.	3.7	3.4	3.2	3.	2.9	2.6	2.5	2.4	2.3	2.2	2.1	2.
80	12.8	10.	8.6	7.3	6.4	5.7	5.1	4.7	4.3	3.9	3.6	3.4	3.2	3.	2.8	2.6	2.6	2.4	2.3	2.2	2.2
85	13.6	10.7	9.2	7.7	6.8	6.	5.4	5.	4.6	4.2	3.8	3.6	3.4	3.4	3.	2.8	2.7	2.6	2.5	2.3	2.3
90	14.4	11.3	9.6	8.2	7.2	6.4	5.8	5.3	4.9	4.4	4.1	3.8	3.6	3.4	3.2	3.	2.9	2.7	2.6	2.5	2.4
95	15.2	12.2	10.2	8.7	7.6	6.8	6.1	5.5	5.1	4.7	4.4	4.1	3.8	3.6	3.4	3.2	3.	2.9	2.8	2.7	2.5
100	16.	12.8	10.7	9.2	8.	7.1	6.4	5.8	5.4	4.9	4.6	4.3	4.	3.8	3.6	3.4	3.2	3.1	2.9	2.8	2.7

Coverages are theoretical, assuming 100 percent utility, no loss, on a perfectly flat, smooth surface.

Source: P. E. Weaver, Industrial Maintenance Painting, p. 137, NACE 1978.

EFFECT OF pH ON CORROSION OF ZINC IN AERATED AQUEOUS SOLUTIONS

Dilute HCl or NaOH at 30°C. CO_2 free

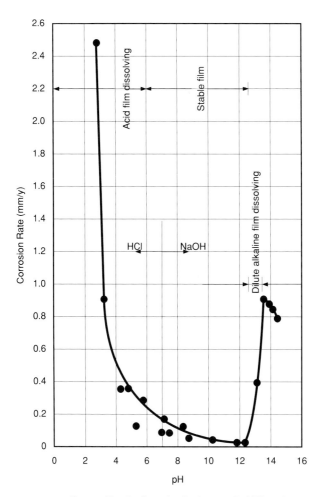

Source: Zinc, Its Corrosion Resistance, 2nd Ed., p. 4,
International Lead Zinc Research Organization, 1983

RUST PREVENTIVES[a]

	Oil Type	Solvent Type	Emulsifiable Type	Wax Type (applied hot)
Metals Coated	Generally applied to ferrous metals; nonferrous metals sometimes with extreme care			
Coating Composition, Structure	Non-setting minerals oils of various weights and viscosities; thin oily layer; thickness depending on viscosity	Petroleum-base film-forming materials and rust inhibitors dissolved in petroleum solvents; soft to hard, depending on composition	Petroleum-base, rust preventives modified to form stable emulsions when mixed with water	Waxy layer; soft to firm, depending on composition
Application Methods	Brushing, spraying, dipping, flushing	Brushing, spraying, dipping, flushing	Brushing, spraying, dipping, flushing	Heating and then dipping, brushing or swabbing; special techniques required (for spraying)
Appearance	Transparent oily film	Transparent to black	Transparent oily to tacky film	Transparent, brown, amber, or black
Thickness, mil	0.2-0.3	0.2-0.4; occasionally up to 2.0	0.2	1.5-3.0
Pretreatment	Alkaline, solvent or emulsion cleaning; scaly surfaces should be freed of all deposits by mechanical cleaning. Emulsifiable coatings can often be applied directly.			
How Coatings Removed	Removal seldom required, solvent rinsing, vapor degreasing, emulsion spraying, or alkaline washing	Removal often unnecessary; solvent rinsing or alkali cleaning	Removal seldom required; solvent rinsing	Solvent rinsing or alkali cleaning

Continued on Next Page

RUST PREVENTIVES[a] (Cont)

Properties	Oil Type	Solvent Type	Emulsifiable Type	Wax Type (applied hot)
Durability	Excellent protection for indoor storage	Excellent indoor protection from 4 months to 2 years; in some cases can also be used outdoors	Excellent indoor protection for 1-2 years	Good protection indoors (up to 3 years) and outdoors (1-2 years)
Adhesion	Good[b]	Good[b]	Good[b]	Good[b]
Abrasion Resistance	Fair	Fair	Fair	Good
Impact Resistance	Very good	Very good	Very good	Very good
Heat Resistance		Up to 120-140°F (50-60°C) (all types)		
Typical Uses	Internal combustion engines, gear cases, hydraulic systems, highly finished auto parts, galvanized products, steel sheet, bar, wire	External surfaces of machinery and tools; highly finished surfaces; steel sheet, bar and wire		Any highly finished part stored for prolonged periods of time, e.g., ball bearings
Relative Cost per ft²	1.0	4	2.	5.-30.

[a]Rust preventives are essentially petroleum-type coatings designed to provide low cost corrosion protection during manufacture, shipment, and storage.

[b]Soft types can be wiped off, but hard types have relatively good adhesion. Degree of adhesion is also influenced by porosity of base metal.

Source: Materials in Design Engineering, p.403, 1964.

PRESSURE LOSS IN HOSE (psi)

Inside Diameter (in.)	Length (ft)	Free Air (cfm)	Line Pressure (psig)					
			60	80	100	120	150	200
3/4	50	60	3.1	2.4	2.0			
		80	5.3	4.2	3.5	2.9	2.4	1.8
		100	8.1	6.4	5.2	4.5	3.6	2.8
		120		9.0	7.4	6.3	5.1	3.9
		140		12.0	9.9	8.4	6.9	5.3
		160			12.7	10.8	8.9	6.8
		180				13.6	11.1	8.5
		200				16.6	13.5	10.4
		220					16.2	12.4
1	50	120	2.7	2.1				
		150	4.1	3.2	2.7	2.3		
		180	5.8	4.6	3.8	3.2	2.6	2.0
		210	7.7	6.1	4.0	4.3	3.5	2.7
		240		7.9	6.5	5.5	4.5	3.4
		270		9.8	8.1	6.9	5.6	4.3
		300		12.0	9.9	8.4	6.9	5.3
		330			11.8	10.0	8.2	6.3
		360			13.9	11.9	9.7	7.4
		390				13.8	11.3	8.7
		420				15.9	13.0	10.0
		450					14.8	11.4
1-1/4	50	200	2.4					
		250	3.7	2.9	2.4	2.0		
		300	5.2	4.1	3.4	2.9	2.3	1.8
		350	7.0	5.5	4.5	3.8	3.1	2.4
		400	8.9	7.0	5.8	4.9	4.0	3.1
		450		8.8	7.3	6.2	5.0	3.9
		500		10.8	8.9	7.6	6.2	4.7
		550			10.7	9.1	7.4	5.7
		600			12.6	10.7	8.7	5.7
		650			14.6	12.4	10.2	7.8
		700				14.3	11.7	9.0
		750					13.3	10.2
		800					15.0	11.5
1-1/2	50	300	2.1					
		400	3.7	2.9	2.4	2.0		
		500	5.6	4.4	3.7	3.1	2.5	1.9
		600	8.0	6.3	5.2	4.4	3.6	2.8
		700		8.5	7.0	5.9	4.9	3.7
		800		10.9	9.0	7.7	6.3	4.8
		900			11.2	9.5	7.8	6.0
		1000			13.6	11.6	9.5	7.3
		1100				14.0	11.4	8.8
		1200					13.6	10.4
		1300					15.8	12.1

Continued on Next Page

PRESSURE LOSS IN HOSE (psi) (Cont)

Inside Diameter (in.)	Length (ft)	Free Air (cfm)	Line Pressure (psig)						
			60	80	100	120	150	200	300
2	50	600	1.9						
		800	3.2	2.5	2.1				
		1000	5.0	3.9	3.2	2.7	2.2	1.7	1.1
		1200	7.0	5.5	4.5	3.8	3.1	2.4	1.6
		1400	9.3	7.4	6.1	5.2	4.2	3.2	2.2
		1600		9.6	7.9	6.7	5.5	4.2	2.8
		1800		12.1	9.9	8.4	6.9	5.3	3.6
		2000			12.2	10.4	8.5	6.5	4.4
		2200			14.6	12.5	10.2	7.8	5.3
		2400				14.7	12.0	9.2	6.3
		2600					14.1	10.8	7.3
		2800					16.2	12.4	8.5
2-1/2	50	1000	1.7						
		1500	3.7	2.9	2.4	2.0			
		2000	6.5	5.1	4.2	3.6	2.9	2.2	1.5
		2500	10.0	7.9	6.5	5.5	4.5	3.4	2.3
		3000		11.2	9.3	7.9	6.4	4.9	3.3
		3500			12.4	10.6	8.7	6.6	4.5
		4000				13.7	11.2	8.6	5.8
		4500					14.0	10.7	7.3
3	50	2000	2.5	2.0					
		2500	3.9	3.0	2.5	2.1			
		3000	5.5	4.4	3.6	3.1	2.5	1.9	1.3
		3500	7.5	5.9	4.9	4.1	3.4	2.6	1.7
		4000	9.8	7.6	6.3	5.3	4.4	3.3	2.3
		4500		9.6	7.9	6.7	5.5	4.2	2.8
		5000		11.7	9.6	8.2	6.7	5.1	3.5
		5500			11.5	9.8	8.0	6.1	4.2
		6000			13.6	11.5	9.4	7.2	4.9
		6500				13.5	11.0	8.4	5.7
		7000				15.6	12.7	9.8	6.6
		7500					14.5	11.1	7.6
4	25	5000	1.9						
		6000	2.7	2.1	1.7				
		7000	3.6	2.8	2.3	2.0		1.2	
		8000	4.7	3.7	3.0	2.6	2.1	1.6	
		9000	5.9	4.6	3.8	3.2	2.6	2.0	
		10000	7.2	5.7	4.7	4.0	3.2	2.5	
		11000	8.7	6.8	5.6	4.8	3.9	3.0	
		12000		8.1	6.7	5.7	4.6	3.5	
		13000		9.4	7.8	6.6	5.4	4.1	
		14000			9.0	7.6	6.2	4.8	
		15000				8.7	7.1	5.4	
		16000				9.8	8.0	6.2	
		17000					9.1	6.9	

Note: Lubrication only at tool, no line lubricator

Source: Journal of Protective Coatings & Linings, Vol. 2, No. 7, p.28 (1985)

APPROXIMATE SQUARE FEET PER LINEAR FOOT AND PER TON FOR DIFFERENT STEEL MEMBERS

Size	Wt.	Sq. Ft./ Lin. Ft.	Sq. Ft. Per Ton	Size	Wt.	Sq. Ft./ Lin. Ft.	Sq. Ft. Per Ton
24 WF (24X14)	160	8.9	110	14 WF (14X16)	426	8.5	40
	145	8.8	121		398	8.5	43
	130	8.7	135		370	8.5	46
24 WF (24X12)	120	8.1	133		342	8.5	50
	110	8.0	144		314	8.5	54
	100	8.0	160		287	8.0	56
					264	8.0	61
24 WF (24X9)	94	7.1	149		246	8.0	65
	84	7.0	167		237	8.0	68
	76	7.0	184		228	8.0	70
21 WF (21X13)	142	7.9	111		219	7.9	72
	127	7.9	124		211	7.9	75
	112	7.8	139		202	7.9	78
					193	7.9	82
21 WF (21X9)	95	6.5	135		184	7.9	86
	82	6.5	159		176	7.7	87
21 WF (21X8 1/4)	73	6.3	173		167	7.7	92
	68	6.3	185		158	7.7	97
	62	6.2	200		150	7.7	103
					142	7.7	108
18 WF (18X11 3/4)	114	7.0	123				
	105	7.0	133	14 WF (14X14 1/2)	136	7.3	107
	96	7.0	146		127	7.3	115
					119	7.3	123
18 WF (18X8 3/4)	85	6.0	141		111	7.3	132
	77	6.0	156		103	7.3	142
	70	5.9	169		95	7.3	154
	64	5.9	184		87	7.3	168
18 WF (18X7 1/2)	60	5.5	183	14 WF (14X12)	84	6.4	152
	55	5.5	200		78	6.3	162
	50	5.5	220	14 WF (14X10)	74	5.7	154
16 WF (16X11 1/2)	96	6.6	137		68	5.7	168
	88	6.5	148		61	5.7	187
16 WF (16X8 1/2)	78	5.6	144	14 WF (14X8)	53	5.0	189
	71	5.5	155		48	5.0	208
	64	5.5	172		43	4.9	228
	58	5.5	190				
16 WF (16X7	50	5.1	204	14 WF (14X6 3/4)	38	4.6	242
	45	5.0	222		34	4.6	271
	40	5.0	250		30	4.6	307
	36	5.0	278				

Continued on Next Page

APPROXIMATE SQUARE FEET PER LINEAR FOOT AND PER TON
FOR DIFFERENT STEEL MEMBERS (Cont)

Size	Wt.	Sq. Ft./ Lin. Ft.	Sq. Ft. Per Ton	Size	Wt.	Sq. Ft./ Lin. Ft.	Sq. Ft. Per Ton
12 WF	190	6.6	69	8 WF	28	3.5	250
(12X12)	161	6.5	81	(8X6 1/2)	24	3.5	292
	133	6.4	96				
	120	6.3	105	8 WF	20	3.1	310
				(8X5 1/4)	17	3.1	365
	106	6.2	117				
	99	6.2	125	6 WF	25	3.1	248
	92	6.2	135	(6X6)	20	3.0	300
	85	6.1	144		15.5	3.0	387
	79	6.1	154				
	72	6.1	169	5 WF	18.5	2.5	270
	65	6.0	185	(5X5)	16	2.5	313
12 WF	58	5.4	186	4 WF	13	2	308
(12X10)	53	5.3	200				
12 WF	50	4.7	188		**I-BEAMS**		
(12X8)	45	4.7	209	24 I	120	6.7	112
	40	4.7	235		106	6.6	125
12 WF	36	4.2	233		100	6.4	128
(12X6 1/2)	31	4.2	271		90	6.4	142
	27	4.2	311		79.9	6.3	158
10 WF	112	5.4	96	20 I	95	5.7	120
(10X10)	100	5.3	106		85	5.7	134
	89	5.2	117				
	77	5.2	135		75	5.5	147
					65.4	5.4	165
	72	5.1	142				
	66	5.1	155	18 I	70	5.1	148
	60	5.1	170		54.7	5.0	183
	54	5.0	185				
	49	5.0	204	15 I	50	4.4	176
					42.9	4.3	200
10 WF	45	4.4	196				
(10X8)	39	4.3	221	12 I	50	3.8	152
	33	4.3	261		40.8	3.8	186
10 WF	29	3.6	248		35	3.7	211
(10X5 3/4)	25	3.6	288		31.8	3.7	233
	21	3.6	343				
				10 I	35	3.3	189
8 WF	67	4.3	128		25.4	3.2	252
(8X8)	58	4.2	145				
	48	4.1	171	8 I	23	2.7	322
					18.4	2.7	402
	40	4.1	205				
	35	4.0	229				
	31	4.0	258	*I-Beams Continued on Next Page*			

**APPROXIMATE SQUARE FEET PER LINEAR FOOT AND PER TON
FOR DIFFERENT STEEL MEMBERS (Cont)**

Size	Wt.	Sq. Ft./ Lin. Ft.	Sq. Ft. Per Ton	Size	Wt.	Sq. Ft./ Lin. Ft.	Sq. Ft. Per Ton
	I-BEAMS (Cont.)				**CHANNELS**		
71	20	2.5	250	10	30	2.7	180
	15.3	2.4	314		25	2.6	208
					20	2.6	260
61	17.25	2.2	255		15.3	2.5	327
	12.5	2.1	336				
				9	20	2.4	240
51	14.75	1.9	258		15	2.3	307
	10	1.8	360		13.4	2.3	343
41	9.5	1.6	337	8	18.75	2.2	235
	7.7	1.6	416		13.75	2.1	305
					11.5	2.1	365
31	7.5	1.3	347				
	5.7	1.3	456	7	14.75	1.9	258
					12.25	1.9	310
					9.8	1.9	388
	CHANNELS						
18	58	4.4	152	6	13.0	1.7	262
	51.9	4.4	172		10.5	1.7	324
	45.8	4.3	188		8.2	1.6	390
	42.7	4.3	201				
				5	9.0	1.5	333
15	50	3.7	148		6.7	1.4	418
	40	3.7	185				
	33.9	3.6	212	4	7.25	1.2	331
					5.4	1.2	444
13	50	3.6	144				
	31.8	3.5	220	3	6.0	1.0	333
					5.0	1.0	400
12	30	3.1	207		4.1	1.0	488
	25	3.0	240				
	20.7	3.0	290				

Continued on Next Page

APPROXIMATE SQUARE FEET PER LINEAR FOOT AND PER TON
FOR DIFFERENT STEEL MEMBERS (Cont)

Size	Wt.	Sq. Ft./ Lin. Ft.	Sq. Ft. Per Ton	Size	Wt.	Sq. Ft./ Lin. Ft.	Sq. Ft. Per Ton
ANGLES EQUAL LEG				**ANGLES UNEQUAL LEG**			
8X8X 1/2	26.4	2.7	205	8X6X 1/2	23.0	2.3	200
6X6X 5/16	12.5	2.0	320	8X4X 1/2	19.6	2.0	204
5X5X 5/16	10.3	1.7	330	7X4X 3/8	13.6	1.8	265
4X4X 1/4	6.6	1.3	394	6X4X 5/16	10.3	1.7	330
3 1/2X 3 1/2X 1/4	5.8	1.2	414	6X3 1/2X5/16	9.8	1.6	327
3X3X 3/16	3.71	1.0	539	5X3 1/2X 5/16	8.7	1.4	322
2 1/2X2 1/4X 3/16	3.07	0.8	521	5X3X 1/4	6.6	1.3	394
2X2X 1/8	1.65	0.7	848	4X3 1/2X 1/4	6.2	1.25	403
1 1/4X1 1/4X 1/8	1.23	0.5	813	4X3X 1/4	5.8	1.17	403
1X1X 1/8	0.80	0.3	750	3 1/2X 3 1/4	5.4	1.08	400
				3X2 1/2X 1/4	4.5	0.92	409
FLAT 1-Surface Only				3X2X 3/16	3.07	0.83	541
1 Foot Wide*				2 1/2X2X 3/16	2.75	0.75	545
				2 1/2X1 1/2X 3/16	2.44	0.67	549
Wt. Lbs./Sq. Ft.				2X1 1/2X 1/8	1.44	0.58	806
				1 1/2X1 11/4X 3/16	1.67	0.31	371
1/16"	2.55	1.0	784	1X3/4X 1/8	0.70	0.15	429
1/8"	5.10	1.0	392	1X 5/8X 1/8	0.64	0.14	438
3/16"	7.65	1.0	261				
1/4"	10.20	1.0	196				
3/8"	15.30	1.0	131				
1/2"	20.40	1.0	98				
5/8"	25.50	1.0	78				
3/4"	30.60	1.0	65				
7/8"	35.70	1.0	56				
1"	40.80	1.0	49				

*If 2 surfaces (top and bottom) are desired, multiply figures in the 2 columns at right above by 2. This is for flat material only, such as plates.

Source: P.E. Weaver, Industrial Maintenance Painting, pp. 129-132, NACE 1978.

SURFACE AREA PER TON OF STEEL FOR VARIOUS TYPES OF CONSTRUCTION

	Sq Ft Per Ton
Light construction	300 to 500
Medium construction	150 to 300
Heavy construction	100 to 150
Extra heavy construction	50 to 100

Note: The average in industrial plants is around 200 to 250 sq ft per ton.

Schedule 40 Steel Pipe
Exterior Surface Area Sq Ft/Ton

Nominal Size Inch	Sq Ft/ Ton	Nominal Size Inch	Sq Ft/ Ton
1	410	8	158
1 1/2	365	10	139
2	341	12	125
2 1/2	260	14	116
3	242	16	101
3 1/2	230	18	90
4	218	20	85
5	199	24	73
6	183		

For weight of pipe not shown use following equation:

$$W = K (D^2 - d^2)$$

where W = weight in lb/1 ft; D = outside diameter; d = inside diameter; K = 2.67 for steel pipe; K = 2.45 for cast iron pipe; K = 2.82 for brass pipe.

Source: P. E. Weaver, Industrial Maintenance Painting, p. 133, NACE 1978.

SQUARE FEET OF AREA AND GALLON CAPACITY
PER FOOT OF DEPTH IN CYLINDRICAL TANKS

Diameter ft.	Circumference ft.	Cross Section Area ft.^{2}	Gallons per ft. of Depth
5.0	15.708	19.635	146.88
5.5	17.279	23.758	177.72
6.0	18.850	28.274	211.51
6.5	20.420	33.183	248.23
7.0	21.991	38.485	287.88
7.5	23.562	44.179	330.48
8.0	25.133	50.265	376.01
8.5	26.704	56.745	424.48
9.0	28.274	63.617	475.89
9.5	29.845	70.882	530.24
10.0	31.416	78.540	587.52
10.5	32.987	86.590	647.74
11.0	34.558	95.033	710.90
11.5	36.128	103.87	776.99
12.0	37.699	113.10	846.03
12.5	39.270	122.72	918.00
13.0	40.841	132.73	992.91
13.5	42.412	143.14	1070.8
14.0	43.982	153.94	1151.5
14.5	45.553	165.13	1235.3
15.0	47.124	176.71	1321.9
15.5	48.695	188.69	1411.5
16.0	50.265	201.06	1504.1
16.5	51.836	213.82	1599.5
17.0	53.407	226.98	1697.9
17.5	54.978	240.53	1799.3
18.0	56.549	254.47	1903.6
18.5	58.119	268.80	2010.8
19.0	59.690	283.53	2120.9
19.5	61.261	298.65	2234.0
20.0	62.832	314.16	2350.1
20.5	64.403	330.06	2469.1
21.0	65.973	346.46	2591.0
21.5	67.544	363.05	2715.8
22.0	69.115	380.13	2843.6
22.5	70.686	397.61	2974.3
23.0	72.257	415.48	3108.0
23.5	73.827	433.74	3244.6
24.0	75.398	452.39	3384.1
24.5	76.969	471.44	3526.6
25.0	78.540	490.87	3672.0
25.5	80.111	510.71	3820.3
26.0	81.681	530.93	3971.6
26.5	83.252	551.55	4125.9
27.0	84.823	572.56	4283.0
27.5	86.394	593.96	4443.1
28.0	87.965	615.75	4606.2
28.5	89.535	637.94	4772.1
29.0	91.106	660.52	4941.0
29.5	92.677	683.49	5112.9
30.0	94.248	706.86	5287.7
30.5	95.819	730.62	5465.4

Total area = circumference × length + area of two ends. (Use area of one end for open top tanks.)

PROPERTIES OF FLAMMABLE LIQUIDS
USED IN PAINTS AND LACQUERS

Name	Flash Point °F Closed Cup	Open Cup	Flash Point °C Closed Cup	Open Cup	Explosive Limits % by Volume in Air Lower	Upper	Vapor Density (Air = 1.00)	Boiling Point °F	
Acetone	0	15	−18	−9	2.15	13.0	2.00	134	
n-Amyl Acetate	76	80	24	27	1.10	—	4.49	300	
n-Amyl Alcohol	91	120	32	49	1.20	—	3.04	280	
Benzene	12	—	−11	—	1.40	8.0	2.77	176	
n-Butyl Acetate	72	90	22	32	1.70	15.0	4.00	260	
n-Butyl Alcohol	84	110	29	43	1.70	—	2.55	243	
Cellosolve	104	120	40	49	2.60	15.7	3.10	275	
Cellosolve Acetate	124	125	51	52	1.71	—	4.72	313	
Cyclohexanone	147	—	64	—	—	—	3.38	313	
Ethyl Acetate	24	30	−5	−1	2.18	11.5	3.04	171	
Ethyl Alcohol	55	—	13	—	3.23	19.0	1.59	173	
Fuel Oil No. 1	100-165	—	38-74	—	—	—	—	—	
Gasoline	−50	—	−45	—	1.30	6.0	3-4	100-400	3
n-Hexane	−7	—	−21	—	1.25	6.9	2.97	156	
Kerosene	100-165	—	38-74	—	—	—	—	—	
Methyl Alcohol	54	60	12	16	6.00	36.5	1.11	147	
Methyl n-Butyl Ketone	—	95	—	35	1.22	8.0	3.45	262	
Methyl Isobutyl Ketone	73	—	23	—	1.34	8.0	3.45	244	
Methyl Ethyl Ketone	30	—	−1	—	1.81	11.5	2.41	176	
Methyl n-Propyl Ketone	—	60	—	16	1.55	8.1	2.96	216	
Mineral Spirits	100-110	—	38-43	—	1.10	6.0	—	300-400	1
Naphtha. Coal Tar	100-110	—	38-43	—	—	—	—	300-400	1
Naphtha. Safety Solvent	100-110	—	38-43	—	1.10	6.0	—	300-400	1
Naphtha, V.M. & P.	20-45	—	−7-7	—	1.20	6.0	—	212-320	1
n-Propyl Alcohol	59	85	15	29	2.50	—	2.07	207	
Isopropyl Alcohol	53	60	11	16	2.50	—	2.07	181	
Stoddard Solvent	100-110	—	38-43	—	1.10	6.0	—	300-400	1
Toluene	40	45	4	7	1.27	7.0	3.14	232	
Turpentine	95	—	35	—	0.80	—	—	300	
o-Xylene	63	75	17	24	1.00	—	3.66	291	

Source: P.E. Weaver, Industrial Maintenance Painting, p.118, NACE, 1978.

Do's and Don'ts for Steel Construction to be Coated

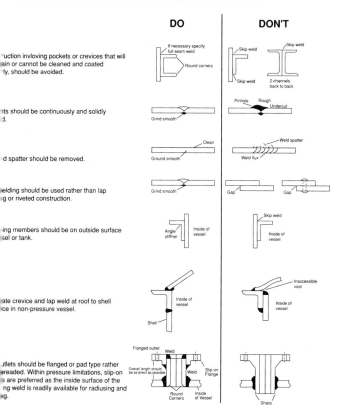

DO	DON'T

...ruction involving pockets or crevices that will ...ain or cannot be cleaned and coated ...ly, should be avoided.

...ts should be continuously and solidly ...d.

...d spatter should be removed.

...elding should be used rather than lap ...g or riveted construction.

...ing members should be on outside surface ...sel or tank.

...ate crevice and lap weld at roof to shell ...ce in non-pressure vessel.

...utlets should be flanged or pad type rather ...readed. Within pressure limitations, slip-on ...s are preferred as the inside surface of the ...ng weld is readily available for radiusing and ...g.

Source:
NACE Standard RP0178
P.E. Weaver, Industrial Maintenance Painting, p. 2, NACE 1973.

SURFACE FINISHING OF WELDS IN PREPARATION FOR LINING

NACE Weld Preparation Designation	Type of Grinding	Butt Weld	Fillet Welded Tee Joint	Lap Weld
A	Ground flush and smooth; free of all defects.	Weld spatter removed and all surface imperfections repaired as necessary. The weld is ground flush with the plate surface.	Not Applicable	Not Applicable
			Not Applicable	Not Applicable
B	Ground flush	Minor imperfections such as porosity and undercutting exist. The weld is ground flush with the plate surface.	Not Applicable	Not Applicable
			Not Applicable	Not Applicable
C	Ground smooth; free of all defects	Weld spatter removed and all surface imperfections repaired as necessary. The weld is ground smooth and blended into the plate surface.	Weld spatter removed and all surface imperfections repaired as necessary. The weld is ground smooth and blended into the plate surfaces.	Full fillet weld between the two plates. Weld spatter removed and all surface imperfections repaired as necessary. The weld is ground smooth and blended into the plate surfaces.
			Inside of Vessel	

SURFACE FINISHING OF WELDS IN PREPARATION FOR LINING (Cont)

NACE Weld Preparation Designation	Type of Grinding	Butt Weld	Fillet Welded Tee Joint	Lap Weld
D	Ground smooth and blended.		Minor imperfections such as porosity and undercutting exist. Weld spatter is removed, welds are ground smooth and blended into the plate surfaces.	Minor imperfections such as porosity and undercutting exist. Weld spatter is removed, welds are ground smooth and blended into the plate surfaces.
		Minor imperfections such as porosity and undercutting exist. Weld spatter is removed, welds are ground smooth and blended into the plate surfaces.	Inside of Vessel	
E	Minimal.		Sharp projections on the weld bead, slag and weld spatter are removed.	Sharp projections on the weld bead, slag, and weld spatter are removed.
		Sharp projections on the weld bead, slag and weld spatter are removed.	Inside of Vessel	
	Weld Condition Prior to Finishing		Inside of Vessel	

Source: NACE Standard RP0178

Standards Organizations

NACE National Association of Corrosion Engineers
P. O. Box 218340
Houston, TX 77218
713/492-0535 Telex 792310 Fax 713/492-8254

API American Petroleum Institute
1220 L. Street NW
Washington, DC 20005
202/682-8375

ASME American Society of Mechanical Engineers
345 East 47th Street
New York, NY 10017
212/644-7722

ASTM American Society for Testing and Materials
1916 Race St.
Philadelphia, PA 19103-1187
215/299-5400

AWS American Welding Society
P.O. Box 251040
Miami, FL 33126
305/443-9353

AWWA American Water Works Association
6666 W. Quincy Ave.
Denver, CO 80235
303/794-7711

BSI British Standards Institution
2 Park St.
London WIA 2BS
England

CSA Canadian Standards Association
178 Rexdale Blvd.
Toronto, Ontario, Canada M9W 1R3
416/747-4000

DIN   Deutsches Institute for Normung
            Postfach 1107
            D-1000 Berlin 30
            Federal Republic of Germany

ISO   International Organization for Standardization
            Case Postale 56
            CH-1121 Geneva 20
            Switzerland

NIST   National Institute of Standards and Technology
            Gaithersburg, MD 20899
            301/975-2000

SAE   Society of Automotive Engineers
            400 Commonwealth Drive
            Warrendale, PA 15096
            412/776-4841

SSPC   Steel Structures Painting Council
            4400 Fifth Avenue
            Pittsburgh, PA 15213
            412/268-3327

Metallic Materials Specifications

<u>API</u>
Pipe
Spec 5CT, Casing and Tubing
Spec 5D, Drill Pipe
Spec 5L, Line Pipe

Valves
Spec 6A, Valves and Wellhead Equipment
Spec 6D, Pipeline Valves, End Closures, Connectors and Swivels

Tanks
Spec 12D, Field Welded Tanks for Storage of Production Liquids
Std 620, Recommended Rules for Design and Construction of Large, Welded, Low-Pressure Storage Tanks
Std 650, Welded Steel Tanks for Oil Storage

<u>ASTM</u>
Plate
A167, Stainless and Heat-Resisting Chromium-Nickel, Steel Plate, Sheet, and Strip
A176, Stainless and Heat-Resisting Chromium Steel Plate, Sheet, and Strip
A240, Heat-Resisting Chromium and Chromium-Nickel Stainless Steel Plate, Sheet, and Strip for Pressure Vessels
A242, High-Strength Low-Alloy Structural Steel
A285, Pressure Vessel Plates, Carbon Steel, Low- and Intermediate-Tensile Strength
A387, Pressure Vessel Plates, Alloy Steel, Chromium-Molybdenum
A457, Hot-Worked, Hot-Cold-Worked, and Cold-Worked Alloy Steel Plate, Sheet, and Strip for High Strength at Elevated Temperatures
A515, Pressure Vessel Plates, Carbon Steel, for Intermediate- and Higher-Temperature Service
A516, Pressure Vessel Plates, Carbon Steel, for Moderate- and Lower-Temperature Service
A517, Pressure Vessel Plates, Alloy Steel, High-Strength, Quenched and Tempered
A542, Pressure Vessel Plates, Alloy Steel, Quenched and Tempered Chromium-Molybdenum-Vanadium-Titanium-Boron
B171, Copper Alloy Condensor Tube Plates
B265, Titanium and Titanium Alloy Strip, Sheet, and Plate
B333, Nickel-Molybdenum Alloy Plate, Sheet, and Strip
B575, Low-Carbon Nickel-Molybdenum-Chromium Alloy Plate, Sheet, and Strip
B582, Nickel-Chromium-Iron-Molybdenum-Copper Alloy Plate, Sheet, and Strip

Pipe/Tube

A53, Pipe, Steel, Black and Hot-Dipped, Zinc-Coated Welded and Seamless

A106, Seamless Carbon Steel Pipe for High-Temperature Service

A199, Seamless Cold-Drawn Intermediate Alloy-Steel Heat-Exchanger and Condenser Tubes

A200, Seamless Intermediate Alloy-Steel Still Tubes for Refinery Service

A209, Seamless Carbon-Molybdenum Alloy-Steel Boiler and Super-heater Tubes

A213, Seamless Ferritic and Austenitic Alloy-Steel Boiler, Superheater, and Heat-Exchanger Tubes

A249, Welded Austenitic Steel Boiler, Superheater, Heat-Exchanger, and Condenser Tubes

A250, Electric-Resistance-Welded Carbon-Molybdenum Alloy Steel Boiler and Superheater Tubes

A268, Seamless and Welded Ferritic and Martensitic Stainless Steel Tubing for General Service

A269, Seamless and Welded Austenitic Stainless Steel Tubing for General Service

A271, Seamless Austenitic Chromium-Nickel Steel Still Tubes for Refinery Service

A312, Seamless and Welded Austenitic Stainless Steel Pipe

A333, Seamless and Welded Steel Pipe for Low-Temperature Service

A335, Seamless Ferritic Alloy Steel Pipe for High-Temperature Service

A358, Electric-Fusion-Welded Austenitic Chromium-Nickel Alloy Steel Pipe for High-Temperature Service

A369, Carbon and Ferritic Steel Forged and Bored Pipe for High-Temperature Service

A376, Seamless Austenitic Steel Pipe for High-Temperature Central-Station Service

A409, Welded Large Diameter Austenitic Steel Pipe for Corrosive or High-Temperature Service

A426, Centrifugally Cast Ferritic Alloy Steel Pipe for High-Temperature Service

A430, Austenitic Steel Forged and Bored Pipe for High-Temperature Service

A511, Seamless Stainless Steel Mechanical Tubing

A672, Electric-Fusion-Welded Steel Pipe for High-Pressure Service at Moderate Temperatures

A691, Carbon and Alloy Steel Pipe, Electric Fusion Welded for High-Pressure Service at High Temperature

A789, Seamless and Welded Ferritic/Austenitic Stainless Steel Tubing for General Service

A790, Seamless and Welded Ferritic/Austenitic Stainless Steel Pipe

A872, Centrifugally Cast Ferritic/Austenitic Stainless Steel Pipe for Corrosive Enviroments

B111, Copper and Copper-Alloy Seamless Condenser Tubes and Ferrule Stock

B135, Seamless Brass Tube

B161, Nickel Seamless Pipe and Tube

B163, Seamless Nickel and Nickel Alloy Condenser and Heat-Exchanger Tubes

B165, Nickel-Copper Alloy (UNS N04400) Seamless Pipe and Tube

B167, Nickel-Chromium-Iron Alloy (UNS N06600 and N06690) Seamless Pipe and Tube

B210, Alumimum Alloy, Drawn Seamless Tubes

B315, Seamless Copper Alloy Pipe and Tube

B338, Seamless and Welded Titanium and Titanium Alloy Tubes for Condensers and Heat Exchangers

B407, Nickel-Iron-Chromium Alloy Seamless Pipe and Tube

B423, Nickel-Iron-Chromium-Molybdenum-Copper Alloy (UNS N08825 and N08221) Seamless Pipe and Tube

B444, Nickel-Chromium-Molybdenum-Columbium Alloy (UNS N06625) Seamless Pipe and Tube

B464, Welded Chromium-Nickel-Iron-Molybdenum-Copper-Columbium Stabilized Alloy (UNS N08020) Pipe

B523, Seamless and Welded Zirconium and Zirconium Alloy Tubes for Condensers and Heat Exchangers

B535, Nickel-Iron-Chromium-Silicon Alloys (UNS N08330 and UNS N08332) Seamless Pipe

B619, Welded Nickel and Nickel-Cobalt Alloy Pipe

B622, Seamless Nickel and Nickel-Cobalt Alloy Pipe and Tube

B668, UNS N08028 Seamless Tubes

B677, UNS N08904 and UNS N08925 Seamless Pipe and Tube

B690, Iron-Nickel-Chromium-Molybdenum Alloys (UNS N08366 and UNS N08367) Seamless Pipe and Tube

B722, Ni-Cr-Mo-Co-W-Fe-Si Alloy (UNS N06333) Seamless Pipe and Tube

B729, Seamless UNS N08020, UNS N08026, and UNS N08024 Nickel-Alloy Pipe and Tube

B751, General Requirements for Nickel and Nickel Alloy Seamless and Welded Tube

B759, Nickel-Chromium-Molybdenum-Tungsten Alloy (UNS N06110) Pipe and Tube

B775, General Requirements for Nickel and Nickel Alloy Seamless and Welded Pipe

Castings

A216, Steel Castings, Carbon Suitable for Fusion Welding for High-Temperature Service

A217, Steel Castings, Martensitic Stainless and Alloy, for Pressure-Containing Parts Suitable for High-Temperature Service

A297, Steel Castings, Iron Chromium and Iron-Chromium-Nickel, Heat Resistant, for General Application

A351, Steel Castings, Austenitic, for High-Temperature Service

A352, Ferritic Steel Castings for Pressure-Containing Parts Suitable for Low-Temperature Service

A395, Ferritic Ductile Iron Pressure-Retaining Castings for Use at Elevated Temperatures

A451, Centrifugally Cast Austenitic Steel Pipe for High-Temperature Service

A452, Centrifugally Cast Austenitic Steel Cold-Wrought Pipe for High-Temperature Service

A487, Steel Castings Suitable for Pressure Service

A518, Corrosion-Resistant High-Silicon Iron Castings

A608, Centrifugally Cast Iron-Chromium-Nickel High-Alloy Tubing for Pressure Application at High Temperature

A743, Corrosion-Resistant Iron-Chromium, Iron-Chromium-Nickel and Nickel-Base Alloy Castings for General Application

A890, Castings, Iron-Chromium-Nickel-Molybdenum Corrosion Resistant Duplex (Austenitic/Ferritic) for General Application

Forgings

A182, Forged or Rolled Alloy-Steel Pipe Flanges, Forged Fittings, and Valves and Parts for High-Temperature Service

A234, Pipe Fittings of Wrought Carbon Steel and Alloy Steel for Moderate and Elevated Temperatures

A336, Steel Forgings, Alloy, for Pressure and High-Temperature Parts

A403, Wrought Austenitic Stainless Steel Piping Fittings

A473, Stainless and Heat-Resisting Steel Forgings

A565, Martensitic Stainless Steel Bars, Forgings, and Forging Stock for High-Temperature Service

A705, Age-Hardening Stainless and Heat-Resisting Steel Forgings

B637, Precipitation Hardening Nickel Alloys Bars, Forgings, and Forging Stock for High-Temperature Service

B639, Precipitation Hardening Cobalt-Containing Alloy Bars, Forgings, and Forging Stock for High-Temperature Service

CSA

Z183, Oil Pipeline Systems

Z184, Gas Pipeline Systems

Z187, Offshore Pipelines

Z245.1, Steel Line Pipe

Z145.11, Steel Fittings

Z145.12, Steel Flanges

Z245.15, Steel Valves

Nonmetallic Materials Specifications

API
Spec 12P, Fiberglass Reinforced Plastic Tanks
Spec 15HR, High Pressure Fiberglass Line Pipe
Spec 15LE, Polyethylene Line Pipe (PE)
Spec 15LP, Thermoplastic Line Pipe (PVC and CPVC)
Spec 15LR, Low Pressure Fiberglass Line Pipe

ASTM
D1527, Acrylonitrile-Butadiene-Styrene(ABS) Plastic Pipe, Schedules 40 and 80
D1785, Poly(Vinyl Chloride) (PVC) Plastic Pipe Schedules 40, 80, and 120
D2104, Polyethylene (PE) Plastic Pipe, Schedule 40
D2239, Polyethylene (PE) Plastic Pipe (SDR-PR) Based on Controlled Inside Diameter
D2241, Poly(Vinyl Chloride) Pressure-Rated Pipe (SDR)
D2282, Acrylonitrile-Butadiene-Styrene (ABS) Plastic Pipe (SDR-PR)
D2447, Polyethylene (PE) Plastic Pipe, Schedules 40 and 80 Based on Outside Diameter
D2513, Thermoplastic Gas Pressure Piping Systems
D2517, Reinforced Epoxy Resin Gas Pressure Pipe and Fittings
D2662, Polybutylene (PB) Plastic Pipe (SDR-PR) Based on Controlled Inside Diameter
D2996, Filament-Wound Reinforced Thermosetting Resin Pipe
D2997, Centrifugally Cast Reinforced Thermosetting Resin Pipe
D3517, "Fiberglass" (Glass-Fiber-Reinforced Thermosetting-Resin) Pressure Pipe
D3754, "Fiberglass" (Glass-Fiber-Reinfoced Thermosetting-Resin) Sewer and Industrial Pressure Piping
D4021, Glass-Fiber-Reinforced Polyester Underground Petroleum Storage Tanks

AWWA
C301, Prestressed Concrete Pressure Pipe, Steel Cylinder Type, for Water, Other Liquids
C900, Polyvinyl Chloride (PVC) Pressure Pipe, 4 Inch Through 12 Inch, for Water
C950, Glass Fiber-Reinforced Thermosetting-Resin Pressure Pipe

Nonmetallic Materials Standards

ASTM
C581, Determining Chemical Resistance of Thermosetting Resins Used in Glass Fiber Reinforced Structures Intended for Liquid Service
D543, Resistance of Plastics to Chemical Reagents
D3681, Chemical Resistance of Reinforced Thermosetting Resin Pipe in a Deflected Condition
D1693, Environmental Stress-Cracking of Ethylene Plastics
G7, Atmospheric Environmental Exposure Testing of Nonmetallic Materials
G21, Determining Resistance of Synthetic Polymeric Materials to Fungi
G22, Determining Resistance of Plastics to Bacteria
G23, Operating Light- and Water-Exposure Apparatus (Carbon-Arc Type) for Exposure of Nonmetallic Materials
G24, Conducting Natural Light Exposures Under Glass
G26, Operating Light-Exposure Apparatus (Xenon-Arc Type)
G29, Determining Algal Resistance of Plastic Films
G53, Operating Light- and Water-Exposure Apparatus (Fluorescent UV-Condensation Type) for Exposure of Nonmetallic Materials
G63, Evaluating Nonmetallic Materials for Oxygen Service
G90, Accelerated Outdoor Weathering of Nonmetallic Materials Using Concentrated Natural Sunlight

Protective Coatings

NACE
TM0170, Visual Standard for Surfaces of New Steel Airblast Cleaned with Sand Abrasive
TM0174, Laboratory Methods for the Evaluation of Protective Coatings Used as Lining Materials in Immersion Service
TM0175, Visual Standard for Surfaces of New Steel Centrifugally Blast Cleaned with Steel Grit and Shot
TM0183, Evaluation of Internal Plastic Coatings for Corrosion Control
TM0184, Accelerated Test Procedures for Screening Atmospheric Surface Coating Systems for Offshore Platforms and Equipment
TM0185, Evaluation of Internal Plastic Coatings for Corrosion Control of Tubular Goods by Autoclave Testing
TM0186, Holiday Detection of Internal Tubular Coatings of 10 to 30 mils (0.25 to 0.76 mm) Dry Film Thickness
TM0375, Abrasion Resistance Testing of Thin Film Baked Coatings and Linings Using the Falling Sand Method
TM0384, Holiday Detection of Internal Tubular Coatings of Less Than 10 mils (0.25 mm) Dry Film Thickness
RP0172, Surface Preparation of Steel and Other Hard Materials by Water Blasting Prior to Coating or Recoating
RP0178, Design, Fabrication, and Surface Finish of Metal Tanks and Vessels to be Lined for Chemical Immersion Service

RP0184, Repair of Lining Systems

RP0188, Discontinuity (Holiday) Testing of Protective Coatings

RP0281, Method for Conducting Coating (Paint) Panel Evaluation Testing in Atmospheric Exposure

RP0287, Field Measurement of Surface Profile of Abrasive Blast Cleaned Steel Surfaces Using a Replica Tape

RP0288, Inspection of Linings on Steel and Concrete

RP0372, Method for Lining Lease Production Tanks with Coal Tar Epoxy

RP0376, Monolithic Organic Corrosion Resistant Floor Surfacings

RP0386, Applications of a Coating System to Interior Surfaces of Covered Railroad Hopper Cars in Plastic, Food and Chemical Service

RP0487, Considerations in the Selection and Evaluation of Interim Petroleum-Based Coatings

ASTM

D609, Preparation of Steel Panels for Testing Paint, Varnish, Lacquer, and Related Products

D610, Evaluating Degree of Rusting on Painted Steel Surfaces

D658, Abrasion Resistance of Organic Coatings by the Air Blast Abrasion Test

D659, Evaluating Degree of Chalking of Exterior Paints

D660, Evaluating Degree of Checking of Exterior Paints

D661, Evaluating Degree of Cracking of Exterior Paints

D662, Evaluating Degree of Erosion of Exterior Paints

D714, Evaluating Degree of Blistering of Paints

D772, Evaluating Degree of Flaking (Scaling) of Exterior Paints

D822, Operating Light- and Water-Exposure Apparatus (Carbon-Arc Type) for Testing Paint, Varnish, Lacquer, and Related Products

D823, Producing Films of Uniform Thickness of Paint, Varnish, Lacquer, and Related Products on Test Panels

D870, Testing Water Resistance of Coatings Using Water Immersion

D968, Abrasion Resistance of Organic Coatings by the Falling Abrasive Tester

D1005, Measurement of Dry Film Thickness of Organic Coatings Using Micrometers

D1014, Conducting Exterior Exposure Tests of Paints on Steel

D1186, Nondestructive Measurement of Dry Film Thickness of Non-magnetic Coatings Applied to a Ferrous Metal Base

D1187, Test Methods for Asphalt-Base Emulsions for Use in Protective Coatings for Metal

D1212, Measurement of Wet Film Thickness of Organic Coatings

D1400, Nondestructive Measurement of Dry Film Thickness of Nonconductive Coatings Applied to a Nonferrous Metal Base

D1653, Water Vapor Permeability of Organic Coating Films

D1654, Evaluation of Painted or Coated Specimens Subjected to Corrosive Environments

D1735, Water Resistance of Coatings Using Water Fog Apparatus

D2197, Adhesion of Organic Coatings by Scrape Adhesion Tester

D2247, Testing Water Resistance of Coatings in 100% Relative Humidity

D2454, Determining the Effect of Overbaking on Organic Coatings

D2485, Test for Coatings Designed to be Resistant to Elevated Temperatures During Their Service Life

D2803, Filiform Corrosion Resistance of Organic Coatings on Metals

D2933, Corrosion Resistance of Coated Steel Specimens (Cyclic Method)

D3258, Porosity of Paint Films

D3276, Guide for Paint Inspectors (Metal Substrates)

D3359, Measuring Adhesion by Tape Test

D3361, Operating Light- and Water-Exposure Apparatus (Unfiltered Carbon-Arc Type) for Testing Paint, Varnish, Lacquer, and Related Products Using the Dew Cycle

D3363, Film Hardness by Pencil Test

E376, Measuring Coating Thickness by Magnetic-Field or Eddy-Current (Electromagnetic) Test Methods

G73, Liquid Impingement Erosion Testing

G85, Modified Salt Spray (Fog) Testing

G87, Conducting Moist SO_2 Tests

SSPC

PA 1, Shop, Field, & Maintenance Painting

PA 2, Measurement of Dry Paint Thickness with Magnetic Gages

PA Guide 3, A Guide To Safety in Paint Application

PA Guide 4, A Guide to Maintenance Repainting with Oil Base or Alkyd Painting System

Guide to Vis 1, Pictorial Surface Preparation Standards for Painting Steel Surfaces

Guide to Vis 2, Standard Method of Evaluating Degree of Rusting on Painted Steel Surfaces

SP 1, Solvent Cleaning

SP 2, Hand Tool Cleaning

SP 3, Power Tool Cleaning

SP 5, White Metal Blast Cleaning

SP 6, Commercial Blast Cleaning

SP 7, Brush-Off Blast Cleaning

SP 8, Pickling

SP 10, Near-White Blast Cleaning

PS Guide 1.00, Guide for Selecting Oil Base Painting Systems

PS 1.04, Three-Coat Oil-Alkyd (Lead and Chromate Free) Painting System for Galvanized or Non-Galvanized Steel (With Zinc Dust-Zinc Oxide Linseed Oil Primer)

PS 1.07, Three-Coat Oil Base Red Lead Painting System

PS 1.08, Four-Coat Oil Base Red Lead Painting System

PS 1.09, Three-Coat Oil Base Zinc Oxide Painting System (Without Lead or Chromate Pigment)

PS 1.10, Four-Coat Oil Base Zinc Oxide Painting System (Without Lead or Chromate Pigment)

PS 1.11, Three-Coat Oil Base Red Lead Painting System

PS 1.12, Three-Coat Oil Base Zinc Chromate Painting System

PS 1.13, One-Coat Oil Base Slow Drying Maintenance Painting System (Without Lead or Chromate Pigment)

PS Guide 2.00, Guide for Selecting Alkyd Painting Systems

PS 2.03, Three-Coat Alkyd Painting System with Red Lead Iron Oxide Primer (For Weather Exposure)

PS 2.05, Three-Coat Alkyd Painting System for Unrusted Galvanized Steel (For Weather Exposure)

PS Guide 3.00, Guide for Selecting Phenolic Painting Systems

PS Guide 4.00, Guide for Selecting Vinyl Painting Systems

PS 4.01, Four-Coat Vinyl Painting System with Red Lead Primer (For Salt Water or Chemical Use)

PS 4.02, Four-Coat Vinyl Painting System (For Fresh Water, Chemical, and Corrosive Atmospheres)

PS 4.03, Three-Coat Vinyl Painting System with Wash Primer (For Salt Water and Weather Exposure)

PS 4.04, Four-Coat White or Colored Vinyl Painting System (For Fresh Water, Chemical, and Corrosive Atmospheres)

PS 4.05, Three-Coat Vinyl Painting System with Wash Primer and Vinyl Alkyd Finish Coat (For Atmospheric Exposure)

PS Guide 7.00, Guide for Selecting One-Coat Shop Painting System

PS 8.01, One-Coat Rust Preventive Painting System with Thick-Film Compounds

PS 9.01, Cold Applied Asphalt Mastic Painting Ssytem with Extra-Thick Film

PS 10.01, Hot Applied Coal Tar Enamel Painting System

PS 10.02, Cold Applied Coal Tar Mastic Painting System

PS 11.01, Black (or Dark Red) Coal Tar Epoxy-Polyamide Painting System

PS Guide 12.00, Guide for Selecting Zinc-Rich Painting System

PS 12.01, One-Coat Zinc-Rich Painting System

PS 13.01, Epoxy-Polyamide Painting System

PS 14.01, Steel Joist Shop Painting System

PS Guide 15.00, Guide for Selecting Chlorinated Rubber Painting Systems

PS 16.01, Silicone Alkyd Painting System for New Steel

PS Guide 17.00, Guide for Selecting Urethane Painting Systems

PS 18.01, Three-Coat Latex Painting System

PS Guide 19.00, Guide for Selecting Painting Systems for Ship Bottoms

PS Guide 20.00, Guide for Selecting Painting Systems for Boottoppings

PS Guide 21.00, Guide for Selecting Painting Systems for Topsides

PS Guide 22.00, Guide for Selecting One-Coat Preconstruction or Prefabrication Painting Systems

Pipeline Coatings

NACE

MR0274, Material Requirements in Prefabricated Plastic Films for Pipeline Coatings

RP0185, Extruded Polyolefin Resin Coating Systems for Underground or Submerged Pipe

RP0274, High Voltage Electrical Inspection of Pipeline Coatings Prior to Installation

RP0275, Application of Organic Coatings to the External Surface of Steel Pipe for Underground Service

RP0276, Extruded Asphalt Mastic Type Protective Coatings for Underground Pipelines

RP0375, Application and Handling of Wax-Type Protective Coatings and Wrapper Systems for Underground Pipelines

ASTM

D427, Shrinkage Factors of Soils

G6, Abrasion Resistance of Pipeline Coatings

G8, Cathodic Disbonding of Pipeline Coatings

G9, Water Penetration into Pipeline Coatings

G10, Specific Bendability of Pipeline Coatings

G11, Effects of Outdoor Weathering on Pipeline Coatings

G12, Nondestructive Measurement of Film Thickness of Pipeline Coatings on Steel

G13, Impact Resistance of Pipeline Coatings (Limestone Drop Test)

G14, Impact Resistance of Pipeline Coatings (Falling Weight Test)

G17, Penetration Resistance of Pipeline Coatings (Blunt Rod)

G18, Test Method for Joints, Fittings, and Patches in Coated Pipelines

G19, Disbonding Characteristics of Pipeline Coatings by Direct Soil Burial

G20, Chemical Resistance of Pipeline Coatings

G42, Cathodic Disbonding of Pipeline Coatings Subjected to Elevated Temperatures

G55, Method for Evaluating Pipeline Coating Patch Materials

G62, Holiday Detection in Pipeline Coatings

G70, Ring Bendability of Pipeline Coatings (Squeeze Test)

G80, Specific Cathodic Disbonding of Pipeline Coatings

G89, Cathodic Disbonding of Pipeline Coatings Subjected to Cyclic Temperatures

G95, Cathodic Disbondment Test of Pipeline Coatings

AWWA

C203, Coal-Tar Protective Coatings and Linings for Steel Water Pipelines - Enamel and Tape - Hot-Applied

C205, Cement - Mortar Protective Lining and Coating for Steel Water Pipe - 4 Inch and Larger - Shop Applied

C209, Cold Applied Tape Coating for the Exterior of Special Sections, Connections, and Fittings for Steel Water Pipe

C213, Fusion Bonded Epoxy Coatings for the Interior and Exterior of Steel Water Pipelines

C214, Tape Coating for the Exterior of Steel Water Pipelines

CSA

Z245.20, External Fusion Bonded Epoxy Coated Steel Pipe

DIN

30 670, Polyethylene Sheathing of Steel Tubes and of Steel Shapes and Fittings

Metallic and Anodic Coatings

ASTM

A143, Safeguarding Against Embrittlement of Hot-Dip Galvanized Structural Steel Products and Procedure for Detecting Embrittlement

B380, Corrosion Testing of Decorative Chromium Electroplating by the Corrodkote Procedure

B457, Measurement of Impedance of Anodic Coatings on Aluminum

B537, Rating of Electroplated Panels Subjected Atmospheric Exposure

B651, Measurement of Corrosion Sites in Nickel Plus Chromium or Copper Plus Nickel Plus Chromium Electroplated Surfaces with the Double-Beam Interference Microscope

B680, Seal Quality of Anodic Coatings on Aluminum by Acid Dissolution

F326, Electronic Hydrogen Embrittlement Test for Cadmium Electroplating Processes

F519, Mechanical Hydrogen Embrittlement Testing of Plating Processes and Aircraft Maintenance Chemicals

G33, Recording Data from Atmospheric Corrosion Tests of Metallic-Coated Steel Specimens

Cathodic Protection

NACE

RP0169, Control of External Corrosion on Underground or Submerged Metallic Piping Systems

RP0572, Design, Installation, Operation, and Maintenance of Impressed Current Deep Groundbeds

RP0174, Corrosion Control of Electric Underground Residential Distribution Systems

RP0575, Design, Installation, Operation, and Maintenance of Internal Cathodic Protection Systems in Oil Treating Vessels

RP0675, Control of Corrosion on Offshore Steel Pipelines

RP0176, Corrosion Control of Steel, Fixed Offshore Platforms Associated with Petroleum Production

RP0177, Mitigation of Alternating Current and Lightning Effects on Metallic Structures and Corrosion Control Systems

RP0180, Cathodic Protection of Pulp and Paper Mill Effluent Clarifiers

RP0285, Control of External Corrosion on Metallic Buried, Partially Buried, or Submerged Liquid Storage Systems

RP0186, Application of Cathodic Protection for Well Casings

RP0286, The Electrical Isolation of Cathodically Protected Pipelines

RP0387, Metallurgical and Inspection Requirements for Cast Sacrificial Anodes for Offshore Applications

RP0388, Impressed Current Cathodic Protection of Internal Submerged Surfaces of Steel Water Storage Tanks

ASTM
B418, Cast and Wrought Galvanic Zinc Anodes
G57, Field Measurement of Soil Resistivity Using the Wenner Four-Electrode Method
G97, Laboratory Evaluation of Magnesium Sacrificial Anodes

Atmospheric Corrosion

ASTM
B117, Salt Spray (Fog) Testing
B368, Copper-Accelerated Acetic Acid-Salt Spray (Fog) Testing (CASS Test)
D2059, Resistance of Zippers to Salt Spray (Fog)
G50, Conducting Atmospheric Corrosion Tests on Metals
G60, Cyclic Humidity Tests
G84, Measurement of Time-of-Wetness on Surfaces Exposed to Wetting Conditions as in Atmospheric Corrosion Testing
G85, Modified Salt Spray (Fog) Testing
G87, Conducting Moist SO2 Tests
G91, Monitoring Atmospheric SO2 Using the Sulfation Plate Technique
G92, Characterization of Atmospheric Test Sites
G101, Guide for Estimating the Atmospheric Corrosion Resistance of Low-Alloy Steels
G104, Test Method for Assessing Galvanic Corrosion Caused by the Atmosphere

Oil Production

NACE
MR0174, Recommendations for Selecting Inhibitors for Use as Sucker Rod Thread Lubricants
MR0175, Sulfide Stress Cracking Resistant Metallic Materials for Oil Field Equipment

MR0176, Metallic Materials for Sucker Rod Pumps for Hydrogen Sulfide Environments

RP0175, Control of Internal Corrosion in Steel Pipelines and Piping Systems

RP0181, Liquid Applied Internal Protective Linings and Coatings for Oil Field Production Equipment

RP0273, Handling and Proper Usage of Inhibited Oilfield Acids (API Bulletin D-15) (Joint API-NACE Project)

RP0278, Design and Operation of Stripping Columns for Removal of Oxygen from Water

RP0475, Selection of Metallic Materials to be Used in All Phases of Water Handling for Injection into Oil Bearing Formations

RP0775, Preparation and Installation of Corrosion Coupons and Interpretation of Test Data in Oil Production Practice

ID182, Wheel Test Method Used for Evaluation of Film Persistent Inhibitors for Oilfield Applications

TM0173, Methods for Determining Water Quality for Subsurface Injection Using Membrane Filters

TM0177, Testing of Metals for Resistance to Sulfide Stress Cracking at Ambient Temperatures

TM0187, Evaluating Elastomeric Materials in Sour Gas Enviroments

TM0275, Performance Testing of Sucker Rods by the Mixed String, Alternate Rod Method

TM0284, Evaluation of Pipeline Steels for Resistance to Stepwise Cracking

TM0374, Laboratory Screening Tests to Determine the Ability of Scale Inhibitors to Prevent Precipitation of $CaSO_4$ and $CaCO_3$ from Solution

API

RP5L2, Recommended Practice for Internal Coating of Line Pipe for Gas Transmission Service

RP5L7, Recommended Practices for Unprimed Internal Fusion Bonded Epoxy Coating of Line Pipe

RP10E, Recommended Practice for Application of Cement Lining to Steel Tubular Goods, Handling and Joining

RP14E, Recommended Practice for Desing and Installation of Offshore Production Platform Piping Systems

RP15A4, Recommended Practice for Care and Use of Reinforced Thermosetting Resin Casing and Tubing

RP15L4, Recommended Practice for Care and Use of Reinforced Thermosetting Resin Line Pipe

Oil Products

ASTM
D130, Detection of Copper Corrosion from Petroleum Products by the Copper Strip Tarnish Test
D665, Rust-Preventing Characteristics of Inhibited Mineral Oil in the Presence of Water
D801, Sampling and Testing Dipentene
D849, Copper Corrosion of Industrial Aromatic Hydrocarbons
D1743, Corrosion Preventive Properties of Lubricating Greases
D1838, Copper Strip Corrosion by Liquified Petroleum (LP) Gases
D2649, Corrosion Characteristics of Solid Film Lubricants
D3603, Rust-Preventing Characteristics of Steam Turbine Oil in the Presence of Water (Horizontal Disk Method)
D4048, Detection of Copper Corrosion from Lubricating Grease by the Copper Strip Tarnish Test

Automotive, Aircraft

ASTM
D930, Total Immersion Corrosion Test of Water-Soluble Aluminum Cleaners
D1280, Total Immersion Corrosion Test of Soak Tank Metal Cleaners
D1374, Aerated Total Immersion Corrosion Test for Metal Cleaners
D1384, Corrosion Test for Engine Coolants in Glassware
D2570, Simulated Service Corrosion Tests of Engine Coolants
D2758, Testing Engine Coolants by Engine Dynamometer
D2809, Cavitation Erosion-Corrosion Characteristics of Aluminum Pumps with Engine Coolants
D2847, Engine Coolants in Car and Light Truck Service
F482, Total Immersion Corrosion Test for Tank-Type Aircraft Maintenance Chemicals
F483, Total Immersion Corrosion Test for Aircraft Maintenance Chemicals

Process and Power Industries

NACE
RP0170, Protection of Austenitic Stainless Steel from Polythionic Acid Stress Corrosion Cracking During Shutdown of Refinery Equipment
RP0173, Collection and Identification of Corrosion Products
RP0182, Initial Conditioning of Cooling Water Equipment
RP0189, On-Line Monitoring of Cooling Waters
RP0272, Direct Calculation of Economic Appraisals of Corrosion Control Measures
RP0472, Methods and Controls to Prevent In-Service Cracking of Carbon Steel (P-1) Welds in Corrosive Petroleum Refining Environments

5A171, Materials for Receiving, Handling, and Storing Hydrofluoric Acid

TM0169, Laboratory Corrosion Testing of Metals for the Process Industries

TM0171, Autoclave Corrosion Testing of Metals for the Process Industries

TM0172, Antirust Properties of Cargoes in Petroleum Product Pipelines

TM0270, Method of Conducting Controlled Velocity Laboratory Corrosion Tests

TM0274, Dynamic Corrosion Testing of Metals in High Temperature Water

TM0286, Cooling Water Test Units Incorporating Heat Transfer Surfaces

3M182, Corrosion Testing of Chemical Cleaning Solvents

ASTM

D807, Assessing the Tendency of Industrial Boiler Waters to Cause Embrittlement (USBM Embrittlement Detector Method)

D2688, Corrosivity of Water in the Absence of Heat Transfer (Weight Loss Methods)

D2776, Corrosivity of Water in the Absence of Heat Transfer (Electrical Methods)

D3263, Corrosivity of Solvent Systems for Removing Water-Formed Deposits

E712, Laboratory Screening of Metallic Containment Materials for Use with Liquids in Solar Heating and Cooling Systems

E745, Simulated Service Testing for Corrosion of Metallic Containment Materials for Use with Heat Transfer Fluids in Solar Heating and Cooling Systems

G2, Corrosion Testing of Samples of Zirconium and Zirconium Alloys

G4, Conducting Corrosion Coupon Tests in Plant Equipment

G54, Simple Static Oxidation Testing

G79, Evaluation of Metals Exposed to Carburization Enviroments

G88, Guide for Designing Systems for Oxygen Service

API

Publ 941, Steels for Hydrogen Service at Elevated Temperatures and Pressures in Petroleum Refineries and Petrochemical Plants

Publ 942, Controlling Weld Hardness of Carbon Steel Refinery Equipment to Prevent Environmental Cracking

General

NACE
RP0187, Design Considerations for Corrosion Control of Reinforcing Steel in Concrete Structures

ASTM
B76, Accelerated Life Test of Nickel-Chromium and Nickel-Chromium-Iron Alloys for Electrical Heating
B78, Accelerated Life Test of Iron-Chromium-Aluminum Alloys for Electrical Heating
B732, Evaluating the Corrosivity of Solder Fluxes for Copper Tubing Systems
C739, Cellulosic Fiber (Wood-Base) Loose-fill Thermal Insulation
D1141, Substitute Ocean Water
D1193, Reagent Water
D1611, Corrosion Produced by Leather in Contact with Metal
D3310, Determining Corrosivity of Adhesive Materials
D3482, Determining Electrolytic Corrosion of Copper by Adhesives
E380, Metric Practice
E937, Corrosion of Steel by Sprayed Fire-Resistive Material Applied to Structural Members
G1, Preparing, Cleaning, and Evaluating Corrosion Test Specimens
G15, Definitions of Terms Relating to Corrosion and Corrosion Testing
G16, Applying Statistics to Analysis of Corrosion Data
G31, Laboratory Immersion Corrosion Testing of Metals
G51, Method for pH of Soil for Use in Corrosion Testing
G52, Exposing and Evaluating Metals and Alloys in Surface Seawater
G68, Liquid Sodium Corrosion Testing of Metals and Alloys
G71, Conducting and Evaluating Galvanic Corrosion Tests in Electrolytes

Electrochemistry

<u>ASTM</u>

B627, Electrolytic Corrosion Testing

C876, Half-Cell Potentials of Reinforcing Steel in Concrete

G3, Conventions Applicable to Electrochemical Measurements in Corrosion Testing

G5, Standard Reference Method for Making Potentiostatic and Potentiodynamic Anodic Polarization Measurements

G59, Potentiodynamic Polarization Resistance Measurements

G61, Cyclic Potentiodynamic Polarization Measurements for Localized Corrosion Susceptibility of Iron-, Nickel-, or Cobalt-Based Alloys

G82, Development and Use of a Galvanic Series for Predicting Galvanic Corrosion Performance

G100, Test Method for Conducting Cyclic Galvanostaircase Polarization

G102, Calculation of Corrosion Rates and Related Information from Electrochemical Measurements

Localized Corrosion

<u>ASTM</u>

F746, Test for Pitting or Crevice Corrosion of Metallic Surgical Implants

G46, Examination and Evaluation of Pitting Corrosion

G48, Pitting and Crevice Corrosion Resistance of Stainless Steels and Related Alloys by the Use of Ferric Chloride Solution

G78, Crevice Corrosion Testing of Iron-Base and Nickel-Base Stainless Alloys in Seawater and Other Chloride Containing Aqueous Environments

Erosion, Wear, and Abrasion

<u>ASTM</u>

G32, Vibratory Cavitation Erosion Test

G40, Terminology Relating to Wear and Erosion

G65, Dry Sand/Rubber Wheel Abrasion Test

G73, Liquid Impingement Erosion Testing

G75, Slurry Abrasivity by Miller Number

G76, Conducting Erosion Tests by Solid Particle Impingement Using Gas Jets

G77, Ranking Resistance of Materials to Sliding Wear Using Block-on-Ring Wear Test

G81, Jaw Crusher Gouging Abrasion Test

G83, Wear Testing with a Crossed-Cylinder Apparatus

Environmental Cracking

ASTM

A262, Detecting Susceptibility to Intergranular Attack in Stainless Steels

A763, Detecting Susceptibility to Intergranular Attack in Ferritic Stainless Steels

B577, Hydrogen Embrittlement of Copper

C692, Evaluating the Influence of Wicking-Type Thermal Insulations on Stress Corrosion Cracking Tendency of Austenitic Stainless Steels

G28, Detecting Susceptibility to Intergranular Attack in Wrought, Nickel-Rich, Chromium Bearing Alloys

G30, Making and Using U-Bend Stress Corrosion Cracking Specimens

G34, Exfoliation Corrosion Susceptibility of 2XXX and 7XXX Series Aluminum Alloys (EXCO Test)

G35, Susceptibility of Stainless Steels and Related Nickel-Chromium-Iron Alloys to Stress Corrosion Cracking in Polythionic Acids

G36, Stress-Corrosion Cracking Tests in a Boiling Magnesium Chloride Solution

G37, Use of Mattsson's Solution of pH 7.2 to Evaluate the Stress Corrosion Cracking Susceptibility of Copper-Zinc Alloys

G38, Making and Using C-Ring Stress-Corrosion Test Specimens

G39, Preparation and Use of Bent-Beam Stress-Corrosion Test Specimens

G41, Determining Cracking Susceptibility of Metals Exposed Under Stress to a Hot Salt Environment

G44, Alternate Immersion Stress Corrosion Testing in 3.5% Sodium Chloride Solution

G47, Determining Susceptibility to Stress-Corrosion Cracking of High-Strength Aluminum Alloy Products

G49, Preparation and Use of Direct Tension Stress-Corrosion Test Specimens

G58, Preparation of Stress Corrosion Test Specimens for Weldments

G64, Classification of the Resistance to Stress-Corrosion Cracking of High-Strength Aluminum Alloys

G66, Visual Assessment of Exfoliation Corrosion Susceptibility of 5XXX Series Aluminum Alloys (Asset Test)

G67, Determining the Susceptbility to Intergranular Corrosion of 5XXX Series Aluminum Alloys by Weight Loss After Exposure to Nitric Acid (NAMLT Test)

G103, Performing a Stress Corrosion Cracking Test of Low Copper Containing Al-Zn-Mg Alloys in Boiling 6% Sodium Chloride Solution